$135.00 JCF

Handbook of Electrical Engineering Calculations

ELECTRICAL ENGINEERING AND ELECTRONICS
A Series of Reference Books and Textbooks

EXECUTIVE EDITORS

Marlin O. Thurston
Department of Electrical Engineering
The Ohio State University
Columbus, Ohio

William Middendorf
Department of Electrical and
Computer Engineering
University of Cincinnati
Cincinnati, Ohio

EDITORIAL BOARD

Maurice Bellanger
Télécommunications, Radioélectriques,
et Téléphoniques (TRT)
Le Plessis-Robinson, France

Norman B. Fuqua
Reliability Analysis Center
Griffiss Air Force Base, New York

Naim A. Kheir
Department of Electrical and
Systems Engineering
Oakland University
Rochester, Michigan

Pradeep Khosla
Carnegie–Mellon University
Pittsburgh, Pennsylvania

Glenn Zelniker
Z-Systems, Inc.
Gainesville, Florida

1. Rational Fault Analysis, *edited by Richard Saeks and S. R. Liberty*
2. Nonparametric Methods in Communications, *edited by P. Papantoni-Kazakos and Dimitri Kazakos*
3. Interactive Pattern Recognition, *Yi-tzuu Chien*
4. Solid-State Electronics, *Lawrence E. Murr*
5. Electronic, Magnetic, and Thermal Properties of Solid Materials, *Klaus Schröder*
6. Magnetic-Bubble Memory Technology, *Hsu Chang*
7. Transformer and Inductor Design Handbook, *Colonel Wm. T. McLyman*
8. Electromagnetics: Classical and Modern Theory and Applications, *Samuel Seely and Alexander D. Poularikas*
9. One-Dimensional Digital Signal Processing, *Chi-Tsong Chen*
10. Interconnected Dynamical Systems, *Raymond A. DeCarlo and Richard Saeks*
11. Modern Digital Control Systems, *Raymond G. Jacquot*

12. Hybrid Circuit Design and Manufacture, *Roydn D. Jones*
13. Magnetic Core Selection for Transformers and Inductors: A User's Guide to Practice and Specification, *Colonel Wm. T. McLyman*
14. Static and Rotating Electromagnetic Devices, *Richard H. Engelmann*
15. Energy-Efficient Electric Motors: Selection and Application, *John C. Andreas*
16. Electromagnetic Compossibility, *Heinz M. Schlicke*
17. Electronics: Models, Analysis, and Systems, *James G. Gottling*
18. Digital Filter Design Handbook, *Fred J. Taylor*
19. Multivariable Control: An Introduction, *P. K. Sinha*
20. Flexible Circuits: Design and Applications, *Steve Gurley, with contributions by Carl A. Edstrom, Jr., Ray D. Greenway, and William P. Kelly*
21. Circuit Interruption: Theory and Techniques, *Thomas E. Browne, Jr.*
22. Switch Mode Power Conversion: Basic Theory and Design, *K. Kit Sum*
23. Pattern Recognition: Applications to Large Data-Set Problems, *Sing-Tze Bow*
24. Custom-Specific Integrated Circuits: Design and Fabrication, *Stanley L. Hurst*
25. Digital Circuits: Logic and Design, *Ronald C. Emery*
26. Large-Scale Control Systems: Theories and Techniques, *Magdi S. Mahmoud, Mohamed F. Hassan, and Mohamed G. Darwish*
27. Microprocessor Software Project Management, *Eli T. Fathi and Cedric V. W. Armstrong (Sponsored by Ontario Centre for Microelectronics)*
28. Low Frequency Electromagnetic Design, *Michael P. Perry*
29. Multidimensional Systems: Techniques and Applications, *edited by Spyros G. Tzafestas*
30. AC Motors for High-Performance Applications: Analysis and Control, *Sakae Yamamura*
31. Ceramic Motors for Electronics: Processing, Properties, and Applications, *edited by Relva C. Buchanan*
32. Microcomputer Bus Structures and Bus Interface Design, *Arthur L. Dexter*
33. End User's Guide to Innovative Flexible Circuit Packaging, *Jay J. Miniet*
34. Reliability Engineering for Electronic Design, *Norman B. Fuqua*
35. Design Fundamentals for Low-Voltage Distribution and Control, *Frank W. Kussy and Jack L. Warren*
36. Encapsulation of Electronic Devices and Components, *Edward R. Salmon*
37. Protective Relaying: Principles and Applications, *J. Lewis Blackburn*
38. Testing Active and Passive Electronic Components, *Richard F. Powell*
39. Adaptive Control Systems: Techniques and Applications, *V. V. Chalam*

40. Computer-Aided Analysis of Power Electronic Systems, *Venkatachari Rajagopalan*
41. Integrated Circuit Quality and Reliability, *Eugene R. Hnatek*
42. Systolic Signal Processing Systems, *edited by Earl E. Swartzlander, Jr.*
43. Adaptive Digital Filters and Signal Analysis, *Maurice G. Bellanger*
44. Electronic Ceramics: Properties, Configuration, and Applications, *edited by Lionel M. Levinson*
45. Computer Systems Engineering Management, *Robert S. Alford*
46. Systems Modeling and Computer Simulation, *edited by Naim A. Kheir*
47. Rigid-Flex Printed Wiring Design for Production Readiness, *Walter S. Rigling*
48. Analog Methods for Computer-Aided Circuit Analysis and Diagnosis, *edited by Takao Ozawa*
49. Transformer and Inductor Design Handbook: Second Edition, Revised and Expanded, *Colonel Wm. T. McLyman*
50. Power System Grounding and Transients: An Introduction, *A. P. Sakis Meliopoulos*
51. Signal Processing Handbook, *edited by C. H. Chen*
52. Electronic Product Design for Automated Manufacturing, *H. Richard Stillwell*
53. Dynamic Models and Discrete Event Simulation, *William Delaney and Erminia Vaccari*
54. FET Technology and Application: An Introduction, *Edwin S. Oxner*
55. Digital Speech Processing, Synthesis, and Recognition, *Sadaoki Furui*
56. VLSI RISC Architecture and Organization, *Stephen B. Furber*
57. Surface Mount and Related Technologies, *Gerald Ginsberg*
58. Uninterruptible Power Supplies: Power Conditioners for Critical Equipment, *David C. Griffith*
59. Polyphase Induction Motors: Analysis, Design, and Application, *Paul L. Cochran*
60. Battery Technology Handbook, *edited by H. A. Kiehne*
61. Network Modeling, Simulation, and Analysis, *edited by Ricardo F. Garzia and Mario R. Garzia*
62. Linear Circuits, Systems, and Signal Processing: Advanced Theory and Applications, *edited by Nobuo Nagai*
63. High-Voltage Engineering: Theory and Practice, *edited by M. Khalifa*
64. Large-Scale Systems Control and Decision Making, *edited by Hiroyuki Tamura and Tsuneo Yoshikawa*
65. Industrial Power Distribution and Illuminating Systems, *Kao Chen*
66. Distributed Computer Control for Industrial Automation, *Dobrivoje Popovic and Vijay P. Bhatkar*
67. Computer-Aided Analysis of Active Circuits, *Adrian Ioinovici*
68. Designing with Analog Switches, *Steve Moore*
69. Contamination Effects on Electronic Products, *Carl J. Tautscher*
70. Computer-Operated Systems Control, *Magdi S. Mahmoud*
71. Integrated Microwave Circuits, *edited by Yoshihiro Konishi*

72. Ceramic Materials for Electronics: Processing, Properties, and Applications, Second Edition, Revised and Expanded, edited by *Relva C. Buchanan*
73. Electromagnetic Compatibility: Principles and Applications, *David A. Weston*
74. Intelligent Robotic Systems, edited by *Spyros G. Tzafestas*
75. Switching Phenomena in High-Voltage Circuit Breakers, edited by *Kunio Nakanishi*
76. Advances in Speech Signal Processing, edited by *Sadaoki Furui and M. Mohan Sondhi*
77. Pattern Recognition and Image Preprocessing, *Sing-Tze Bow*
78. Energy-Efficient Electric Motors: Selection and Application, Second Edition, *John C. Andreas*
79. Stochastic Large-Scale Engineering Systems, edited by *Spyros G. Tzafestas and Keigo Watanabe*
80. Two-Dimensional Digital Filters, *Wu-Sheng Lu and Andreas Antoniou*
81. Computer-Aided Analysis and Design of Switch-Mode Power Supplies, *Yim-Shu Lee*
82. Placement and Routing of Electronic Modules, edited by *Michael Pecht*
83. Applied Control: Current Trends and Modern Methodologies, edited by *Spyros G. Tzafestas*
84. Algorithms for Computer-Aided Design of Multivariable Control Systems, *Stanoje Bingulac and Hugh F. VanLandingham*
85. Symmetrical Components for Power Systems Engineering, *J. Lewis Blackburn*
86. Advanced Digital Signal Processing: Theory and Applications, *Glenn Zelniker and Fred J. Taylor*
87. Neural Networks and Simulation Methods, *Jian-Kang Wu*
88. Power Distribution Engineering: Fundamentals and Applications, *James J. Burke*
89. Modern Digital Control Systems: Second Edition, *Raymond G. Jacquot*
90. Adaptive IIR Filtering in Signal Processing and Control, *Phillip A. Regalia*
91. Integrated Circuit Quality and Reliability: Second Edition, Revised and Expanded, *Eugene R. Hnatek*
92. Handbook of Electric Motors, edited by *Richard H. Engelmann and William H. Middendorf*
93. Power-Switching Converters, *Simon S. Ang*
94. Systems Modeling and Computer Simulation: Second Edition, *Naim A. Kheir*
95. EMI Filter Design, *Richard Lee Ozenbaugh*
96. Power Hybrid Circuit Design and Manufacture, *Haim Taraseiskey*
97. Robust Control System Design: Advanced State Space Techniques, *Chia-Chi Tsui*
98. Spatial Electric Load Forecasting, *H. Lee Willis*

99. Permanent Magnet Motor Technology: Design and Applications, *Jacek F. Gieras and Mitchell Wing*
100. High Voltage Circuit Breakers: Design and Applications, *Ruben D. Garzon*
101. Integrating Electrical Heating Elements in Appliance Design, *Thor Hegbom*
102. Magnetic Core Selection for Transformers and Inductors: A User's Guide to Practice and Specification, Second Edition, *Colonel Wm. T. McLyman*
103. Statistical Methods in Control and Signal Processing, *edited by Tohru Katayama and Sueo Sugimoto*
104. Radio Receiver Design, *Robert C. Dixon*
105. Electrical Contacts: Principles and Applications, *edited by Paul G. Slade*
106. Handbook of Electrical Engineering Calculations, *edited by Arun G. Phadke*
107. Reliability Control for Electronic Systems, *Donald J. LaCombe*

Additional Volumes in Preparation

Embedded Systems Design with 8051 Microcontrollers: Hardware and Software, *Zdravko Karakehayov, Knud Smed Christensen, and Ole Winther*

Protective Relaying in Communications Technology, *edited by Walter A. Elmore*

Handbook of Electrical Engineering Calculations

edited by
Arun G. Phadke
*Virginia Polytechnic Institute and State University
Blacksburg, Virginia*

MARCEL DEKKER, INC.　　　　　　　　　NEW YORK · BASEL

Library of Congress Cataloging-in-Publication Data

Handbook of electrical engineering calculations / edited by Arun G. Phadke.
    p. cm. -- (Electrical engineering and electronics ; 106)
    ISBN 0-8247-1955-7 (alk. paper)
    1. Electrical engineering Problems, exercises, etc. 2. Electric engineering--Mathematics. I. Phadke, Arun G. II. Series.
TS168.H36 1999
621.3'076--dc21
                        99-14706
                              CIP

This book is printed on acid-free paper.

Headquarters
Marcel Dekker, Inc.
270 Madison Avenue, New York, NY 10016
tel: 212-696-9000; fax: 212-685-4540

Eastern Hemisphere Distribution
Marcel Dekker AG
Hutgasse 4, Postfach 812, CH-4001 Basel, Switzerland
tel: 41-61-261-8482; fax: 41-61-261-8896

World Wide Web
http://www.dekker.com

The publisher offers discounts on this book when ordered in bulk quantities. For more information, write to Special Sales/Professional Marketing at the headquarters address above.

Copyright © 1999 by Marcel Dekker, Inc. All Rights Reserved.

Neither this book nor any part may be reproduced or transmitted in any form or by any means, electronic or mechanical, including photocopying, microfilming, and recording, or by any information storage and retrieval system, without permission in writing from the publisher.

Current printing (last digit):
10 9 8 7 6 5 4 3 2 1

PRINTED IN THE UNITED STATES OF AMERICA

Preface

The field of engineering differs from pure sciences in one special aspect: as an engineer one must be able to measure, calculate, design, or build something in accordance with the underlying theory. There is no question that U.S. engineering colleges train engineering students well. However, in recent years the field of electrical engineering has grown to include a vast array of subjects. The variety of subjects that must be included in the electrical engineering syllabus has inevitably led to a reduced emphasis on many important disciplines, among them the art of engineering calculations.

Calculations give substance to the theory. In engineering, it is not enough to understand the theory—one must be able to quantify. Calculations teach the student what is important and under which circumstances, what effects are negligible and why, and what the sensitivities of the answers are to the parameters of a problem. Not all calculations are necessarily deep and insightful. However, in every branch of engineering there are seminal problems that bring out the essence of the underlying theory. In solving these problems, several key ideas are introduced and used, and when the calculations are completed, the procedure becomes a model for a whole class of problems. Experienced and talented teachers know the importance of such problems.

One of the ironies of the computer age is that many engineering problems are solved using prepackaged software. Although this provides an excellent problem-solving tool for the practicing engineer, it is a mixed blessing in the hands of a student. While the inner workings of a calculation are exposed, the student is not prepared to deal with new and unforeseen problems that may arise.

This book identifies and solves the seminal problems in the principal branches of electrical engineering. It is designed to bring to the widest audience the insight that a detailed and documented solution procedure provides. The book is to be used as a supplementary reference in undergraduate and graduate electrical engineering curricula. It should also be useful to practicing engineers who wish to revisit the details of calculations and obtain a new understanding of the underlying theory.

Modern electrical engineering can be divided into several subdisciplines. The following subjects, which reflect each of the main subdisciplines, are covered in this book: electric power engineering, electromagnetic fields, signal analysis, communication systems, control systems, and computer engineering. Problems that are central to each of these areas have been solved, and the authors of each chapter provide illuminating comments at appropriate places where key ideas are introduced. Where commercial software packages are available for solving a class of problems, appropriate comments about their capabilities are provided. The authors of each of these chapters are individuals who not only have made original contributions to the field, but also are experienced teachers of undergraduate and graduate electrical engineering students.

Arun G. Phadke

Contents

Preface		*iii*
Contributors		*vii*
1	Electric Power Engineering *Arun G. Phadke*	1
2	Electromagnetics *Edmund K. Miller*	61
3	Algorithms Used in Signal Analysis *Hugh F. VanLandingham*	123
4	Communication Systems *Timothy Pratt*	157
5	Algorithms Used in Control Systems *Hugh F. VanLandingham*	225
6	Computer Engineering *Peter Athanas and Yosef Tirat-Gefen*	259
Index		*309*

Contributors

Peter Athanas, Ph.D. Associate Professor, The Bradley Department of Electrical and Computer Engineering, Virginia Polytechnic Institute and State University, Blacksburg, Virginia

Edmund K. Miller, Ph.D.[*] Consultant, Santa Fe, New Mexico.

Arun G. Phadke, Ph.D. Professor and Director, Center for Power Engineering, The Bradley Department of Electrical and Computer Engineering, Virginia Polytechnic Institute and State University, Blacksburg, Virginia

Timothy Pratt, Ph.D. Professor, The Bradley Department of Electrical and Computer Engineering, Virginia Polytechnic Institute and State University, Blacksburg, Virginia

Yosef Tirat-Gefen, Ph.D. Assistant Professor, The Bradley Department of Electrical and Computer Engineering, Virginia Polytechnic Institute and State University, Blacksburg, Virginia

Hugh F. VanLandingham, Ph.D. Professor, The Bradley Department of Electrical and Computer Engineering, Virginia Polytechnic Institute and State University, Blacksburg, Virginia

---

[*] Retired from Los Alamos National Laboratory, Los Alamos, New Mexico.

1
Electric Power Engineering

Arun G. Phadke
Virginia Polytechnic Institute and State University, Blacksburg, Virginia

1.1 SYMMETRICAL COMPONENTS

1.1.1 Introduction

Symmetrical components were introduced in 1918 by C. L. Fortescue as an analytical tool for analyzing the operation of three-phase electric machines under unbalanced operating conditions. Over the years, symmetrical components have become invaluable in electric power system engineering, and are routinely used in calculating the performance of a power system under balanced and unbalanced operating conditions. One of the main applications of symmetrical components is in analyzing unbalanced faults on three-phase a.c. networks.

In the terminology of matrix theory, the symmetrical components of voltages and currents are obtained by a linear transformation from phase quantities, and the main benefit of symmetrical component analysis derives from the fact that the symmetrical component transformation matrix is made up of eigenvectors of the impedance matrix of a balanced three-phase power apparatus.

1.1.2 Definitions and Basic Equations

Symmetrical components of three-phase voltages (E_a, E_b, E_c) or currents (I_a, I_b, I_c) are defined as zero sequence (E_0 or I_0), positive sequence (E_1 or I_1), and negative sequence (E_2 or I_2) components. Let X stand for voltages or currents. Symmetrical components of phase quantities are calculated by the transformation:

$$\begin{bmatrix} X_0 \\ X_1 \\ X_2 \end{bmatrix} = \frac{1}{3} \begin{bmatrix} 1 & 1 & 1 \\ 1 & -\frac{1}{2}+j\frac{\sqrt{3}}{2} & -\frac{1}{2}-j\frac{\sqrt{3}}{2} \\ 1 & -\frac{1}{2}-j\frac{\sqrt{3}}{2} & -\frac{1}{2}+j\frac{\sqrt{3}}{2} \end{bmatrix} \begin{bmatrix} X_a \\ X_b \\ X_c \end{bmatrix} \equiv [S] \begin{bmatrix} X_a \\ X_b \\ X_c \end{bmatrix} \quad (1.1)$$

The complex numbers appearing in the matrix above are the two cube roots of 1, the third cube root being 1 itself. In polar coordinates the three cube roots are given by

$$1 = \varepsilon^{j0}$$
$$-\frac{1}{2}+j\frac{\sqrt{3}}{2} = \varepsilon^{j2\pi/3} \quad (1.2)$$
$$-\frac{1}{2}-j\frac{\sqrt{3}}{2} = \varepsilon^{j4\pi/3}$$

S is the symmetrical component transformation matrix. Its inverse is S^{-1}, and transforms the symmetrical components of X back into phase quantities:

$$\begin{bmatrix} X_a \\ X_b \\ X_c \end{bmatrix} = [S]^{-1} \begin{bmatrix} X_0 \\ X_1 \\ X_2 \end{bmatrix} = \begin{bmatrix} 1 & 1 & 1 \\ 1 & -\frac{1}{2}-j\frac{\sqrt{3}}{2} & -\frac{1}{2}+j\frac{\sqrt{3}}{2} \\ 1 & -\frac{1}{2}+j\frac{\sqrt{3}}{2} & -\frac{1}{2}-j\frac{\sqrt{3}}{2} \end{bmatrix} \begin{bmatrix} X_0 \\ X_1 \\ X_2 \end{bmatrix} \quad (1.3)$$

The impedance matrix $\mathbf{Z}_\phi$ of a three-phase power system element such as a line, load, or a generator is transformed into its sequence impedances (symmetrical component impedances) $\mathbf{Z}_s$ by the transformation

$$[\mathbf{Z}_s] = [S][\mathbf{Z}_\phi][S]^{-1} \quad (1.4)$$

The following calculations explain the use of these equations.

Electric Power Engineering

1.1.3 Calculation of Symmetrical Components of Voltages and Currents

Given

$$\begin{bmatrix} E_a \\ E_b \\ E_c \end{bmatrix} = \begin{bmatrix} 10683 + j5428 \\ -3230 - j12180 \\ -7453 + j6752 \end{bmatrix} \text{volts}$$

$$\begin{bmatrix} I_a \\ I_b \\ I_c \end{bmatrix} = \begin{bmatrix} 8660.3 + j5000 \\ -4773.8 - j7629.6 \\ -5433.6 + j9564.3 \end{bmatrix} \text{amps}$$

The symmetrical components are given by

$$\begin{bmatrix} E_0 \\ E_1 \\ E_2 \end{bmatrix} = \frac{1}{3} \begin{bmatrix} 1 & 1 & 1 \\ 1 & -\frac{1}{2}+j\frac{\sqrt{3}}{2} & -\frac{1}{2}-j\frac{\sqrt{3}}{2} \\ 1 & -\frac{1}{2}-j\frac{\sqrt{3}}{2} & -\frac{1}{2}+j\frac{\sqrt{3}}{2} \end{bmatrix} \begin{bmatrix} 10683 + j5428 \\ -3230 - j12180 \\ -7453 + j6752 \end{bmatrix}$$

$$= \begin{bmatrix} 0 \\ 10806 + j3933 \\ -124 + j1495 \end{bmatrix} \text{volts}$$

$$\begin{bmatrix} I_0 \\ I_1 \\ I_2 \end{bmatrix} = \frac{1}{3} \begin{bmatrix} 1 & 1 & 1 \\ 1 & -\frac{1}{2}+j\frac{\sqrt{3}}{2} & -\frac{1}{2}-j\frac{\sqrt{3}}{2} \\ 1 & -\frac{1}{2}-j\frac{\sqrt{3}}{2} & -\frac{1}{2}+j\frac{\sqrt{3}}{2} \end{bmatrix} \begin{bmatrix} 8660.3 + j5000 \\ -4773.8 - j7629.6 \\ -5433.6 - j9564.3 \end{bmatrix}$$

$$= \begin{bmatrix} -515.7 + j2311.6 \\ 9551.4 + j1534.7 \\ -375.5 + j1153.7 \end{bmatrix} \text{amps}$$

As in this example, it is sometimes the case that one of the components – in this case the zero sequence voltage – is zero.

1.1.4 Calculation of Power

The power in the three-phase system can be calculated using the phase quantities or symmetrical components. First in terms of the phase powers:

$$S_\phi = [E_a \quad E_b \quad E_c] \begin{bmatrix} I_a^* \\ I_b^* \\ I_c^* \end{bmatrix}$$

$$= [(10683 + j5428) \quad (-3230 - j1280) \quad (-7453 + j6752)] \begin{bmatrix} 8660.3 - j5000 \\ -4773.8 + j7629.6 \\ -5433.6 - j9564.3 \end{bmatrix}$$

$$= (333.07 + j61.695) \times 10^6 \quad \text{volt} \quad \text{amps}$$

The power in symmetrical components is given by

$$S_s = 3 \times [E_0 \quad E_1 \quad E_2] \begin{bmatrix} I_0^* \\ I_1^* \\ I_2^* \end{bmatrix}$$

$$= [(0) \quad (10806 + j3933) \quad (-124 + j1495)] \begin{bmatrix} -515.7 - j2311.6 \\ 9551.4 - j1534.7 \\ -375.5 - j1153.7 \end{bmatrix}$$

$$= (333.07 + j61.695) \times 10^6 \quad \text{volt} \quad \text{amps}$$

Of course, the power calculated by the two methods is the same. The factor 3 in the formula for calculating the power in symmetrical components arrives from the fact that the symmetrical component transformation is not power invariant. It should also be remembered that if the voltages and currents are expressed in per-unit, the symmetrical component power calculation is not multiplied by 3. On the other hand, the three-phase power calculated by adding the individual phase powers must be divided by 3.

1.1.5 Power Calculation in Per-Unit

Assume the base voltage and power for the system being considered to be 20 kV and 300 MVA respectively. The base voltage is then ($E_{\text{base}} = 20000/\sqrt{3} = 11547$ volts), and the base current is ($I_{\text{base}} = (100 \times 10^6/11547 = 8660.3$ amps). The phase voltages and currents in per-unit are given by E/E_{base} and I/I_{base}, respectively.

$$\begin{bmatrix} E_a \\ E_b \\ E_c \end{bmatrix} = \begin{bmatrix} 0.9251 + j0.4701 \\ -0.2797 - j1.0548 \\ -0.6454 + j0.5847 \end{bmatrix} \text{pu} \quad \begin{bmatrix} I_a \\ I_b \\ I_c \end{bmatrix} = \begin{bmatrix} 1.0 + j0.5774 \\ -0.5512 - j0.8810 \\ -0.6274 + j1.1044 \end{bmatrix} \text{pu}$$

The symmetrical components in per-unit are obtained in the same manner:

$$\begin{bmatrix} E_0 \\ E_1 \\ E_2 \end{bmatrix} = \begin{bmatrix} 0 \\ 0.9359 + j0.3406 \\ -0.0107 + j0.1295 \end{bmatrix} \text{pu} \quad \begin{bmatrix} I_0 \\ I_1 \\ I_2 \end{bmatrix} = \begin{bmatrix} -0.0595 + j0.2669 \\ 1.1029 + j0.1772 \\ -0.0434 + j0.1332 \end{bmatrix} \text{pu}$$

The power in per-unit is $(333.07 + j61.69)/300 = 1.1102 + j0.20565$ pu. If we calculate the power using the phase voltages and currents in per-unit:

$$S_{pu} = \frac{1}{3}[(0.9251 + j0.4701) \quad (-0.2797 - j1.0548) \quad (-0.6454 + j0.5847)]$$
$$\begin{bmatrix} 1.0 - j0.574 \\ -0.5512 + j0.8810 \\ -0.6274 - j1.1044 \end{bmatrix}$$
$$= 1.1102 + j0.2057 \text{pu}$$

Also, calculating the power with per-unit symmetrical component voltages and currents:

$$S_{pu} = [(0) \quad (0.9359 + j0.3406) \quad (-0.0107 + j0.1295)]$$
$$\begin{bmatrix} -0.595 + j0.2669 \\ 1.1029 + j0.1772 \\ -0.0434 + j0.1332 \end{bmatrix}$$
$$= 1.1102 + j0.2057 \text{ pu}$$

1.1.6 Sequence Impedances

In the following it is assumed that the impedances are in per-unit. This is not very important. If the impedances are in physical units, the same computation procedure is still valid.

A Transposed Transmission Line

A typical phase impedance matrix for a transposed transmission line is given by

$$\mathbf{Z}_\phi = \begin{bmatrix} 0.1 + j2.0 & 0.05 + j0.5 & 0.05 + j0.5 \\ 0.05 + j0.5 & 0.1 + j2.0 & 0.05 + j0.5 \\ 0.05 + j0.5 & 0.05 + j0.5 & 0.1 + j2.0 \end{bmatrix}$$

Using Eq. (1.4), the symmetrical component (sequence) impedances of the transmission line can be calculated:

$$\mathbf{Z}_s = \frac{1}{3}\begin{bmatrix} 1 & 1 & 1 \\ 1 & -\frac{1}{2}+j\frac{\sqrt{3}}{2} & -\frac{1}{2}-j\frac{\sqrt{3}}{2} \\ 1 & -\frac{1}{2}-j\frac{\sqrt{3}}{2} & -\frac{1}{2}+j\frac{\sqrt{3}}{2} \end{bmatrix}$$

$$\begin{bmatrix} 0.1+j2.0 & 0.05+j0.5 & 0.05+j0.5 \\ 0.05+j0.5 & 0.1+j2.0 & 0.05+j0.5 \\ 0.05+j0.5 & 0.05+j0.5 & 0.1+j2.0 \end{bmatrix}$$

$$\begin{bmatrix} 1 & 1 & 1 \\ 1 & -\frac{1}{2}+j\frac{\sqrt{3}}{2} & -\frac{1}{2}+j\frac{\sqrt{3}}{2} \\ 1 & -\frac{1}{2}+j\frac{\sqrt{3}}{2} & -\frac{1}{2}-j\frac{\sqrt{3}}{2} \end{bmatrix}$$

$$= \begin{bmatrix} 0.2+j3.0 & 0 & 0 \\ 0 & 0.05+j1.5 & 0 \\ 0 & 0 & 0.05+j1.5 \end{bmatrix}$$

An Untransposed Transmission Line

An untransposed transmission line will have unequal mutual impedances between a–b and b–c phases. A typical phase impedance matrix for an untransposed transmission line with a flat conductor configuration is

$$\mathbf{Z}_\phi = \begin{bmatrix} 0.1+j2.0 & 0.07+j0.7 & 0.05+j0.5 \\ 0.07+j0.7 & 0.1+j2.0 & 0.07+j0.7 \\ 0.05+j0.5 & 0.07+j0.7 & 0.1+j2.0 \end{bmatrix}$$

Repeating the calculation of the sequence impedances:

$$\mathbf{Z}_s = \frac{1}{3}\begin{bmatrix} 1 & 1 & 1 \\ 1 & -\frac{1}{2}+j\frac{\sqrt{3}}{2} & -\frac{1}{2}-j\frac{\sqrt{3}}{2} \\ 1 & -\frac{1}{2}-j\frac{\sqrt{3}}{2} & -\frac{1}{2}+j\frac{\sqrt{3}}{2} \end{bmatrix}$$

$$\begin{bmatrix} 0.1+j2.0 & 0.07+j0.7 & 0.05+j0.5 \\ 0.07+j0.7 & 0.1+j2.0 & 0.07+j0.7 \\ 0.05+j0.5 & 0.07+j0.7 & 0.1+j2.0 \end{bmatrix}$$

$$\begin{bmatrix} 1 & 1 & 1 \\ 1 & -\dfrac{1}{2}-j\dfrac{\sqrt{3}}{2} & -\dfrac{1}{2}+j\dfrac{\sqrt{3}}{2} \\ 1 & -\dfrac{1}{2}+j\dfrac{\sqrt{3}}{2} & -\dfrac{1}{2}-j\dfrac{\sqrt{3}}{2} \end{bmatrix}$$

$$= \begin{bmatrix} 0.2267 + j3.2667 & 0.0544 - j0.0391 & -0.0611 - j0.0276 \\ -0.0611 - j0.0276 & 0.0367 + j1.3667 & -0.1088 + j0.0782 \\ 0.0544 - j0.0391 & 0.1221 + j0.0551 & 0.0367 + j1.3667 \end{bmatrix}$$

Note that with untransposed transmission lines, the sequence impedance matrix is no longer diagonal.

Rotating Machine

A rotating machine is characterized by the fact that the mutual impedances between the phases depend upon whether the phase windings in question are contiguous in the sense of the direction of rotation of the machine rotor. Thus, if the rotor moves from the winding centers in the normal order of a–b–c–a, ..., the mutual impedance between phases a–b, b–c, and c–a would be equal to each other, but different from the mutual impedance between phases b–a, c–b, and a–c. A typical synchronous generator phase impedance matrix is

$$\mathbf{Z}_\phi = \begin{bmatrix} 0 + j0.4167 & -0.2309 - j0.1833 & 0.2309 - j0.1833 \\ 0.2309 - j0.1833 & 0 + j0.4167 & -0.2309 - j0.1833 \\ -0.2309 - j0.1833 & 0.2309 - j0.1833 & 0 + j0.4167 \end{bmatrix}$$

The symmetrical component (sequence) impedances of the machine are calculated as before:

$$\mathbf{Z}_s = \dfrac{1}{3} \begin{bmatrix} 1 & 1 & 1 \\ 1 & -\dfrac{1}{2}+j\dfrac{\sqrt{3}}{2} & -\dfrac{1}{2}-j\dfrac{\sqrt{3}}{2} \\ 1 & -\dfrac{1}{2}-j\dfrac{\sqrt{3}}{2} & -\dfrac{1}{2}+j\dfrac{\sqrt{3}}{2} \end{bmatrix}$$

$$\begin{bmatrix} 0 + j0.4167 & -0.2309 - j0.1833 & 0.2309 - j0.1833 \\ 0.2309 - j0.1833 & 0 + j0.4167 & -0.2309 - j0.1833 \\ -0.2309 - j0.1833 & 0.2309 - j0.1833 & 0 + j0.4167 \end{bmatrix}$$

$$\begin{bmatrix} 1 & 1 & 1 \\ 1 & -\frac{1}{2}-j\frac{\sqrt{3}}{2} & -\frac{1}{2}+j\frac{\sqrt{3}}{2} \\ 1 & -\frac{1}{2}+j\frac{\sqrt{3}}{2} & -\frac{1}{2}-j\frac{\sqrt{3}}{2} \end{bmatrix}$$

or

$$\mathbf{Z}_s = \begin{bmatrix} 0+j0.05 & 0 & 0 \\ 0 & 0+j1.0 & 0 \\ & & 0+j0.2 \end{bmatrix}$$

Note once again that the sequence impedance matrix of the machine is diagonal.

1.1.7 Some Observations About Symmetrical Components

- Symmetrical components are one of the many linear transformations on three-phase quantities that are possible. Other well-known transformations are Clarke components, Park transformation, and Kimbark components. Symmetrical components produce uncoupled (diagonal) impedance matrices for all unbalanced symmetrical power apparatus.
- In a perfectly balanced three-phase system, only positive sequence voltages and currents exist.
- In a practical power system, unbalanced voltages and currents (negative and zero sequence) of the order of 2–10% of the positive sequence components may be found.
- Negative sequence currents create excessive heating in rotating machines. Unbalanced voltages at rotating machine terminals may lead to excessive vibration, currents, and possible stalling.
- The presence of large zero sequence currents and voltages is usually an indication of a ground fault on the system.
- The presence of a large negative sequence current may be caused by one phase of a three-phase device being open-circuit. This condition is often associated with an open circuit breaker pole.

1.2 FAULT CALCULATIONS

1.2.1 Introduction

Faults on a power network occur due to equipment failure or other natural catastrophes, such as tornadoes, fires, and floods. Although a fault may

Electric Power Engineering

create an open circuit in the network, the short circuit is the fault which causes the greatest concern. About 80% of all short circuits are single line to ground and are temporary in nature. The other likely short circuits are three-phase, phase-to-phase, or phase-to-phase-to ground. Faults are of great concern to power system engineers, as they stress the equipment directly involved in the short circuit and may cause extensive damage to it, and if left unattended, may cause instabilities in the power system. The design of protective equipment such as relays and circuit breakers is also determined by the conditions accompanying short circuits. Short circuits produce transient components in the current and voltage wave forms, which are of interest to designers of protective equipment. These transients last a relatively short time. Of considerable importance are the steady state voltages and currents which last several cycles after the occurrence of the short circuit, and this is the subject we will consider here.

It is common to make several simplifying assumptions before calculating short circuit currents and voltages: (1) The network carries no load before the fault occurs. (2) The network branches are purely reactive (i.e. they have zero resistance). (3) The transmission system is completely balanced. (4) The generator impedance can be correctly represented by a single reactance, usually the transient reactance in the direct axis.

Bus impedance matrices using generator internal nodes as references are used in short-circuit calculations. Positive and negative sequence matrices are usually assumed to be equal, while the zero sequence matrix is expected to be different. The bus voltages calculated in the positive sequence short-circuit network are modified by adding the pre-fault voltage (which is the same as the voltage of all the generators in a short-circuit study).

1.2.2 Calculation of Three-Phase Short-Circuit Currents and Voltages

Only the positive sequence short-circuit bus impedance matrix is needed. Consider the network shown in Fig. 1.1(a), wherein the assumptions stated above have been made. The short-circuit network representation is shown in Fig. 1.1(b). The positive sequence admittances of the branches are given in Table 1.1. All admittances and voltages are in per-unit. The bus impedance matrix may be obtained by inverting the bus admittance matrix. In large system studies, this would not be a practical technique for finding the bus impedance matrix. Efficient programs using sparse matrix and vector techniques are available for finding impedance matrices of large networks.

The corresponding positive sequence admittance matrix and its inverse, the positive sequence impedance matrix are given next.

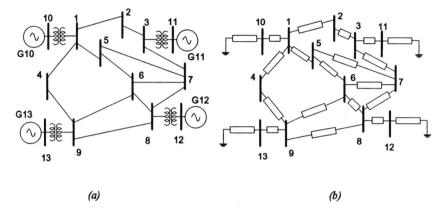

Figure 1.1 (a) System one line diagram. (b) Positive sequence short-circuit diagram. The admittance values in per unit are given in Table 1.1.

Table 1.1 Positive and Zero Sequence Susceptance Data for the Network in Fig. 1.1(b)

Element	B_1	B_0
G10	$-j200$	$-j1$
G11	$-j40$	$-j1$
G12	$-j100$	$-j1$
G13	$-j100$	$-j1$
1-2	$-j20$	$-j8$
1-4	$-j25$	$-j8$
1-5	$-j30$	$-j10$
1-10	$-j40$	$-j15$
2-3	$-j50$	$-j20$
3-7	$-j20$	$-j10$
3-11	$-j10$	$-j4$
4-9	$-j30$	$-j10$
5-6	$-j30$	$-j10$
5-7	$-j45$	$-j15$
6-7	$-j50$	$-j20$
6-8	$-j20$	$-j10$
6-9	$-j30$	$-j10$
7-8	$-j20$	$-j8$
8-9	$-j40$	$-j15$
8-12	$-j30$	$-j15$
9-13	$-j15$	$-j5$

$$\mathbf{Y}_{\mathrm{BUS}(1)} = j \times \begin{bmatrix}
-115 & 20 & 0 & 25 & 30 & 0 & 0 & 0 & 0 & 40 & 0 & 0 & 0 \\
20 & -70 & 50 & 0 & 0 & 0 & 0 & 0 & 0 & 0 & 0 & 0 & 0 \\
0 & 50 & -80 & 0 & 0 & 0 & 20 & 0 & 0 & 0 & 10 & 0 & 0 \\
25 & 0 & 0 & -55 & 0 & 0 & 0 & 0 & 30 & 0 & 0 & 0 & 0 \\
30 & 0 & 0 & 0 & -105 & 30 & 45 & 0 & 0 & 0 & 0 & 0 & 0 \\
0 & 0 & 0 & 0 & 30 & -130 & 50 & 20 & 30 & 0 & 0 & 0 & 0 \\
0 & 0 & 20 & 0 & 45 & 50 & -135 & 20 & 0 & 0 & 0 & 0 & 0 \\
0 & 0 & 0 & 0 & 0 & 20 & 20 & -110 & 40 & 0 & 0 & 30 & 0 \\
0 & 0 & 0 & 30 & 0 & 30 & 0 & 40 & -115 & 0 & 0 & 0 & 15 \\
40 & 0 & 0 & 0 & 0 & 0 & 0 & 0 & 0 & -240 & 0 & 0 & 0 \\
0 & 0 & 10 & 0 & 0 & 0 & 0 & 0 & 0 & 0 & -50 & 0 & 0 \\
0 & 0 & 0 & 0 & 0 & 0 & 0 & 30 & 0 & 0 & 0 & -130 & 0 \\
0 & 0 & 0 & 0 & 0 & 0 & 0 & 0 & 15 & 0 & 0 & 0 & -115
\end{bmatrix}$$

$$\mathbf{Z}_{\mathrm{BUS}(1)} = j \times \begin{bmatrix}
0.0184 & 0.0134 & 0.0114 & 0.0133 & 0.0130 & 0.0105 & 0.0111 \\
0.0134 & 0.0384 & 0.0283 & 0.0112 & 0.0136 & 0.0123 & 0.0146 \\
0.0114 & 0.0283 & 0.0351 & 0.0104 & 0.0138 & 0.0129 & 0.0160 \\
0.0133 & 0.0112 & 0.0104 & 0.0331 & 0.0129 & 0.0133 & 0.0124 \\
0.0130 & 0.0136 & 0.0138 & 0.0129 & 0.0271 & 0.0188 & 0.0199 \\
0.0105 & 0.0123 & 0.0129 & 0.0133 & 0.0188 & 0.0255 & 0.0198 \\
0.0111 & 0.0146 & 0.0160 & 0.0124 & 0.0199 & 0.0198 & 0.0259 \\
0.0077 & 0.0088 & 0.0093 & 0.0113 & 0.0125 & 0.0148 & 0.0141 \\
0.0090 & 0.0094 & 0.0095 & 0.0163 & 0.0128 & 0.0156 & 0.0136 \\
0.0031 & 0.0022 & 0.0019 & 0.0022 & 0.0022 & 0.0018 & 0.0018 \\
0.0023 & 0.0057 & 0.0070 & 0.0021 & 0.0028 & 0.0026 & 0.0032 \\
0.0018 & 0.0020 & 0.0021 & 0.0026 & 0.0029 & 0.0034 & 0.0033 \\
0.0012 & 0.0012 & 0.0012 & 0.0021 & 0.0017 & 0.0020 & 0.0018
\end{bmatrix}$$

$$\begin{bmatrix} 0.0077 & 0.0090 & 0.0031 & 0.0023 & 0.0018 & 0.0012 \\ 0.0088 & 0.0094 & 0.0022 & 0.0057 & 0.0020 & 0.0012 \\ 0.0093 & 0.0095 & 0.0019 & 0.0070 & 0.0021 & 0.0012 \\ 0.0113 & 0.0163 & 0.0022 & 0.0021 & 0.0026 & 0.0021 \\ 0.0125 & 0.0128 & 0.0022 & 0.0028 & 0.0029 & 0.0017 \\ 0.0148 & 0.0156 & 0.0018 & 0.0026 & 0.0034 & 0.0020 \\ 0.0141 & 0.0136 & 0.0018 & 0.0032 & 0.0033 & 0.0018 \\ 0.0209 & 0.0143 & 0.0013 & 0.0019 & 0.0048 & 0.0019 \\ 0.0143 & 0.0224 & 0.0015 & 0.0019 & 0.0033 & 0.0029 \\ 0.0013 & 0.0015 & 0.0047 & 0.0004 & 0.0003 & 0.0002 \\ 0.0019 & 0.0019 & 0.0004 & 0.0214 & 0.0004 & 0.0002 \\ 0.0048 & 0.0033 & 0.0003 & 0.0004 & 0.0088 & 0.0004 \\ 0.0019 & 0.0029 & 0.0002 & 0.0002 & 0.0004 & 0.0091 \end{bmatrix}$$

Three-phase fault current at any of the buses can now be calculated by dividing the pre-fault voltage (1.0 pu) by the diagonal entry of the bus impedance matrix at the bus. For example, three-phase fault current due to a fault at bus 4 is

$$I_{3-\varphi(4)} = 1/(j0.0331) = -j30.1906 \text{ pu}.$$

This current flows *into* the fault, i.e. away from the bus. To calculate the voltages at all the network buses, we multiply the 4th column (corresponding to the faulted bus) of the bus impedance matrix $\mathbf{Z}_{BUS,4}$, by the injected current into bus 4 (I_4), which therefore is the fault current with its sign changed. By the principle of superposition, the voltages calculated in this fashion will be the changes in the bus voltages from their pre-fault values. The actual bus voltages will be obtained by adding 1.0 to each bus voltage change calculated in this fashion.

$$\mathbf{E}_{BUS} = [\mathbf{1}] + \mathbf{Z}_{BUS,4}\mathbf{I}_4$$

where **1** represents a column of 1s.

Thus,

$$\mathbf{E}_{BUS} = \begin{bmatrix} 1 \\ 1 \\ 1 \\ 1 \\ 1 \\ 1 \\ 1 \\ 1 \\ 1 \\ 1 \\ 1 \\ 1 \\ 1 \end{bmatrix} + (+j30.1906) \times j \begin{bmatrix} 0.0133 \\ 0.0112 \\ 0.0104 \\ 0.0331 \\ 0.0129 \\ 0.0133 \\ 0.0124 \\ 0.0113 \\ 0.0163 \\ 0.0022 \\ 0.0021 \\ 0.0026 \\ 0.0021 \end{bmatrix} = \begin{bmatrix} 0.5988 \\ 0.6617 \\ 0.6869 \\ 0.0000 \\ 0.6100 \\ 0.5994 \\ 0.6246 \\ 0.6583 \\ 0.5074 \\ 0.9331 \\ 0.9374 \\ 0.9211 \\ 0.9357 \end{bmatrix}$$

Note that the voltage at bus 4 is 0, as it should be because there is a fault at that bus. Also, voltages at buses 10, 11, 12, and 13 are close to 1.0, as they are at the terminals of the generators. The contribution to the fault from any of the lines can be found by taking the voltage difference across that line, and multiplying the voltage difference by the line admittance. Thus, the current in line 1–4 is

$$I_{1(1-4)} = [E_{BUS}(1) - E_{BUS}(4)] \times y_{14}$$
$$= [0.5988 - 0.0000] \times [0 - j25] = -j14.97 \text{ pu.}$$

Current contributions from all other lines can be calculated in this fashion. The ratio of the current contribution from a line to the total fault current is known as a *distribution factor*. The distribution factor for line 1–4 for a fault at bus 4 is thus

$$D_{1-4,4} = (-j14.97)/(-j30.1906) = 0.4958.$$

Distribution factors for a linear network are independent of actual current flows, depending only upon the network structure and line impedances. Distribution factors are extensively used in contingency analysis of power systems.

1.2.3 Calculation of Ground Fault Currents and Voltages

The zero sequence network is usually different from the positive sequence network because of the wye–delta transformers, and because the zero sequence admittances of network elements are generally different from those of the positive sequence network. Consider the zero sequence diagram of the network shown in Fig. 1.1. All transformers are assumed to be delta connected on the generator side, and wye connected on the system side, with their neutrals solidly grounded (grounding impedance 0 ohms). The generator neutrals are assumed to be ungrounded. (Although we assume the generator neutrals to be ungrounded for this example, in general modern generators are grounded through high resistances in order to eliminate the likelihood of arcing grounds involving the stray capacitance associated with the network on the generator side of the transformer. When a neutral is grounded through an impedance $\mathbf{Z}_g$, it appears as an impedance of $3\mathbf{Z}_g$ between the neutral and the zero sequence reference bus in the zero sequence diagram.)

Figure 1.2 is the zero sequence impedance diagram of the network of Fig. 1.1, where the delta side of the transformers and the generators is isolated from the rest of the network. The zero sequence admittance and impedance matrices are calculated using the zero sequence admittance data from Table 1.1.

$$\mathbf{Y}_{BUS(0)} = j \times \begin{bmatrix} -66 & 8 & 0 & 8 & 10 & 0 & 0 & 0 & 0 \\ 8 & -28 & 20 & 0 & 0 & 0 & 0 & 0 & 0 \\ 0 & 20 & -40 & 0 & 0 & 0 & 10 & 0 & 0 \\ 8 & 0 & 0 & -18 & 0 & 0 & 0 & 0 & 10 \\ 10 & 0 & 0 & 0 & -35 & 10 & 15 & 0 & 0 \\ 0 & 0 & 0 & 0 & 10 & -50 & 20 & 10 & 10 \\ 0 & 0 & 10 & 0 & 15 & 20 & -53 & 8 & 0 \\ 0 & 0 & 0 & 0 & 0 & 10 & 8 & -63 & 15 \\ 0 & 0 & 0 & 10 & 0 & 10 & 0 & 15 & -50 \end{bmatrix}$$

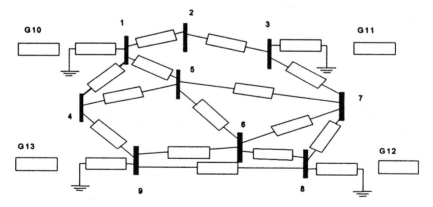

Figure 1.2 System impedance diagram for the zero sequence network of Fig. 1.1. The generator impedances are shown isolated, as the generator neutrals are ungrounded.

$$Z_{BUS(0)} = j \times \begin{bmatrix} 0.0194 & 0.0106 & 0.0071 & 0.0110 & 0.0104 & 0.0064 & 0.0072 & 0.0030 & 0.0044 \\ 0.0106 & 0.0644 & 0.0359 & 0.0075 & 0.0124 & 0.0104 & 0.0149 & 0.0047 & 0.0050 \\ 0.0071 & 0.0359 & 0.0475 & 0.0061 & 0.0132 & 0.0120 & 0.0180 & 0.0054 & 0.0052 \\ 0.0110 & 0.0075 & 0.0061 & 0.0708 & 0.0103 & 0.0109 & 0.0093 & 0.0073 & 0.0185 \\ 0.0104 & 0.0124 & 0.0132 & 0.0103 & 0.0508 & 0.0253 & 0.0279 & 0.0100 & 0.0101 \\ 0.0064 & 0.0104 & 0.0120 & 0.0109 & 0.0253 & 0.0415 & 0.0271 & 0.0135 & 0.0145 \\ 0.0072 & 0.0149 & 0.0180 & 0.0093 & 0.0279 & 0.0271 & 0.0423 & 0.0123 & 0.0110 \\ 0.0030 & 0.0047 & 0.0054 & 0.0073 & 0.0100 & 0.0135 & 0.0123 & 0.0222 & 0.0108 \\ 0.0044 & 0.0050 & 0.0052 & 0.0185 & 0.0101 & 0.0145 & 0.0110 & 0.0108 & 0.0299 \end{bmatrix}$$

The line-to-ground (phase a) fault is calculated by first calculating the positive sequence fault current, and then multiplying it by 3 to find the phase a fault current.

$$I_{a,F} = 3I_{1,F} = \frac{3}{2Z_{FF(1)} + Z_{FF(0)}}$$

The Zs in the formula are the diagonal entries of the bus impedance matrices (positive and zero sequence respectively) corresponding to the fault bus. We will now illustrate this procedure for a phase-a-to-ground fault at bus 4. The positive, negative, and zero sequence currents *into* the fault are given by

$$I_{a,4} = 3I_{1,4} = 3/(2Z_{44(1)} + Z_{44(0)})$$
$$= 3/j(2 \times 0.0331 + 0.0708) = 3 \times (-j7.2993) = -j21.8978$$

The currents calculated above flow *into* the fault, or out of the bus. Thus the *injected currents into* the positive, negative, and zero sequence networks are obtained by reversing the sign of the currents calculated above:

$$I_{injected,(4)} = j7.2993 = I_{4(1)} = I_{4(2)} = I_{4(0)}$$

The resulting bus voltages in the positive, negative, and zero sequence networks are calculated as in the case of the three-phase fault. Of course, the negative and zero sequence networks do not have any pre-fault voltages, hence they do not have the pre-fault value of 1.0 added to them:

$$\mathbf{E}_{BUS(1)} = [1] + \mathbf{Z}_{BUS,4(1)}\mathbf{I}_{4(1)}$$
$$\mathbf{E}_{BUS(2)} = \mathbf{Z}_{BUS,4(1)}\mathbf{I}_{4(1)}$$
$$\mathbf{E}_{BUS(0)} = \mathbf{Z}_{BUS,4(0)}\mathbf{I}_{4(0)}$$

Using the values calculated above,

$$\mathbf{E}_{\text{BUS}(1)} = \begin{bmatrix} 0.9030 \\ 0.9182 \\ 0.9243 \\ 0.7583 \\ 0.9057 \\ 0.9031 \\ 0.9092 \\ 0.9174 \\ 0.8809 \\ 0.9838 \\ 0.9849 \\ 0.9809 \\ 0.9845 \end{bmatrix}; \mathbf{E}_{\text{BUS}(2)} = \begin{bmatrix} -0.0970 \\ -0.0818 \\ -0.0757 \\ -0.2418 \\ -0.0943 \\ -0.0969 \\ -0.0908 \\ -0.0826 \\ -0.1191 \\ -0.0162 \\ -0.0151 \\ -0.0191 \\ -0.0155 \end{bmatrix}; \mathbf{E}_{\text{BUS}(0)} = \begin{bmatrix} -0.0806 \\ -0.0546 \\ -0.0442 \\ -0.5165 \\ -0.0748 \\ -0.0798 \\ -0.0677 \\ -0.0535 \\ -0.1353 \end{bmatrix}$$

Note that there are no zero sequence voltages at buses 10, 11, 12, and 13, as these buses are not connected to the rest of the network. The phase a voltage at any of the buses can be obtained by adding the corresponding symmetrical components. For example, the phase voltage at bus 4 is given by adding the 4th row of each of the voltage vectors above:

$E_{a(4)} = 0.7583 - 0.2418 - 0.5165 = 0$

This is of course correct, as there is a phase-a-to-ground fault on bus 4. Phase b and c voltages can be calculated by applying appropriate symmetrical component transformations. Voltages at any of the other buses are calculated in the same manner.

The contribution to the fault current can be calculated by first calculating the sequence currents in the corresponding line. For example, the current in line 1–4 is calculated as follows:

$I_{1,(1-4)} = (E_{1,1} - E_{1,4})y_{1,1-4} = (0.9030 - 0.7583) \times (-j25) = -j3.6200$
$I_{2,(1-4)} = (E_{2,1} - E_{2,4})y_{1,1-4} = (-0.0970 + 0.2418) \times (-j25) = -j3.6200$
$I_{0,(1-4)} = (E_{0,1} - E_{0,4})y_{0,1-4} = (-0.0806 - 0.5165) \times (-j8) = -j3.4872$

and

$I_{a,(1-4)} = I_{1,(1-4)} + I_{2,(1-4)} + I_{0,(1-4)} = -j3.62 - j3.62 - j3.4872 = -j10.7272$

Note that the positive and negative sequence currents in the line are equal. This is generally true, as we have assumed that the positive and negative sequence networks are identical. The zero sequence current is different, as that network is substantially different from the other two. A consequence of the zero sequence current being different is that phase b and c currents in this line will not be zero, even though the fault is on phase a. Applying the formula for calculating phase b and c currents:

$$I_{b,(1-4)} = \alpha^2 I_{1,(1-4)} + \alpha I_{2,(1-4)} + I_{0,(1-4)}$$
$$= -j\alpha^2 3.62 - j\alpha 3.62 - j3.4872 = j0.1328$$
$$I_{c,(1-4)} = \alpha I_{1,(1-4)} + \alpha^2 I_{2,(1-4)} + I_{0,(1-4)}$$
$$= -j\alpha 3.62 - j\alpha^2 3.62 - j3.4872 = j0.1328$$

Contributions to the fault from all the lines can be calculated in this fashion. As before, distribution factors could be used to calculate these contributions. The distribution factors for the positive and negative sequence networks are identical, whereas the distribution factors is different for the zero sequence network.

1.2.4 Some Observations About Fault Calculations

Fault calculations are necessary in order to determine the rating of equipment, the settings of protective equipment (relaying), and the strength of equipment to withstand destructive forces created by the heavy fault currents. Since it is customary to use a safety margin in all designs, it is usually not necessary to determine fault currents very precisely. To that end, many simplifying assumptions are commonly made in fault calculations. These are: all generators are assumed to produce normal voltages and their phase angles are all assumed to be equal ($0°$); resistance of network elements is generally neglected; load currents are generally neglected; all network elements are assumed to be balanced and symmetric; and all positive and negative sequence impedances are assumed to be equal.

The three-phase fault is generally most severe, although in some cases, especially near a generator, the ground fault current may be higher. All other fault currents are bracketed by these fault currents. The ground fault is the most common – occurring about 80% of the fault incidences.

The transient components in fault current and voltages are treated separately. The most important transient component in a fault current is the d.c. offset, which may be as high as the peak value of the a.c. fault current.

1.3 LOAD FLOW

1.3.1 Introduction

Load flow is one of the most important programs used in power system engineering. It determines the prevailing voltages and currents on the network, when the connected generators supply the connected loads. The input statement of the load flow problem describes the loads (active and reactive power) and the generator active power, and their terminal voltages. The turbine governor and excitation controllers hold these parameters constant for all the generators according to their set schedules. Since the losses in the network are not known from the input data (since it is not known how the currents will distribute among the available transmission lines), the real power output of one of the generators must be left undetermined. This generator is known as the swing bus (or slack bus), and only its terminal voltage is specified. It is customary to assign the voltage of the swing bus a phase angle of $0°$, since all the phasor angles are indeterminate to an additive constant. The swing bus voltage is the reference phasor for all the voltage and current phasors of the load flow.

It is usual to assume that the entire network is balanced symmetric, so that a positive sequence network is assumed. Furthermore, the resistance of network elements and charging capacitance of transmission lines is important, since they contribute to network losses and reactive power consumption.

Since the terminal voltage of the generators is controlled by the voltage regulators, it is not necessary to model the internal impedance of the generators. If there are n load buses, m generator buses, and 1 swing bus, or a total of $n + m + 1$ buses in a load flow problem, the swing bus voltage is completely known. At the load buses (n), neither the magnitude nor the phase angle of the voltages is known, and they must be calculated. At the generator buses (m), the voltage magnitude is known (since it is specified in the input statement), and only the phase angles need to be determined. Thus in the present load flow problem, there are n voltage magnitudes and $n + m$ phase angles to be determined. Thus, there is a total of $2n + m$ unknown quantities. In the input data, there are P and Q specified for the load buses, and P specified for the generators, or a total of $2n + m$ input quantities. The load flow problem consists in finding the $2n + m$ unknown voltage magnitudes and angles from the $2n + m$ specified P and Q injections in the network. The normal convention is for P and Q flowing into the network to be considered positive. Thus, the load buses will in general have negative P and Q (if the loads are lagging power factor loads), while the generator Ps will be positive.

Electric Power Engineering

The most powerful method for solving load flow problem is known as the Newton–Raphson method. The Newton–Raphson method consists of expressing the P and Q injected into a bus as a non-linear function of all voltage magnitudes and angles. The non-linear equations are solved by an iterative technique based upon determining the linear approximation to the non-linear equations. The coefficient matrix of the linearized equations is known as the Jacobian of the system.

As in all non-linear iterative processes, an initial guess is made about the answer. In load flow work, it is common to assume that the initial guess for all voltages is the voltage of the swing bus. This is known as the flat start. From the initial guess, the Jacobian is estimated, and the mismatch between the given P, Q and the actual P, Q produced by the guessed solution is used to drive the solution closer to the true value. The iterations continue until the mismatch becomes close to zero. In most load flow problems, no more than 5–10 iterations are needed to obtain the load flow solution to a very high degree of accuracy.

A very useful variant of the Newton–Raphson method is known as the "Fast Decoupled Load Flow". It consists in calculating an *approximate* Jacobian matrix, which is easy to calculate and remains constant from one iteration to the next. The approximate Jacobian matrix is made up of the negative imaginary part of the bus admittance matrix. This will be illustrated in the numerical example to follow.

1.3.2 Equations

The injected currents and voltages of a network with admittance matrix **Y** are related by

$$\mathbf{I} = \mathbf{YE}$$

where all matrices and vectors are complex. The complex power at any bus p is given by (where the '*' stands for complex conjugate)

$$\begin{aligned}
P_p - jQ_p &= E_p^* I_p = E_p^* \sum Y_{pq} E_q \\
&= |E_p|(\cos \delta_p - j \sin \delta_p) \sum (G_{pq} + jB_{pq})|E_q|(\cos \delta_q + j \sin \delta_q) \\
&= |E_p| \sum (G_{pq} + jB_{pq})|E_q|[\cos (\delta_q - \delta_p) + j \sin (\delta_q - \delta_p)]
\end{aligned}$$

The summation is carried over all $n + m + 1$ buses of the network. Separating the above equations in real and imaginary parts:

$$\begin{aligned}
P_p &= |E_p||E_q| \sum [G_{pq} \cos (\delta_q - \delta_p) + B_{pq} j \sin (\delta_q - \delta_p)] \\
Q_p &= |E_p||E_q| \sum [B_{pq} \cos (\delta_q - \delta_p) + G_{pq} j \sin (\delta_q - \delta_p)]
\end{aligned} \quad (1.5)$$

There are $n+m$ equations for P and n equations for Q. The unknowns are $|E|$ and δ for the n and $n+m$ buses respectively. Equations (1.5) can be expressed in matrix form as

$$\begin{bmatrix} \mathbf{P} \\ \mathbf{Q} \end{bmatrix} = \begin{bmatrix} \mathbf{P}(|\mathbf{E}|, \delta) \\ \mathbf{Q}(|\mathbf{E}|, \delta) \end{bmatrix}$$

where the formal dependence of P and Q on $|E|$ and δ is expressed by parentheses, and represents the equations (1.5). The equation is linearized around a starting point identified by the subscript '0':

$$\begin{bmatrix} \mathbf{P} \\ \mathbf{Q} \end{bmatrix} = \begin{bmatrix} \mathbf{P}_0 \\ \mathbf{Q}_0 \end{bmatrix} + \begin{bmatrix} \Delta \mathbf{P} \\ \Delta \mathbf{Q} \end{bmatrix} = \begin{bmatrix} \mathbf{P}_0 \\ \mathbf{Q}_0 \end{bmatrix} + \begin{bmatrix} \dfrac{\partial \mathbf{P}}{\partial \delta} & \dfrac{\partial \mathbf{P}}{\partial |\mathbf{E}|} \\ \dfrac{\partial \mathbf{Q}}{\partial \delta} & \dfrac{\partial \mathbf{Q}}{\partial |\mathbf{E}|} \end{bmatrix} \begin{bmatrix} \Delta \delta \\ \Delta |\mathbf{E}| \end{bmatrix}$$

The matrix of partial derivatives is known as the Jacobian **J**. In terms of the Jacobian, the above equation can be written as:

$$\Delta \mathbf{S} = \begin{bmatrix} \Delta \mathbf{P} \\ \Delta \mathbf{Q} \end{bmatrix} = \begin{bmatrix} \mathbf{P} \\ \mathbf{Q} \end{bmatrix} - \begin{bmatrix} \mathbf{P}_0 \\ \mathbf{Q}_0 \end{bmatrix} = \begin{bmatrix} \dfrac{\partial \mathbf{P}}{\partial \delta} & \dfrac{\partial \mathbf{P}}{\partial |\mathbf{E}|} \\ \dfrac{\partial \mathbf{Q}}{\partial \delta} & \dfrac{\partial \mathbf{Q}}{\partial |\mathbf{E}|} \end{bmatrix} \begin{bmatrix} \Delta \delta \\ \Delta |\mathbf{E}| \end{bmatrix} = j \begin{bmatrix} \Delta \delta \\ \Delta |\mathbf{E}| \end{bmatrix} \quad (1.6)$$

$\Delta \mathbf{S}$ is known as the mismatch vector. The Jacobian **J** is evaluated at the assumed values of the unknown vector $|\mathbf{E}_0|, \delta_0$. The powers **P** and **Q** are the input data powers (specified), while $\mathbf{P}_0, \mathbf{Q}_0$ are the powers calculated at $|\mathbf{E}_0|, \delta_0$. The improved estimate of the unknown vector is $|\mathbf{E}_0| + \Delta |\mathbf{E}|$, $\delta_0 + \Delta \delta$, where the increments are calculated by

$$\begin{bmatrix} \Delta \delta \\ \Delta |\mathbf{E}| \end{bmatrix} = \mathbf{J}^{-1} \Delta \mathbf{S}$$

As mentioned before, the iterations are continued until the mismatch $\Delta \mathbf{S}$ becomes very small. The Jacobian is not actually inverted as indicated in the above expression, but a very efficient sparse matrix inversion technique is used to solve the above equations.

The Fast Decoupled Load Flow approximation consists in setting the off-diagonal blocks of the Jacobian matrix equal to zero:

$$\dfrac{\partial \mathbf{P}}{\partial |\mathbf{E}|} = 0, \quad \dfrac{\partial \mathbf{Q}}{\partial \delta} = 0$$

Also, the main diagonal blocks are approximated by negation of the imaginary part of the admittance matrix, taking care to note the dimension of the matrix in question.

Electric Power Engineering

$$\frac{\partial \mathbf{P}}{\partial \delta} = \mathbf{B}'; \mathbf{E}\frac{\partial \mathbf{Q}}{\partial |\mathbf{E}|} = \mathbf{B}''$$

where $\mathbf{B}'$ is the negative of Imag($\mathbf{Y}_{BUS,1}$) and $\mathbf{B}''$ is the negative of Imag($\mathbf{Y}_{BUS,2}$). $\mathbf{Y}_{BUS,1}$ and $\mathbf{Y}_{BUS,2}$ are obtained from $\mathbf{Y}_{BUS}$ by eliminating the swing bus row and column and the swing bus and generator bus rows and columns respectively. An additional modification is to change the corrections to the unknown voltage magnitude from $\Delta \mathbf{E}$ to $\Delta \mathbf{E}/\mathbf{E}$. This last variation is necessary in order to normalize the real and reactive power equations. Thus, for the Fast Decoupled Load Flow method, Eq. (1.6) is modified to

$$\Delta \mathbf{S} = \begin{bmatrix} \mathbf{B}' & 0 \\ 0 & \mathbf{B}'' \end{bmatrix} \begin{bmatrix} \Delta \delta \\ \frac{\Delta |\mathbf{E}|}{|\mathbf{E}|} \end{bmatrix}$$

1.3.3 Numerical Example

We will use the same system for load flow calculation as that used for short-circuit calculation, except that we will now include the resistance and charging capacitance of the network elements. Note that half the charging susceptance is given, and it is assumed to be connected between each terminal of the line and ground. The values are given in Table 1.2.

Table 1.2 Line Admittance and Charging Current Data for Load Flow Calculation

Element	$G_1 + jB_1$	$jB_c/2$
1–2	2–j20	j0.0100
1–4	1.5–j25	j0.0120
1–5	2–j30	j0.0120
1–10	3–j40	j0.0100
2-3	4–j50	j0.0000
3–7	1.5–j20	j0.0015
3–11	0.5–j10	j0.0000
4–9	3–j30	j0.0060
5–6	3–j30	j0.0060
5–7	4–j45	j0.0000
6–7	3–j50	j0.0020
6–8	1–j20	j0.0000
6–9	3–j30	j0.0020
7–8	1.5–j20	j0.0010
8–9	3–j40	j0.0015
8–12	3–j30	j0.0060
9–13	1.0–j15	j0.0000

The bus admittance matrix for this system is given by:

$$\mathbf{Y}_{\text{BUS}} = 10^2 \times \begin{bmatrix}
0.085 - j1.1496 & -0.020 + j0.2 & 0 & -0.015 + j0.25 & -0.020 + j0.3 & 0 \\
-0.020 + j0.20 & 0.060 - j0.6999 & -0.040 + j0.50 & 0 & 0 & 0 \\
0 & -0.040 + j0.5000 & 0.060 - j0.8000 & 0 & 0 & 0 \\
-0.015 + j0.25 & 0 & 0 & 0.045 - j0.5498 & 0 & 0 \\
-0.020 + j0.30 & 0 & 0 & 0 & 0.090 - j1.0498 & -0.03 + j0.30 \\
0 & 0 & 0 & 0 & -0.03 + j0.30 & 0.10 - j1.299 \\
0 & 0 & -0.015 + j0.20 & 0 & -0.040 + j0.45 & -0.03 + j0.50 \\
0 & 0 & 0 & 0 & 0 & -0.010 + j0.200 \\
0 & 0 & 0 & -0.030 + j0.30 & 0 & -0.030 + j0.3 \\
-0.030 + j0.40 & 0 & 0 & 0 & 0 & 0 \\
0 & 0 & -0.005 + j0.10 & 0 & 0 & 0 \\
0 & 0 & 0 & 0 & 0 & 0 \\
0 & 0 & 0 & 0 & 0 & 0
\end{bmatrix}$$

$$\begin{bmatrix}
0 & 0 & 0 & -0.0300 + j0.40 & 0 & 0 & 0 \\
0 & 0 & 0 & 0 & 0 & 0 & 0 \\
-0.015 + j0.2 & 0 & 0 & 0 & -0.0050 + j0.10 & 0 & 0 \\
0 & 0 & -0.030 + j0.3 & 0 & 0 & 0 & 0 \\
-0.040 + j0.45 & 0 & 0 & 0 & 0 & 0 & 0 \\
-0.03 + j0.50 & -0.010 + j0.20 & -0.030 + j0.3 & 0 & 0 & 0 & 0 \\
0.10 - j1.350 & -0.015 + j0.20 & 0 & 0 & 0 & 0 & 0 \\
-0.015 + j0.200 & 0.085 - j1.0999 & -0.030 + j0.4 & 0 & 0 & -0.030 + j0.3 & 0 \\
0 & -0.030 + j0.40 & 0.100 - j1.1499 & 0 & 0 & 0 & -0.01 + j0.15 \\
0 & 0 & 0 & 0.0300 - j0.3999 & 0 & 0 & 0 \\
0 & 0 & 0 & 0 & 0.0050 - j0.1 & 0 & 0 \\
0 & -0.030 + j0.30 & 0 & 0 & 0 & 0.0300 - j0.2999 & 0 \\
0 & 0 & -0.010 + j0.15 & 0 & 0 & 0 & 0.01 - j0.15
\end{bmatrix}$$

To obtain the approximate Jacobian, the imaginary part of the admittance matrix is used. The first block diagonal of the approximate Jacobian is obtained by eliminating the swing bus row and column (13), and changing its sign. The lower block diagonal is obtained by eliminating buses 10, 11, and 12 (the generator buses) as well, and reversing the sign. The approximate Jacobian is given below:

Electric Power Engineering

$$J(\text{approximate}) = \begin{bmatrix}
115.0 & -20.0 & 0 & -25.0 & -30.0 & 0 & 0 & 0 & 0 & 0 & 0 & 0 & 0 & 0 & 0 & 0 & 0 & 0 & 0 & 0 & 0 & 0 \\
-20.0 & 70.0 & -50.0 & 0 & 0 & 0 & 0 & 0 & 0 & 0 & 0 & 0 & 0 & 0 & 0 & 0 & 0 & 0 & 0 & 0 & 0 & 0 \\
0 & -50.0 & 80.0 & 0 & 0 & 0 & 0 & 0 & 0 & 0 & 0 & 0 & 0 & 0 & 0 & 0 & 0 & 0 & 0 & 0 & 0 & 0 \\
-25.0 & 0 & 0 & 54.98 & 0 & 0 & 0 & 0 & 0 & 0 & 0 & 0 & 0 & 0 & 0 & 0 & 0 & 0 & 0 & 0 & 0 & 0 \\
-30.0 & 0 & 0 & 0 & 105.0 & -30 & -45.0 & 0 & 0 & 0 & 0 & 0 & 0 & 0 & 0 & 0 & 0 & 0 & 0 & 0 & 0 & 0 \\
0 & 0 & 0 & 0 & -30 & 130.0 & -50 & -20.0 & -30.0 & 0 & 0 & 0 & 0 & 0 & 0 & 0 & 0 & 0 & 0 & 0 & 0 & 0 \\
0 & 0 & 0 & 0 & -45.0 & -50 & 135.0 & -20.0 & 0 & -20.0 & 0 & 0 & 0 & 0 & 0 & 0 & 0 & 0 & 0 & 0 & 0 & 0 \\
0 & 0 & 0 & 0 & 0 & -20.0 & -20.0 & 110.0 & -40.0 & 0 & -30.0 & 0 & 0 & 0 & 0 & 0 & 0 & 0 & 0 & 0 & 0 & 0 \\
0 & 0 & 0 & -30.0 & 0 & -30.0 & 0 & -40.0 & 115.0 & 0 & 0 & 0 & 0 & 0 & 0 & 0 & 0 & 0 & 0 & 0 & 0 & 0 \\
0 & 0 & 0 & 0 & 0 & 0 & 0 & 0 & 0 & -40.0 & 0 & -10.0 & 0 & 0 & 0 & 0 & 0 & 0 & 0 & 0 & 0 & 0 \\
-40.0 & 0 & 0 & 0 & 0 & 0 & 0 & 0 & 0 & 0 & 40.0 & 0 & -30.0 & 0 & 0 & 0 & 0 & 0 & 0 & 0 & 0 & 0 \\
0 & 0 & 0 & 0 & 0 & 0 & 0 & 0 & 0 & -10.0 & 0 & 10.0 & 0 & 0 & 0 & 0 & 0 & 0 & 0 & 0 & 0 & 0 \\
0 & 0 & 0 & 0 & 0 & 0 & 0 & 0 & -30.0 & 0 & 0 & 0 & 30.0 & 0 & 0 & 0 & 0 & 0 & 0 & 0 & 0 & 0 \\
0 & 0 & 0 & 0 & 0 & 0 & 0 & 0 & 0 & 0 & 0 & 0 & 0 & 115.0 & -20.0 & 0 & -25.0 & -30.0 & 0 & 0 & 0 & 0 \\
0 & 0 & 0 & 0 & 0 & 0 & 0 & 0 & 0 & 0 & 0 & 0 & 0 & -20.0 & 70.0 & -50.0 & 0 & 0 & 0 & 0 & 0 & 0 \\
0 & 0 & 0 & 0 & 0 & 0 & 0 & 0 & 0 & 0 & 0 & 0 & 0 & 0 & -50.0 & 80.0 & 0 & 0 & 0 & 0 & 0 & 0 \\
0 & 0 & 0 & 0 & 0 & 0 & 0 & 0 & 0 & 0 & 0 & 0 & 0 & -25.0 & 0 & 0 & 54.98 & 0 & 0 & 0 & 0 & 0 \\
0 & 0 & 0 & 0 & 0 & 0 & 0 & 0 & 0 & 0 & 0 & 0 & 0 & -30.0 & 0 & 0 & 0 & 105.0 & -30 & -45.0 & 0 & 0 \\
0 & 0 & 0 & 0 & 0 & 0 & 0 & 0 & 0 & 0 & 0 & 0 & 0 & 0 & 0 & 0 & 0 & -30 & 130.0 & -50 & -20.0 & -30.0 \\
0 & 0 & 0 & 0 & 0 & 0 & 0 & 0 & 0 & 0 & 0 & 0 & 0 & 0 & -20.0 & 0 & 0 & -45.0 & -50 & 135.0 & -20.0 & 0 \\
0 & 0 & 0 & 0 & 0 & 0 & 0 & 0 & 0 & 0 & 0 & 0 & 0 & 0 & 0 & 0 & 0 & 0 & -20.0 & -20.0 & 110.0 & -40.0 \\
0 & 0 & 0 & 0 & 0 & 0 & 0 & 0 & 0 & 0 & 0 & 0 & 0 & 0 & 0 & 0 & 0 & -30.0 & 0 & -30.0 & -40.0 & 115.0
\end{bmatrix}$$

The bus data is given in the following table. Note that at the load buses, the active and reactive powers are given, while at the generator buses the active power and voltage magnitudes are given. At the swing bus, only the voltage magnitude and the voltage phase angle (0°) are specified. Also note that the convention for power is positive flowing into the bus. Thus, loads have negative **P** and **Q** as appropriate.

Let the starting values assumed for bus voltages magnitudes and angles be as shown below:

Bus No.	|E|	δ
1	1.0	0.0
2	1.0	0.0
3	1.0	0.0
4	1.0	0.0
5	1.0	0.0
6	1.0	0.0
7	1.0	0.0
8	1.0	0.0
9	1.0	0.0
10	1.05	0.0
11	1.03	0.0
12	1.07	0.0
13	1.000	0.0

The power flow at each bus resulting from these assumed voltages can be calculated by first calculating the injected current:

$$\mathbf{I}_{BUS} = \mathbf{Y}_{BUS}\mathbf{E}_{BUS}$$
$$P_p + jQ_p = \mathbf{E}_p \mathbf{I}_p^*$$

for each of the load and generator buses. The real power at the load and generator buses and the reactive power at the load buses are tabulated in a column, and subtracted from the specified real and reactive power given in Table 1.3.

Electric Power Engineering

Table 1.3 Bus Data Specification for Load Flow

Bus Type	Bus No.	P	Q	|E|	δ
Load buses	1	−1.0206	−0.4081		
	2	−3.8633	−0.9027		
	3	−0.6810	1.2621		
	4	−2.1960	−0.3385		
	5	−2.0398	0.7346		
	6	−0.4853	0.6220		
	7	−1.1507	−1.5003		
	8	−2.4488	−0.5109		
	9	−4.5796	0.7984		
Gen. buses	10	7.4986		1.05	
	11	2.5225		1.03	
	12	7.5279		1.07	
Swing	13			1.000	0.0

Bus Type	Bus No.	S_{sp}	S_0	ΔS
Load	1	−1.0206	−0.1500	−0.8706
	2	−3.8633	0	−3.8633
	3	−0.6810	−0.0150	−0.6660
	4	−2.1960	0	−2.1960
P	5	−2.0398	0	−2.0398
	6	−0.4853	0	−0.4853
	7	−1.1507	0	−1.1507
	8	−2.4488	−0.2100	−2.2388
	9	−4.5796	0	−4.5796
Gen	10	7.4986	0.1575	7.3411
P	11	2.5225	0.0155	2.5071
	12	7.5279	0.2247	7.3032
Load	1	−0.4081	−2.0440	1.6359
	2	−0.9027	−0.0100	−0.8927
	3	1.2621	−0.3015	1.5636
Q	4	−0.3385	−0.0180	−0.3205
	5	0.7346	−0.0180	0.7526
	6	0.6220	−0.0100	0.6320
	7	−1.5003	−0.0045	−1.4958
	8	−0.5109	−2.1085	1.5976
	9	0.7984	−0.0095	0.8079

The mismatch vector $\Delta \mathbf{S}$ is pre-multiplied by the inverse of the (approximate) Jacobian, in order to produce the correction vector $[\Delta\delta, \Delta|\mathbf{E}|/|\mathbf{E}|]'$. The voltage corrections are multiplied by appropriate bus voltages to obtain the changes to be applied to the voltage magnitudes. The corrected values of bus voltage angles and magnitudes at the end of the first iteration are tabulated below:

Bus No.	Bus Type	$[\Delta\delta\Delta|\mathbf{E}|/|\mathbf{E}|]'$	$[\Delta\delta\Delta|\mathbf{E}|/|\mathbf{E}|]'$	$[\delta|\mathbf{E}|]'$
1		0.0303	0.0303	1.7388
2		−0.0739	−0.0739	−4.2354
3		−0.0383	−0.0383	−2.1972
4	Load	−0.0602	−0.0602	−3.4517
5		−0.0404	−0.0404	−2.3176
6		−0.0417	−0.0417	−2.3895
7		−0.0415	−0.0415	−2.3757
8		0.0113	0.0113	0.6465
9		−0.0625	−0.0625	−3.5809
10		2.139	0.2139	12.2571
11	Gen	0.2124	0.2124	12.1674
12		0.2548	0.2548	14.5975
13	Swing			0
1		0.0417	0.0417	1.0417
2		0.0360	0.0360	1.0360
3		0.0515	0.0515	1.0515
4	Load	0.0383	0.0383	1.0383
5		0.0493	0.0493	1.0493
6		0.0487	0.0487	1.0487
7		0.0380	0.0380	1.0380
8		0.0471	0.0471	1.0471
9		0.0461	0.0461	1.0461
10				1.05
11	Gen			1.03
12				1.07
13	Swing			1.00

Note that, in this first iteration, there is no difference between the third and fourth columns of the above table, as the assumed voltage magnitudes of all the buses were 1.0. Furthermore, the swing bus voltage (magnitude and angle) as well as the generator bus voltage magnitudes are not calculated. They are set equal to the originally specified values. The iterations are continued in this fashion until the mismatch vector becomes small enough. The results of seven iterations are tabulated in Table 1.4.

As can be seen, the mismatch vector has essentially vanished by the seventh iteration. The result after 12 iterations can be taken as the final answer. It is

Bus No.	$|E|$	δ
1	1.0000	0.0000
2	0.9800	−6.0000
3	1.0000	−4.0000
4	0.9900	−5.0000
5	1.0000	−4.0000
6	1.0000	−4.0000
7	0.9880	−4.0000
8	1.0000	−1.0000
9	1.0000	−5.0000
10	1.0500	10.0000
11	1.0300	10.0000
12	1.0700	12.0000
13	1.0000	0

The flow in each line can now be calculated. For example, the current in line 1-2 at its two terminals is calculated as follows:

$$I_{1-2} = (\mathbf{E}_1 - \mathbf{E}_2)y_{12} + \mathbf{E}_1(y_o/2)$$
$$= (1.0\angle 0 - 0.98\angle -6°)(2 - j20) + 1.0\angle 0(j0.01) = 2.0995 - j0.2925$$
$$I_{2-1} = (\mathbf{E}_2 - \mathbf{E}_1)y_{12} + \mathbf{E}_2(y_o/2)$$
$$= (0.98\angle -6° - 1.0\angle 0)(2 - j20) + 0.98\angle -6°(j0.01)$$
$$= -2.0985 + j0.3122$$

The corresponding line flows are:

$$\mathbf{S}_{12} = \mathbf{E}_1 \mathbf{I}_{12}^* = 1.0\angle 0(2.0995 + j0.2925) = 2.0995 + j0.2925$$
$$\mathbf{S}_{21} = \mathbf{E}_2 \mathbf{I}_{21}^* = 0.98\angle -6°(2.0985 - j0.3122) = -2.0772 - j0.0894$$

The power flows in all the lines are calculated in the same manner and are given as follows.

Table 1.4 The Mismatch Vector and the Corresponding Bus Voltages for Seven Iterations

Bus No.	Iter. No. = 1		2		3		4		5		6		7	
	ΔS	$[\delta, |E|]'$	ΔS	$[\delta, |E|]'$	ΔS	$[\delta, |E|]'$	ΔS	$[\delta, |E|]'$	ΔS	$[\delta, |E|]'$	ΔS	$[\delta, |E|]'$	ΔS	$[\delta, |E|]'$
1	−0.8706	1.7388	−0.1001	−0.6936	0.4008	0.1929	−0.0673	−0.0532	0.0303	0.0156	−0.0060	−0.0042	0.0023	0.0012
2	−3.8633	−4.2354	0.3821	−6.1774	−0.4028	−5.8663	0.0724	−6.0160	−0.0314	−5.9900	0.0065	−6.0013	−0.0025	−5.9992
3	−0.6660	−2.1972	−0.1170	−4.3805	0.0989	−3.8379	−0.0208	−4.0320	0.0084	−3.9872	−0.0020	−4.0026	0.0007	−3.9990
4	−2.1960	−3.4517	0.1996	−5.2231	−0.2089	−4.8800	0.0386	−5.0187	−0.0157	−4.9908	0.0033	−5.0015	−0.0012	−4.9993
5	−2.0398	−2.3176	0.1179	−4.3441	−0.1908	−3.8549	0.0316	−4.0278	−0.0138	−3.9887	0.0027	−4.0022	−0.0011	−3.9991
6	−0.4853	−2.3895	0.0039	−4.3293	−0.0442	−3.8600	0.0051	−4.0264	−0.0026	−3.9892	0.0004	−4.0021	−0.0002	−3.9992
7	−1.1507	−2.3757	0.2131	−4.3389	−0.1334	−3.8585	0.0326	−4.0270	−0.0105	−3.9890	0.0027	−4.0021	−0.0008	−3.9991
8	−2.2388	0.6465	0.0602	−1.6229	0.2947	−0.8145	−0.0534	−1.0461	0.0208	−0.9856	−0.0045	−1.0036	0.0016	−0.9989
9	−4.5796	−3.5809	0.3148	−5.1817	−0.4098	−4.8927	0.0749	−5.0156	−0.0288	−4.9919	0.0061	−5.0012	−0.0022	−4.9994
10	7.3411	12.257	−0.5695	9.0083	0.2443	10.245	−0.0514	9.9252	0.0181	10.019	−0.0042	9.9942	0.0014	10.001
11	2.5071	12.167	−0.1703	9.0083	0.1247	10.265	−0.0268	9.9178	0.0104	10.022	−0.0023	9.9934	0.0008	10.001
12	7.3032	14.597	−0.7482	10.898	0.3005	12.281	−0.0672	11.920	0.0210	12.021	−0.0051	11.993	0.0016	12.001
1	1.6359	1.0417	−1.1014	0.9934	0.2130	1.0031	−0.0690	0.9995	0.0158	1.0002	−0.0054	1.0000	0.0013	1.0000
2	−0.8927	1.0360	−0.4988	0.9724	0.0113	0.9844	−0.0388	0.9793	0.0010	0.9803	−0.0029	0.9799	0.0002	0.9800
3	1.5636	1.0515	−0.4125	0.9911	0.2038	1.0043	−0.0521	0.9992	0.0181	1.0003	−0.0047	0.9999	0.0014	0.9999
4	−0.3205	1.0383	−0.2374	0.9826	0.0265	0.9938	−0.0223	0.9894	0.0030	0.9903	−0.0017	0.9899	0.0003	0.9900
5	0.7526	1.0493	−0.2645	0.9913	0/1344	1.0039	−0.0404	0.9993	0.0116	1.0003	−0.0034	0.9999	0.0009	1.0000
6	0.6320	1.0487	−0.0629	0.9913	0.0789	1.0038	−0.0186	0.9993	0.0067	1.0003	−0.0016	0.9999	0.0005	1.0000
7	−1.4958	1.0380	−0.0680	0.9803	−0.1313	0.9918	0.0140	0.9874	−0.0097	0.9883	0.0015	0.9879	−0.0007	0.9880
8	1.5976	1.0471	−1.5716	0.9914	0.3693	1.0033	−0.0848	0.9994	0.0211	1.0003	−0.0065	1.0000	0.0016	1.0000
9	0.8079	1.0461	−0.5772	0.9919	0.1571	1.0036	−0.0681	0.9993	0.0164	1.0003	−0.0052	0.9999	0.0013	1.0000

Line	P	Q
1-2	2.0995	+0.2925
2-1	−2.0772	−0.0894
1-4	2.1778	+0.2028
4-1	−2.1663	−0.0357
1-5	2.0976	−0.0784
5-1	−2.0878	+0.2006
1-10	−7.3954	−0.8249
10-1	7.4986	+2.1801
2-3	−1.7861	−0.8133
3-2	1.7925	+0.8930
3-7	0.0180	+0.2385
7-3	−0.0178	−0.2386
3-11	−2.4915	+0.1305
11-3	2.5225	+0.4904
4-9	−0.0297	−0.3029
9-4	0.0300	+0.2940
5-6	0.0000	−0.0060
6-5	0.0000	−0.0060
5-7	0.0480	+0.5400
7-5	−0.0474	−0.5335
6-7	0.0360	+0.5980
7-6	−0.0356	−0.5948
6-8	−1.0453	+0.0797
8-6	1.0481	−0.0249
6-9	0.5240	−0.0498
9-6	−0.5231	+0.0549
7-8	−1.0499	−0.1335
8-7	1.0542	+0.1885
8-9	2.7976	−0.1133
9-8	−2.7830	+0.3052
8-12	−7.3487	−0.5612
12-8	7.5279	+2.3408
9-13	−1.3035	+0.1442
13-9	1.3111	−0.0301

Line flow calculation completes the load flow solution. The difference between the flows at the two ends of the line represents line loss. The real power loss is always positive, whereas the reactive power loss may be negative if the line has significant charging capacitance which generates reactive power. The network condition – such as under- or over-voltage at buses or line overloads – can now be determined. If the conditions on the network are not satisfactory, appropriate changes or controls can be applied.

1.3.4 Some Observations About Load Flow Calculations

- For given input data, load flow has multiple solution possibilities. Thus, it is not always possible to guarantee that the converged solution is the preferred solution. However, in most practical cases, the solution reached with a flat start is also the preferred solution.
- The fast decoupled load flow technique provided here may diverge in some extreme cases. In that case, it is advisable to calculate the correct Jacobian matrix. However, the fast decoupled load flow method is the preferred technique in most situations.
- The results of the load flow are used to determine the condition of the network: namely, whether the load flows in the lines are within safe limits, and whether the bus voltages are within acceptable limits. To control these conditions, generator voltages and output powers are changed, and new load flows are run until desirable conditions are obtained.
- Load flows are also the building blocks of many other programs, such as stability and economic dispatch. Techniques similar to load flow are also applicable in static state estimation.

1.4 PROTECTION WITH OVERCURRENT RELAYS

1.4.1 Introduction

Distribution and subtransmission systems (systems below 138 kV) frequently consist of transmission lines which are connected radially from the sources to the loads. While the simplest protective devices are fuses, most distribution systems are generally protected with *overcurrent* relays. Overcurrent relays can also be used for distribution systems with simple loops, as well as for protecting transmission lines against ground faults. Overcurrent relays are relatively inexpensive, and are easily set for radial systems. The protection system design consists of selecting current transformer ratios, pick-up settings for the overcurrent relays, and their time dial settings, so that the relays on a feeder provide well-coordinated protection. Overcurrent relays are often supplemented with *instantaneous* relays, and in a looped system *directional* overcurrent relays are required.

It often happens that as the power system grows and becomes more complex overcurrent relays may no longer be appropriate, in which case distance relays may be required. These are considered in a later example.

Electric Power Engineering 31

1.4.2 Time Overcurrent Relay Characteristics

A typical time overcurrent relay characteristic is shown in Fig. 1.3. Time dial settings may be adjusted in the discrete settings shown in Fig. 1.3, as well as at intermediate values between these settings. The horizontal axis is in terms of the pick-up setting of the relay, whereas the vertical axis may be in cycles or seconds. The actual curves are provided by relay manufacturer. Relay

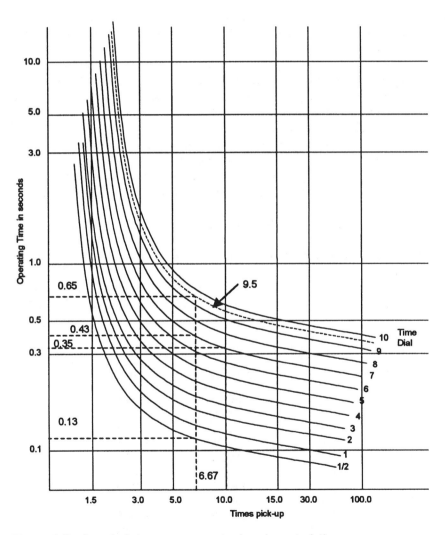

Figure 1.3 A typical time overcurrent relay characteristic.

pick-up current can be selected from a given range of values (taps). We will assume that the tap settings available for this relay are [5, 7.5, 10, 12.5, 15, 20, 25, 30] amperes.

Standard multi-ratio (MR) current transformer ratios available include those shown in Table 1.5.

1.4.3 Principles of Relay Coordination

The following are general rules to be followed in coordinating time overcurrent relays. These principles are further explained in the numerical example below.

The CT ratios for each location are selected so that the steady state secondary current does not exceed 5 amperes (for 5 amp CTs).

The pick-up setting is chosen so that it is greater than twice the maximum load current, and less than one-third of the minimum fault current for which the relay must operate.

The time dial setting for the last relay on a feeder chain is selected at the fastest possible setting – which usually corresponds to $\frac{1}{2}$. Other time dials are chosen so that for the maximum fault current for which the two relays under consideration may operate, the back-up relay must be slower than the primary relay by 0.3 s.

For a given fault, the back-up relay should pick up for a somewhat greater current than the primary delay.

Ground fault protection is usually considered separately from phase fault protection.

For phase faults, the maximum fault current is produced by a three-phase fault, while minimum fault is produced by a phase–phase fault. (This

Table 1.5 Standard Multi-Ratio Current Transformers

600:5 MR	122:5 MR	2000:5 MR	3000:5 MR
50:5	100:5	300:5	300:5
100:5	200:5	400:5	500:5
150:5	300:5	500:5	800:5
200:5	400:5	800:5	1000:5
250:5	500:5	1100:5	1200:5
300:5	600:5	1200:5	1500:5
400:5	800:5	1500:5	2000:5
450:5	900:5	1600:5	2200:5
500:5	1000:5	2000:5	2500:5
600:5	1200:5	–	3000:5

Electric Power Engineering

should be about 87% of the three-phase fault current at the same location.) Furthermore, the maximum fault current is calculated with the strongest source (least source impedance), while the minimum fault current is calculated with the weakest source (greatest source impedance).

1.4.4 Numerical Example

The load and fault current data at various locations on the above system are given in Table 1.6.

Settings for R_b

Since the maximum load current seen by this relay is 300 amps, select a **CT ratio of 300:5**.

The secondary maximum load current is 300 × CT ratio = 300 × 5/300 = 5 amps.

The minimum fault curernt for which this relay must operate is the minimum at bus A: 1200 amps. In terms of secondary winding, this is 1200 × 5/300 = 40 amps.

Therefore we may select a pick-up setting between 2 × 5 and 40/3 amps. Select a **pick-up setting of 10 amps**.

Since this is the last relay in the string to be coordinated, select the fastest **time-dial setting of $\frac{1}{2}$**.

Settings for R_c

Maximum load current is 400 amps. Select a **CT ratio of 400:5**.

The corresponding secondary current is 5 amps.

The minimum fault current for which this relay must operate is at bus A (in the back-up mode), 1200 amps. The corresponding secondary current is 1200 × 5/400 = 15 amps. The ratio between the maximum load current and

Table 1.6 Maximum Load, and Minimum and Maximum Fault Currents for the System in Fig. 1.4

Bus, Relay	Max Load Current (Amps)	Minimum Fault Current (Amps)	Maximum Fault Current (Amps)
D, R_d	600	–	–
C, R_c	400	3200	6000
B, R_b	300	2000	4000
A	–	1200	3000

minimum fault current is only 3, hence an optimum pick-up setting is not possible. A compromise **pick-up setting of 7.5 amps** can therefore be selected.

This relay must coordinate with R_b at the maximum fault both relays will see. This will be the maximum fault current at bus B: 4000 amps. Relay R_b will see this current as $4000 \times 5/300 = 66.7$ amps, which is $66.7/10 =$ **6.67 times pick-up**. The relay R_c sees the same current as $4000 \times 5/(400 \times 7.5) =$ **6.67 times pick-up**. (The fact that these two numbers are identical is an accident). Relay R_b has a time-dial setting of $\frac{1}{2}$, hence from the characteristic corresponding to TD $\frac{1}{2}$ at 6.67 times pick-up, we see that it will operate in 0.13 s for this fault. The back-up relay R_b must operate in $0.13 + 0.3 = 0.43$ s, 0.3 being the coordinating margin. From Figure 1.3, 0.43 s operating time at 6.67 times pick-up will require a **time-dial setting of 7** for R_c.

Settings for R_d

Maximum load current is 600 amps. Select a **CT ratio of 600:5**.

The corresponding secondary current is 5 amps.

The minimum fault current for which this relay must operate is at bus B (in the back-up mode), 2000 amps. The corresponding secondary current is $2000 \times 5/600 = 16.67$ amps. The ratio between the maximum load current and minimum fault current is only 3.33, hence an optimum pick-up setting is not possible. A compromise **pick-up setting of 7.5 amps** can therefore be selected.

This relay must coordinate with R_c at the maximum fault both relays will see. This will be the maximum fault current at bus C: 6000 amps. Relay R_c will see this current as $6000 \times 5/400 = 75$ amps, which is $75/7.5 =$ **10 times pick-up**. The relay R_d sees the same current as $6000 \times 5/(600 \times 7.5) =$ **6.67 times pick-up**. Relay R_c has a time-dial setting of 7; hence from the characteristic corresponding to TD of 7 at 10 times pick-up, we see that it will operate in 0.35 s for this fault. The back-up relay R_d must operate in

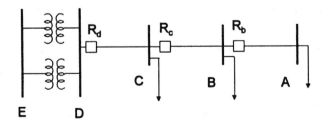

Figure 1.4 System for coordination of time overcurrent relays.

$0.35 + 0.3 = 0.65$ s, 0.3 being the coordinating margin. From Figure 1.3, 0.65 s operating time at 6.67 times pick-up will require a **time-dial setting of 9.5** for R_d.

These settings are summarized as follows:

Relay	CT Ratio	Pick-up (Amps)	Time-dial
R_b	300:5	10	1/2
R_c	400:5	7.5	7
R_d	600:5	7.5	9.5

1.4.5 Coordination for System With One Loop

Consider the power system shown in Fig. 1.5. It consists of a single loop, which now requires that each feeder segment has a circuit breaker at both ends. In addition to the time overcurrent relays, each location will now require a directional element as well. The directional element of each relay will look into the line segment.

The protection system design for a looped system proceeds exactly as before, except that it must be done twice: once for relays with subscript 1, and again for relays with subscript 2 in Fig. 1.3. For relays with subscript 1, the last relay in the feeder chain is R_{a1}, so that its protection settings must be determined first. R_{b1} will be set next to coordinate with it, R_{c1} set to coordinate with R_{b1}, etc. For relays with subscript 2, the last relay in the chain is

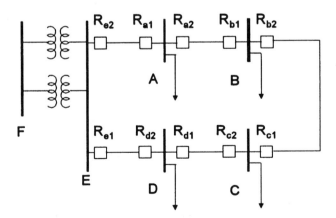

Figure 1.5 A distribution system with a single loop.

R_{d2} and it will be set first. Design of R_{c2}, R_{b2}, R_{a2}, and R_{e2} will follow to complete the protection system design for this looped system.

As mentioned before, for each of these cases, a set of ground relays must also be designed. Maximum and minimum ground fault currents must be calculated at each of the system buses. It is generally assumed that the maximum steady state ground current is 10% of the maximum load current. This may be caused by unbalanced loads between the phases of the feeder system. Consequently the pick-up setting for ground relays can be set very sensitive (i.e., the pick-up setting only needs to be greater than $2 \times 10 = 20\%$ of the maximum load current.)

1.4.6 Some Observations on Overcurrent Relaying

- Overcurrent relays are most effective for protecting radial feeders, which is the common feature of most distribution networks. In the case of looped distribution systems, it becomes necessary to supervise the overcurrent relays with directional elements.
- Overcurrent relays are generally coordinated with the upstream and downstream elements. The coordination is with time delays. It is also good practice to make sure that the back-up zones of two neighboring circuits do not overlap.
- Instantaneous relays are an excellent supplement to overcurrent relays, as they allow fast clearing of faults with heavy currents. It may not always be possible to set instantaneous relays so that they do not overreach the end of the feeder.
- Overcurrent protection against ground faults can be set to be far more sensitive than that for phase faults.

1.5 DISTANCE RELAYING

1.5.1 Introduction

As transmission systems become more complex – deviating from radial or single loop systems as in a distribution network – and become integrated networks, it becomes increasingly difficult, if not impossible, to rely on overcurrent relaying for protection. The minimum fault current and maximum load current, which are necessary to calculate the pick-up setting, become increasingly difficult to determine, as the network configuration may change unpredictably during the normal course of operation, thereby changing both the maximum load current and the minimum fault current with which the protection system must contend. Consequently the protection system of choice for networked systems is distance relaying. Other options require communication systems, and are often employed as a second set of protection systems on important transmission lines in a network. Distance relaying, since it is not dependent upon communication systems, remains the workhorse of high-voltage transmission line protection.

1.5.2 Definitions and Basic Equations

Determining the settings for distance relays is a relatively simple task compared with that required for overcurrent relays. Distance relay characteristics are defined in terms of the R–X diagram in secondary ohms. Secondary ohms are obtained from the actual impedance of transmission lines (primary impedance) by using an impedance conversion factor. If the current transformation ratio is n_i, and the voltage transformation ratio is n_e, the impedance in secondary ohms Z_s is obtained from the impedance in primary ohms Z_p by the formula:

$$Z_s = Z_p \frac{n_i}{n_c} \tag{1.7}$$

The setting of a distance relay is generally in terms of the impedance of the transmission line being protected and that of the lines in its immediate neighborhood. The exact procedure will be explained in the following section. The impedance characteristic of the protective zone of a relay could be a circle, a polygon, or a rectangle. In any case, these shapes are usually centered along an axis, which is known as the 'maximum torque angle' axis. This terminology follows the tradition of electromechanical relays, although there is no torque in electronic and computer relays. When the characteristic is a circle, the setting involves the specification of its diameter only. On the other hand, when the characteristic is a polygon, its vertices must be specified by their (R, X) coordinates. In all cases, the most important feature of the setting is the point where the characteristic intersects the line impedance axis. A popular variation is the 'mho' characteristic, which uses a circular characteristic for the third zone setting and reactance characteristic for the first and second zones. These characteristics are illustrated in Fig. 1.6.

1.5.3 Setting Rules for Distance Relays

Phase Distance Relays

All zone settings are expressed in terms of the positive sequence impedance of the transmission lines. Zone 1 of a distance relay is usually set at between 80% and 90% of the impedance of the line being protected. Zone 2 is set at about 120% of the line impedance, whereas Zone 3 is set equal to the sum of

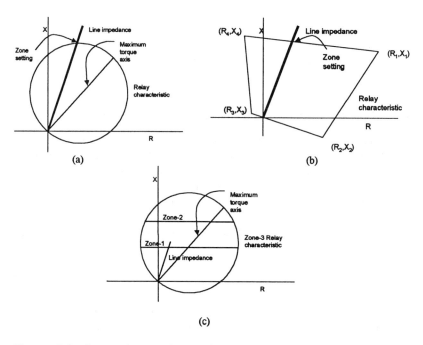

Figure 1.6 Some characteristics of an impedance relay. (a) A circular characteristic, (b) a quadrilateral characteristic, and (c) a mho-supervised reactance characteristic. All the impedances are expressed in secondary ohms. The intersection of the characteristic with the line impedance is the relay zone setting. In (a) and (b), the zones shown are very likely the Zone 1 settings.

the impedance of the line being protected, and 150% of the impedance of the longest line on the next bus for which back-up protection is desired.

The Zone 1 element will operate instantaneously, which means that there are no intentional delays added to the relay operating time. The relay time will be between $\frac{1}{2}$ cycle and 2 cycles in modern distance relays, depending upon the manufacturer. Zone 2 will generally have a time delay of about 0.30 ms (18 cycles on the 60 Hz power system). The Zone 3 timer delay is usually set at about 1.0 s.

Ground Distance Relays

The setting rules for the ground distance relays are the same as for the phase distance relays. Because the ground faults must deal with fault resistance, usually the quadrilateral characteristic is preferred to circular relay characteristic. Often ground fault protection is left to the overcurrent relays because of the uncertainty of the ground fault impedance.

1.5.4 Numerical Example

Consider the system shown in Fig. 1.7. The transmission line to be protected is designated by the terminals A–B, while the neighboring lines emanating from bus B are B–C, B–D, and B–E. The relay design at terminal A of the line A–B is under consideration. Line B–D is the longest of the lines from bus B; hence it will govern the setting of Zone 3 of the distance relay at A. The settings are determined as points along the line of maximum torque of the relay. As mentioned earlier, this is literally a torque axis only for the electromechanical relay. For such relays, the maximum torque angle axis can be adjusted in fixed increments. Typical settings may be 70°, 80°, or 85°. This is the angle between the X-axis and the maximum torque axis. With electronic or digital relay, continuous settings of the maximum torque angle may be available.

The CTs (current transformers) used are 500:5, or a ratio n_i of 100. The VT (voltage transformer) ratio is 138 kV primary and 120 volts secondary (on a line-to-line voltage basis). The VT ratio is $n_e = 138 \times 10^3 / 120 = 1150{:}1$.

Thus the impedance conversion factor is $100/1150 = 0.086\,96$.

The positive sequence impedances of the four lines are given in Table 1.7. The secondary impedance values are obtained by multiplying the primary impedances by 0.086 96.

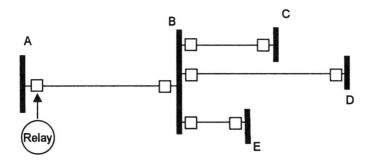

Figure 1.7 System for the numerical example. The current and voltage transformers for the relay are connected to the line at bus A. The current transformers are assumed to be in the phase and ground circuit, whereas the voltage transformers are assumed to be connected phase to neutral to the system bus.

Table 1.7 Positive Sequence Line Impedances in Primary and Secondary Ohms

Line	Primary Impedance	Secondary Impedance
A-B	3 + j40	0.2609 + j3.4784
B-C	2 + j30	0.1739 + j2.6088
B-D	5 + j50	0.4348 + j4.3480
B-E	3 + j40	0.2609 + j3.4784

The impedance relay settings are generally available at fixed intervals. For example, they may be available in the range 1–100 ohms in steps of 0.1 ohm. The range of settings available will vary among different relay manufacturers.

Circular Characteristics

The line impedance angle is arctan $(40/3) = 85.71°$. We will select the nearest available maximum torque angle, say 85°. This is so close to the line impedance angle that we may assume relay settings made along the maximum torque axis to be directly applicable for the line.

1. Zone 1 setting: $0.85 \times |(0.2609 + j3.4784)| = 2.9649$ ohms. We select the nearest value of 3.0 ohms.
2. Zone 2 setting: $1.2 \times |(0.2609 + j3.4784)| = 4.1858$ ohms. We select the nearest value of 4.2 ohms.
3. Zone 3 setting: $|0.2609 + j3.4784 + 1.5 \times (0.4348 + j4.3480)| = 10.042$ ohms. We select the nearest value of 10.0 ohms.

Reactance Characteristics Supervised by Zone 3 Mho Characteristic

The settings calculated above are directly useable for this characteristic.

Quadrilateral Characteristic

We will illustrate the principle of quadrilateral zone setting with the help of a rectangular shape. In many computer relays this may be the preferred shape. The procedure for setting the relay involves computing the coordinates (R, X) for the corners of the zone. As coordinates may have to be transferred to a set of axes rotated with respect to a different set of axes, we will illustrate this procedure in this example.

A reasonable setting for phase distance relays is to use a rectangular characteristic, which has a width of say 40% of the height, and which is centered on the line impedance axis. A rectangle of height Y and width $0.4Y$, centered on the reactance axis, would have for its vertices coordinates as follows:

1. $R_1, X_1 = (0.2Y, Y)$
2. $R_2, X_2 = (0.2Y, 0)$
3. $R_3, X_3 = (-0.2Y, 0)$
4. $R_4, X_4 = (-0.2Y, Y)$

If the coordinate axes are rotated by θ in the positive (counter-clockwise) direction, the coordinates (p, q) will be transformed by the rotation into new coordinates (p', q') as follows:

$$p' = p\cos\theta + q\sin\theta$$
$$q' = -p\sin\theta + q\cos\theta$$

This is illustrated in Fig. 1.8. Applying this formula to the four co-ordinates given above, we get the setting coordinates when the line angle is 85.71° (consequently the axes are rotated by an angle of $90° - 85.71° = 4.29°$ in the clockwise direction):

$$R_1', X_1' = (0.2\cos 4.29° + \sin 4.29°)Y, (-0.2\sin 4.29° + \cos 4.29°)Y$$
$$= 0.2742Y, 0.9822Y$$
$$R_2', X_2' = (0.2\cos 4.29°)Y, (-0.2\sin 4.29°)Y = 0.1994Y, -0.015Y$$
$$R_3', X_3' = (-0.2\cos 4.29°)Y, (0.2\sin 4.29°)Y = -0.1994Y, 0.015Y$$
$$R_4', X_4' = (-0.2\cos 4.29° + \sin 4.29°)Y, (0.2\sin 4.29° + \cos 4.29°)Y$$
$$= -0.1246Y, 1.0122Y$$

Using the zone settings determined above, i.e., $Y = 3.0, 4.2, 10.4$ ohms for the three zones, the coordinates of the corners of the rectangle are found, and then the nearest values selected for the actual zone settings. These values are summarized in Table 1.8.

Other Zone Shapes

The principle of setting zones of any shape can be deduced from the considerations given above. The main exercise is in determining the coordinates of zone boundaries so that the intercept on the line impedance is at the desired level. As mentioned earlier, ground distance relays will have a greater extension in the R direction in order to accommodate fault path resistance. In general, fault arc resistance depends upon line voltage and short-circuit capacity at the fault point. In most overhead transmission lines the fault resistance can be expected to be less than 100 ohms primary.

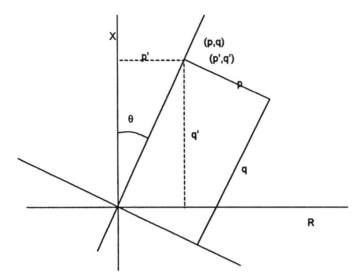

Figure 1.8 Effect of rotation on coordinates (p, q) of a point. The new coordinate system is obtained by rotating the axes in the counter-clockwise direction by θ, thereby changing the coordinates (p, q) to (p', q').

Table 1.8 Coordinates of Zone Vertices for Rectangular Zone Shapes

Zone	Coordinates	Desired Setting	Nearest Available Setting
Zone 1 $Y = 3$	(R_1, X_1)	0.8226, 2.9466	0.83, 3.0
	(R_2, X_2)	0.5982, −0.0450	0.6, 0.0
	(R_3, X_3)	−0.5982, 0.0450	−0.6, 0
	(R_4, X_4)	−0.3738, 3.0366	−0.4, 3.0
Zone 2 $Y = 4.2$	(R_1, X_1)	1.1516, 4.1252	1.1, 4.1
	(R_2, X_2)	0.8375, −0.0630	0.8, −0.1
	(R_3, X_3)	−0.8375, 0.0630	−0.8, 0.1
	(R_4, X_4)	−0.5233, 4.2512	−0.5, 4.2
Zone 3 $Y = 10.0$	(R_1, X_1)	2.7420, 9.8220	2.7, 9.8
	(R_2, X_2)	1.9940, −0.1500	2.0, −0.1
	(R_3, X_3)	−1.9940, 0.1500	−2.0, 0.1
	(R_4, X_4)	−1.2460, 10.122	−1.2, 10.1

Electric Power Engineering

1.5.5 Some Observations on Distance Relays

- Distance relays are the most common form of protection for high-voltage overhead transmission lines. They are relatively immune to levels of load flow and short-circuit currents.
- Ground distance relaying is not as common as phase distance relaying. This is because of the difficulty of accommodating ground fault resistance in the presence of pre-fault load flow. With computer-based relays, this problem may not be as severe as with electronic or electromechanical relays.
- Another difficulty with ground distance relays is the effect of mutual impedance between parallel circuits. The mutual impedance generally creates an error in distance estimation, and may lead to overreaching of the zone under certain conditions.
- Electromechanical relay characteristics are generally limited to circular shapes or straight lines. With electronic and computer relays, other arbitrary shapes are possible. The most common shape for relay zones with the latter type of relays is a quadrilateral.
- Distance relays may also be used as elements of other types of protection system: for example, they may be used in pilot relaying. They may also be used in loss-of-field relays for generators, and for out-of-step relaying on generators and systems.

1.6 TRANSIENT STABILITY

1.6.1 Introduction

During normal system conditions, the power system supplies its connected load, and the resulting conditions on the network are determined with the help of load flow calculations, described in Section 1.3. Of course, the load demand changes during the day through a cyclic variation, and for each load requirement a new steady state is reached by automatic adjustment of generation and other controls. However when faults occur on the network, they disrupt the quasi-steady state prevailing during normal conditions. A major consequence of faults is the mismatch created between the mechanical power input to the generators and the electrical output, which are held in balance during normal system conditions. Because of the mismatch between the input and output powers of the generator, the rotors accelerate or decelerate depending upon the power balance, and set up large oscillations of the rotors around their steady state position. When the faults are particularly severe, the resulting oscillations may be so large that the rotors cannot recover their synchronous speed, thus leading to transient instability in the network. Transient instability is one of the major concerns in power system planning and operation, and the calculation of rotor 'swings' during these disturbances occupies a central role in power system computations.

The generators are fitted with automatic controls – particularly speed governors and excitation controllers – which play a role during transient stability swings. Similarly, the load behavior during a stability swing, when voltages at the load buses, as well as the system frequency, may be different from their normal values, could be quite complex. In programs used for practical system studies, it is customary to consider these effects in stability calculations. However, it has been found that the essence of the stability oscillations could be simulated by considering the generators to be modeled by constant voltages behind their transient reactances, and the loads to be modeled as constant impedances. We will use these approximations in this section. We will also neglect the damping which exists in all generators. The system model which uses the assumptions described above is generally known as a 'classical' model.

1.6.2 Rotor Equations of Motion

The rotor of each generator is assumed to be governed by the following equation of motion:

$$M\frac{d^2\delta}{dt^2} = P_m - P_e; \omega = \frac{d\delta}{dt} \tag{1.8}$$

where ω is the rotor velocity relative to the synchronous speed and δ is the rotor angle measured with respect to its steady state value. M is the inertia constant of the rotor, and P_m and P_e are the mechanical input power and the electrical output power of the generator. The inertia constant of the generator M is usually defined as:

$$M = \frac{2H}{\omega_s} \tag{1.9}$$

where ω_s is the synchronous speed in radians/s of the rotor, and H is the per-unit inertia constant of the rotor system which is usually supplied by the generator manufacturer. The per-unit inertia constants for most common types of generator are given in Table 1.9.

The mechanical power of the generator is equal to the electrical power output before the onset of the fault, and the electrical output power is determined by solving the network equations. Before the disturbance (fault) the electrical and mechanical powers of all the generators are

Table 1.9 Representative Inertia Constants of Generators

Type of Machine	Inertia Constant H in Megajoules/MVA
Turbine generator	3–9
Synchronous motors	1–5
Hydro-generators	2–4
Synchronous condensers	1–1.25

equal, and consequently the right-hand side of the first equation in (1.8) is zero. Thus, the relative speed of the rotors remains equal to the synchronous speed, and the rotor angles remain constant at their pre-disturbance values. When the fault occurs the network equations change, thus changing the electrical power output of the generators, and this sets the rotors in motion. The network equations change once again when the fault is removed by the protection system, and the rotor motions continue to be governed by Eqs. (1.8) with apropriately changing expressions for the electrical output power. The procedure for setting up and solving the transient stability equations is illustrated by the following example.

1.6.3 The Study System

We will use the sample power system used earlier in load flow and short-circuit studies. For convenience, the system diagram is reproduced in Fig. 1.9. Note that the generators are modeled as constant voltages behind their respective transient reactances. The network performance in the unfaulted state is described by the bus impedance matrix $Z_{\text{pre-fault}}$, in the faulted state by Z_{fault}, and in the post fault state by $Z_{\text{post-fault}}$. Note that every load has been converted to constant admittance calculated by the formula:

$$G_{\text{load}} + jB_{\text{load}} = Y_{\text{load}} = -(P + jQ)^*/|E|^2 \tag{1.10}$$

where E is the voltage at the load bus and $(P + jQ)$ is the load in MVA. There is a minus sign in front of $(P + jQ)$ because the bus power is defined as power injected into the bus, whereas the load power flows out of the bus into the load.

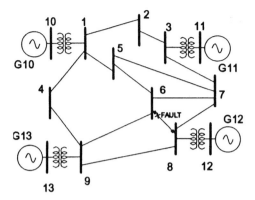

Figure 1.9 System for transient stability study.

Using the load flow solution of Section 1.3, the load admittances are calculated as follows:

Bus No.	|E|	P	Q	G_{load}	B_{load}
1	1.0000	−1.0206	−0.4081	1.0206	−0.4081
2	0.9800	−3.8633	−0.9027	4.0226	−0.9399
3	1.0000	−0.6810	1.2621	0.6810	1.2621
4	0.9900	−2.1960	−0.3385	2.2406	−0.3454
5	1.0000	−2.0398	0.7346	2.0398	0.7346
6	1.0000	−0.4853	0.6220	0.4853	0.6220
7	0.9880	−1.1507	−1.5003	1.1788	−1.5370
8	1.0000	−2.4488	−0.5109	2.4488	−0.5109
9	1.0000	−4.5796	0.7984	4.5796	0.7984

These admittances are added to the diagonal entries of the bus admittance matrix used in the load flow calculation to provide a matrix for the network which includes the load impedances:

$\mathbf{Y}_{BUS}$ (with loads) = $\mathbf{Y}_{BUS}$ (without loads) + diagonal matrix [load admittances]

Recall from Section 1.3 that the $\mathbf{Y}_{BUS}$ (without loads) is

Electric Power Engineering

$$\mathbf{Y}_{\text{BUS}} \text{ (without loads)} = 10^2 \times \begin{bmatrix}
0.085 - j1.1496 & -0.020 + j0.2 & 0 & -0.015 + j0.25 & -0.020 + j0.3 & 0 \\
-0.020 + j0.20 & 0.060 - j0.6999 & -0.040 + j0.50 & 0 & 0 & 0 \\
0 & -0.040 + j0.5000 & 0.060 - j0.8000 & 0 & 0 & 0 \\
-0.015 + j0.25 & 0 & 0 & 0.045 - j0.5498 & 0 & 0 \\
-0.020 + j0.30 & 0 & 0 & 0 & 0.90 - j1.0498 & -0.03 + j0.30 \\
0 & 0 & 0 & 0 & -0.03 + j0.30 & 0.10 - j1.299 \\
0 & 0 & -0.015 + j0.20 & 0 & -0.040 + j0.45 & -0.03 + j0.50 \\
0 & 0 & 0 & 0 & 0 & -0.010 + j0.200 \\
0 & 0 & 0 & -0.030 + j0.30 & 0 & -0.030 + j0.3 \\
-0.030 + j0.40 & 0 & 0 & 0 & 0 & 0 \\
0 & 0 & -0.005 + j0.10 & 0 & 0 & 0 \\
0 & 0 & 0 & 0 & 0 & 0 \\
0 & 0 & 0 & 0 & 0 & 0
\end{bmatrix}$$

$$\begin{bmatrix}
0 & 0 & 0 & -0.0300 + j0.40 & 0 & 0 & 0 \\
0 & 0 & 0 & 0 & 0 & 0 & 0 \\
-0.015 + j0.2 & 0 & 0 & 0 & -0.0050 + j0.10 & 0 & 0 \\
0 & 0 & -0.030 + j0.3 & 0 & 0 & 0 & 0 \\
-0.040 + j0.45 & 0 & 0 & 0 & 0 & 0 & 0 \\
-0.03 + j0.50 & -0.010 + j0.20 & -0.030 + j0.3 & 0 & 0 & 0 & 0 \\
0.10 - j1.350 & -0.015 + j0.20 & 0 & 0 & 0 & 0 & 0 \\
-0.015 + j0.200 & 0.085 - j1.0999 & -0.030 + j0.3 & 0 & 0 & -0.030 + j0.3 & 0 \\
0 & -0.030 + j0.40 & 0.100 - j1.1499 & 0 & 0 & 0 & -0.01 + j0.15 \\
0 & 0 & 0 & 0.0300 - j0.3999 & 0 & 0 & 0 \\
0 & 0 & 0 & 0 & 0.0050 - j0.1 & 0 & 0 \\
0 & -0.030 + j0.30 & 0 & 0 & 0 & 0.0300 - j0.2999 & 0 \\
0 & 0 & -0.010 + j0.15 & 0 & 0 & 0 & 0.01 - j0.15
\end{bmatrix}$$

Adding the load admittances to the corresponding diagonal terms of the above matrix produces the bus admittance matrix including the load admittances:

$$\mathbf{Y}_{\text{BUS}} \text{ (with loads)} = 10^2 \times \begin{bmatrix}
0.0952 - j1.1536 & -0.020 + j0.2 & 0 & -0.015 + j0.25 & -0.020 + j0.3 & 0 \\
-0.020 + j0.20 & 0.1002 - j0.7093 & -0.040 + j0.50 & 0 & 0 & 0 \\
0 & -0.040 + j0.5000 & 0.0668 - j0.7874 & 0 & 0 & 0 \\
-0.015 + j0.25 & 0 & 0 & 0.0674 - j0.5533 & 0 & 0 \\
-0.020 + j0.30 & 0 & 0 & 0 & 0.1104 - j1.0425 & -0.03 + j0.30 \\
0 & 0 & 0 & 0 & -0.03 + j0.30 & 0.10 - j1.299 \\
0 & 0 & -0.015 + j0.20 & 0 & -0.040 + j0.45 & -0.03 + j0.50 \\
0 & 0 & 0 & 0 & 0 & -0.010 + j0.200 \\
0 & 0 & 0 & -0.030 + j0.30 & 0 & -0.030 + j0.3 \\
-0.030 + j0.40 & 0 & 0 & 0 & 0 & 0 \\
0 & 0 & -0.005 + j0.10 & 0 & 0 & 0 \\
0 & 0 & 0 & 0 & 0 & 0 \\
0 & 0 & 0 & 0 & 0 & 0
\end{bmatrix}$$

$$\begin{bmatrix}
0 & 0 & 0 & -0.0300 + j0.40 & 0 & 0 & 0 \\
0 & 0 & 0 & 0 & 0 & 0 & 0 \\
-0.015 + j0.2 & 0 & 0 & 0 & -0.0050 + j0.10 & 0 & 0 \\
0 & 0 & -0.030 + j0.3 & 0 & 0 & 0 & 0 \\
-0.040 + j0.45 & 0 & 0 & 0 & 0 & 0 & 0 \\
-0.03 + j0.50 & -0.010 + j0.20 & -0.030 + j03 & 0 & 0 & 0 & 0 \\
0.1118 - j1.3653 & -0.015 + j0.20 & 0 & 0 & 0 & 0 & 0 \\
-0.015 + j0.200 & 0.1095 - j1.1050 & -0.030 + j0.4 & 0 & 0 & -0.030 + j0.3 & 0 \\
0 & -0.030 + j0.40 & 0.1458 - j1.1419 & 0 & 0 & 0 & -0.01 + j0.15 \\
0 & 0 & 0 & 0.0300 - j0.3999 & 0 & 0 & 0 \\
0 & 0 & 0 & 0 & 0.0050 - j0.1 & 0 & 0 \\
0 & -0.030 + j0.30 & 0 & 0 & 0 & 0.0300 - j0.2999 & 0 \\
0 & 0 & -0.010 + j0.15 & 0 & 0 & 0 & 0.01 - j0.15
\end{bmatrix}$$

Next, the generator internal bus voltages and angles are calculated. The generator terminal currents are obtained by multiplying the admittance matrix by the bus voltage vector:

$$\mathbf{I}_{BUS} = \mathbf{Y}_{BUS}\mathbf{E}_{BUS} \tag{1.11}$$

The last four entries of the current injection vector are the pre-fault generator currents:

$$\mathbf{I}_{GEN0} = \mathbf{I}_{BUS}(10, 11, 12, 13) \tag{1.12}$$

Using the bus voltage vector from the pre-fault load flow result of Section 1.3,

$|E|$	δ (deg)
1.0000	0.0000
0.9800	−6.0000
1.0000	−4.0000
0.9900	−5.0000
1.0000	−4.0000
1.0000	−4.0000
0.9880	−4.0000
1.0000	−1.0000
1.0000	−5.0000
1.0500	10.0000
1.0300	10.0000
1.0700	12.0000
1.0000	0

Performing the matrix multiplication indicated in Eq. (1.11), the generator currents are given by

$$\begin{bmatrix} 7.3935 - j0.8046 \\ 2.4945 - j0.0436 \\ 7.3365 - j0.6771 \\ 1.3111 + j0.0301 \end{bmatrix}$$

Next, the voltage drop in the generator transient reactances due to these currents is added to the generator terminal voltages to obtain the generator internal voltages:

$$\begin{bmatrix} E_{14} \\ E_{15} \\ E_{16} \\ E_{17} \end{bmatrix} = \begin{bmatrix} E_{10} \\ E_{11} \\ E_{12} \\ E_{13} \end{bmatrix} + j \begin{bmatrix} X_{14} & & & \\ & X_{15} & & \\ & & X_{16} & \\ & & & X_{17} \end{bmatrix} \begin{bmatrix} I_{10} \\ I_{11} \\ I_{12} \\ I_{13} \end{bmatrix} \quad (1.13)$$

where the internal buses of the four generators are assumed to be numbered 14, 15, 16, and 17 respectively. Assume the generator transient reactances to be $[j0.005, j0.025, j0.01, j0.01]$. Equation (1.13) then becomes:

$$\begin{bmatrix} E_{\text{GEN0}}(14) \\ E_{\text{GEN0}}(15) \\ E_{\text{GEN0}}(16) \\ E_{\text{GEN0}}(17) \end{bmatrix} = \begin{bmatrix} E_{\text{BUS}}(10) \\ E_{\text{BUS}}(11) \\ E_{\text{BUS}}(12) \\ E_{\text{BUS}}(13) \end{bmatrix} + \begin{bmatrix} j0.005 & & & \\ & j0.025 & & \\ & & j0.01 & \\ & & & j0.01 \end{bmatrix} \begin{bmatrix} 7.3935 - j0.8046 \\ 2.4945 - j0.0436 \\ 7.3365 - j0.6771 \\ 1.3111 + j0.0301 \end{bmatrix}$$

or

$$\begin{bmatrix} E_{\text{GEN0}}(14) \\ E_{\text{GEN0}}(15) \\ E_{\text{GEN0}}(16) \\ E_{\text{GEN0}}(17) \end{bmatrix} = \begin{bmatrix} 1.0381 + j0.2193 \\ 1.0154 + j0.2412 \\ 1.0534 + j0.2958 \\ 0.9997 + j0.0131 \end{bmatrix}$$

During the stability calculation, it will be assumed that the generator internal voltage magnitude will remain constant, while the angles will vary as governed by the differential equations of motion.

1.6.4 Pre-Fault, Fault, and Post-Fault Systems

The admittance matrix of the system given above is now modified to include the generator impedances and the generator buses. This increases the size of the admittance matrix by 4 in our cases. We have assigned bus numbers 14, 15, 16, and 17 to the internal buses of generators connected to buses 10, 11, 12, and 13 respectively. The new admittance matrix is created simply by the rules for constructing the admittance matrix. The negative of the generator admittance is entered at the off-diagonal location corresponding the two buses corresponding to the generator terminal and internal buses, and the admittance is added to the diagonal entries corresponding to the two buses. The resulting admittance matrix is inverted to produce the impedance matrix, and from the bus impedance matrix a 4×4 matrix is extracted. This is the equivalent bus impedance matrix for the entire network as seen from the generator internal buses. The procedure is summarized in the flow-chart of Fig. 1.10.

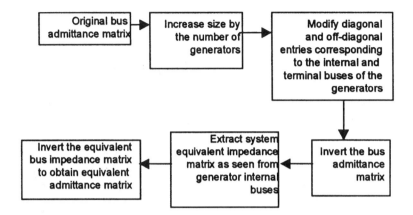

Figure 1.10 Flow-chart for constructing equivalent admittance matrix.

The system has different equivalent admittance matrices depending upon the fault occurrence and the effect of the resulting circuit breaker operation. In most cases, there are three minimum states of the network: pre-fault state, in-fault, and post-fault. Consider a fault on line 6-8 near bus 6 of the system shown in Fig. 1.9. It is assumed that the fault is cleared by opening of the circuit breakers at the two ends of the transmission line. Thus the in-fault system has bus 6 grounded, and the post-fault system has no fault, but the line 6-8 is removed from the network. For each of these three states, there is an equivalent admittance matrix obtained by following the procedure described above and illustrated in Fig. 1.10, except that in each case one starts from a different system admittance matrix. The end result is that there are three (or more, if there are more complex circuit breaker operations) equivalent admittance matrices as seen from the generator internal buses. For the fault and breaker operations described above, the following equivalent admittance matrices are obtained.

Pre-Fault Equivalent Admittance Matrix

The admittance matrix of dimension 17 created by adding the generator admittances as described above is given by:

Electric Power Engineering

$$\mathbf{Y}_{\text{BUS}} \text{ (pre-fault)} = 10^2 \times$$

$$\begin{bmatrix}
0.0952 - j1.153 & -0.020 + j0.2 & 0 & -0.015 + j0.25 & -0.020 + j0.3 & 0 \\
-0.020 + j0.20 & 0.1002 - j0.7093 & -0.040 + j0.50 & 0 & 0 & 0 \\
0 & -0.040 + j0.500 & 0.0668 - j0.787 & 0 & 0 & 0 \\
-0.015 + j0.25 & 0 & 0 & 0.0674 - j0.553 & 0 & 0 \\
-0.020 + j0.30 & 0 & 0 & 0 & 0.1104 - j1.042 & -0.030 + j0.30 \\
0 & 0 & 0 & 0 & -0.03 + j0.30 & 0.10 - j1.299 \\
0 & 0 & -0.015 + j0.20 & 0 & -0.040 + j0.45 & -0.03 + j0.50 \\
0 & 0 & 0 & 0 & 0 & -0.010 + j0.200 \\
0 & 0 & 0 & -0.030 + j0.30 & 0 & -0.030 + j0.3 \\
-0.030 + j0.40 & 0 & 0 & 0 & 0 & 0 \\
0 & 0 & -0.005 + j0.10 & 0 & 0 & 0 \\
0 & 0 & 0 & 0 & 0 & 0 \\
0 & 0 & 0 & 0 & 0 & 0 \\
0 & 0 & 0 & 0 & 0 & 0 \\
0 & 0 & 0 & 0 & 0 & 0 \\
0 & 0 & 0 & 0 & 0 & 0 \\
0 & 0 & 0 & 0 & 0 & 0
\end{bmatrix}$$

$$\begin{bmatrix}
0 & 0 & 0 & -0.0300 + j0.40 & 0 & 0 \\
0 & 0 & 0 & 0 & 0 & 0 \\
-0.015 + j0.2 & 0 & 0 & 0 & -0.0050 + j0.10 & 0 \\
0 & 0 & -0.030 + j0.3 & 0 & 0 & 0 \\
-0.040 + j0.45 & 0 & 0 & 0 & 0 & 0 \\
-0.03 + j0.50 & -0.010 + j0.20 & -0.030 + j0.3 & 0 & 0 & 0 \\
0.1118 - j1.365 & -0.015 + j0.20 & 0 & 0 & 0 & 0 \\
-0.015 + j0.200 & 0.1095 - j1.105 & -0.030 + j0.4 & 0 & 0 & -0.030 + j0.3 \\
0 & -0.030 + j0.40 & 0.1458 - j1.141 & 0 & 0 & 0 \\
0 & 0 & 0 & 0.0300 - j2.399 & 0 & 0 \\
0 & 0 & 0 & 0 & 0.0050 - j0.5 & 0 \\
0 & -0.030 + j0.30 & 0 & 0 & 0 & 0.0300 - j1.299 \\
0 & 0 & -0.010 + j0.15 & 0 & 0 & 0 \\
0 & 0 & 0 & 0 + j2.0 & 0 & 0 \\
0 & 0 & 0 & 0 & 0 + j0.4 & 0 \\
0 & 0 & 0 & 0 & 0 & 0 + j1.0 \\
0 & 0 & 0 & 0 & 0 & 0
\end{bmatrix}$$

$$\begin{bmatrix}
0 & 0 & 0 & 0 & 0 \\
0 & 0 & 0 & 0 & 0 \\
0 & 0 & 0 & 0 & 0 \\
0 & 0 & 0 & 0 & 0 \\
0 & 0 & 0 & 0 & 0 \\
0 & 0 & 0 & 0 & 0 \\
0 & 0 & 0 & 0 & 0 \\
0 & 0 & 0 & 0 & 0 \\
-0.01 + j0.15 & 0 & 0 & 0 & 0 \\
0 & 0 + j2.0 & 0 & 0 & 0 \\
0 & 0 & 0 + j0.4 & 0 & 0 \\
0 & 0 & 0 & 0 + j1.0 & 0 \\
0.01 - j1.15 & 0 & 0 & 0 & 0 + j1.0 \\
0 & 0 - j2.0 & 0 & 0 & 0 \\
0 & 0 & 0 - j0.4 & 0 & 0 \\
0 & 0 & 0 & 0 - j1.0 & 0 \\
0 + j1.0 & 0 & 0 & 0 & 0 - j1.0
\end{bmatrix}$$

The pre-fault bus impedance matrix $\mathbf{Z}_{\text{BUS(pre-fault)}}$ is the inverse of the above matrix, and is given by:

$$\begin{bmatrix}
0.0543 + j0.0149 & 0.0531 + j0.0015 & 0.0539 + j0.0004 & 0.0534 + j0.0027 & 0.0537 + j0.0027 & 0.0532 - j0.0011 \\
0.0531 + j0.0015 & 0.0572 + j0.0209 & 0.0570 + j0.0138 & 0.0516 - j0.0062 & 0.0528 - j0.0032 & 0.0523 - j0.0052 \\
0.0539 + j0.0004 & 0.0570 + j0.0138 & 0.0584 + j0.0240 & 0.0527 - j0.0060 & 0.0540 - j0.0015 & 0.0537 - j0.0030 \\
0.0534 + j0.0027 & 0.0516 - j0.0062 & 0.0527 - j0.0060 & 0.0554 + j0.0178 & 0.0533 - j0.0020 & 0.0533 - j0.0017 \\
0.0537 + j0.0027 & 0.0528 - j0.0032 & 0.0540 - j0.0015 & 0.0533 - j0.0020 & 0.0553 + j0.0124 & 0.0544 + j0.0040 \\
0.0532 - j0.0011 & 0.0523 - j0.0052 & 0.0537 + j0.0030 & 0.0533 - j0.0017 & 0.0544 + j0.0040 & 0.0551 + j0.0112 \\
0.0527 - j0.0004 & 0.0523 - j0.0025 & 0.0537 + j0.0007 & 0.0525 - j0.0028 & 0.0539 + j0.0049 & 0.0540 + j0.0050 \\
0.0524 - j0.0036 & 0.0512 - j0.0079 & 0.0526 - j0.0058 & 0.0530 - j0.0019 & 0.0535 - j0.0004 & 0.0542 + j0.0032 \\
0.0530 - j0.0034 & 0.0514 - j0.0090 & 0.0528 - j0.0075 & 0.0540 + j0.0016 & 0.0538 - j0.0021 & 0.0545 + j0.0016 \\
0.0543 + j0.0149 & 0.0531 + j0.0015 & 0.0539 + j0.0004 & 0.0534 + j0.0027 & 0.0538 + j0.0027 & 0.0532 - j0.0011 \\
0.0539 + j0.0004 & 0.0570 + j0.0138 & 0.0584 + j0.0240 & 0.0527 - j0.0060 & 0.0540 - j0.0015 & 0.0537 - j0.0030 \\
0.0524 - j0.0036 & 0.0512 - j0.0079 & 0.0526 - j0.0058 & 0.0530 - j0.0019 & 0.0536 - j0.0004 & 0.0542 + j0.0032 \\
0.0530 - j0.0034 & 0.0514 - j0.0090 & 0.0528 - j0.0075 & 0.0540 + j0.0016 & 0.0538 - j0.0021 & 0.0545 + j0.0016 \\
0.0543 + j0.0149 & 0.0531 + j0.0015 & 0.0539 + j0.0004 & 0.0534 + j0.0027 & 0.0538 + j0.0027 & 0.0532 - j0.0011 \\
0.0539 + j0.0004 & 0.0570 + j0.0138 & 0.0584 + j0.0240 & 0.0527 - j0.0060 & 0.0540 - j0.0015 & 0.0537 - j0.0030 \\
0.0524 - j0.0036 & 0.0512 - j0.0079 & 0.0526 - j0.0058 & 0.0530 - j0.0019 & 0.0536 - j0.0004 & 0.0542 + j0.0032 \\
0.0530 - j0.0034 & 0.0514 - j0.0090 & 0.0528 - j0.0075 & 0.0540 + j0.0016 & 0.0538 - j0.0021 & 0.0545 + j0.0016
\end{bmatrix}$$

$$\begin{matrix}
0.0527 - j0.0004 & 0.0524 - j0.0036 & 0.0530 - j0.0034 & 0.0543 + j0.0149 & 0.0539 + j0.0004 & 0.0524 - j0.0036 \\
0.0523 - j0.0025 & 0.0512 - j0.0079 & 0.0514 - j0.0090 & 0.0531 + j0.0015 & 0.0570 + j0.0138 & 0.0512 - j0.0079 \\
0.0537 + j0.0007 & 0.0526 - j0.0058 & 0.0528 - j0.0075 & 0.0539 + j0.0004 & 0.0584 + j0.0240 & 0.0526 - j0.0058 \\
0.0525 - j0.0028 & 0.0530 - j0.0019 & 0.0540 + j0.0016 & 0.0534 + j0.0027 & 0.0527 - j0.0060 & 0.0530 - j0.0019 \\
0.0539 + j0.0049 & 0.0535 - j0.0004 & 0.0538 - j0.0021 & 0.0538 + j0.0027 & 0.0540 - j0.0015 & 0.0536 - j0.0004 \\
0.0540 + j0.0050 & 0.0542 + j0.0032 & 0.0545 + j0.0016 & 0.0532 - j0.0011 & 0.0537 - j0.0030 & 0.0542 + j0.0032 \\
0.0538 + j0.0107 & 0.0533 + j0.0018 & 0.0534 - j0.0011 & 0.0527 - j0.0004 & 0.0537 + j0.0007 & 0.0533 + j0.0018 \\
0.0533 + j0.0018 & 0.0550 + j0.0136 & 0.0546 + j0.0033 & 0.0524 - j0.0036 & 0.0526 - j0.0058 & 0.0550 + j0.0136 \\
0.0534 - j0.0011 & 0.0546 + j0.0033 & 0.0560 + j0.0097 & 0.0530 - j0.0034 & 0.0528 - j0.0075 & 0.0546 + j0.0033 \\
0.0527 - j0.0004 & 0.0524 - j0.0036 & 0.0530 - j0.0034 & 0.0562 + j0.0397 & 0.0539 + j0.0004 & 0.0524 - j0.0036 \\
0.0537 + j0.0007 & 0.0526 - j0.0058 & 0.0528 - j0.0075 & 0.0539 + j0.0004 & 0.0634 + j0.1238 & 0.0526 - j0.0058 \\
0.0533 + j0.0018 & 0.0550 + j0.0136 & 0.0546 + j0.0033 & 0.0524 - j0.0036 & 0.0526 - j0.0058 & 0.0583 + j0.0466 \\
0.0534 - j0.0011 & 0.0546 + j0.0033 & 0.0560 + j0.0097 & 0.0530 - j0.0034 & 0.0528 - j0.0075 & 0.0546 + j0.0033 \\
0.0527 - j0.0004 & 0.0524 - j0.0036 & 0.0530 - j0.0034 & 0.0562 + j0.0397 & 0.0539 + j0.0004 & 0.0524 - j0.0036 \\
0.0537 + j0.0007 & 0.0526 - j0.0058 & 0.0528 - j0.0075 & 0.0539 + j0.0004 & 0.0634 + j0.1238 & 0.0526 - j0.0058 \\
0.0533 + j0.0018 & 0.0550 + j0.0136 & 0.0546 + j0.0033 & 0.0524 - j0.0036 & 0.0526 - j0.0058 & 0.0583 + j0.0466 \\
0.0534 - j0.0011 & 0.0546 + j0.0033 & 0.0560 + j0.0097 & 0.0530 - j0.0034 & 0.0528 - j0.0075 & 0.0546 + j0.0033
\end{matrix}$$

$$\begin{matrix}
0.0530 - j0.0034 & 0.0543 + j0.0149 & 0.0539 + j0.0004 & 0.0524 - j0.0036 & 0.0530 - j0.0034 \\
0.0514 - j0.0090 & 0.0531 + j0.0015 & 0.0570 + j0.0138 & 0.0512 - j0.0079 & 0.0514 - j0.009 \\
0.0528 - j0.0075 & 0.0539 + j0.0004 & 0.0584 + j0.0240 & 0.0526 - j0.0058 & 0.0528 - j0.007 \\
0.0540 + j0.0016 & 0.0534 + j0.0027 & 0.0527 - j0.0060 & 0.0530 - j0.0019 & 0.0540 + j0.0016 \\
0.0538 - j0.0021 & 0.0538 + j0.0027 & 0.0540 - j0.0015 & 0.0536 - j0.0004 & 0.0538 - j0.0021 \\
0.0545 + j0.0016 & 0.0532 - j0.0011 & 0.0537 - j0.0030 & 0.0542 + j0.0032 & 0.0545 + j0.0016 \\
0.0534 - j0.0011 & 0.0527 - j0.0004 & 0.0537 + j0.0007 & 0.0533 + j0.0018 & 0.0534 - j0.0011 \\
0.0546 + j0.0033 & 0.0524 - j0.0036 & 0.0526 - j0.0058 & 0.0550 + j0.0136 & 0.0546 + j0.0033 \\
0.0560 + j0.0097 & 0.0530 - j0.0034 & 0.0528 - j0.0075 & 0.0546 + j0.0033 & 0.0560 + j0.0097 \\
0.0530 - j0.0034 & 0.0562 + j0.0397 & 0.0539 + j0.0004 & 0.0524 - j0.0036 & 0.0530 - j0.0034 \\
0.0528 - j0.0075 & 0.0539 + j0.0004 & 0.0634 + j0.1238 & 0.0526 - j0.0058 & 0.0528 - j0.0075 \\
0.0546 + j0.0033 & 0.0524 - j0.0036 & 0.0526 - j0.0058 & 0.0583 + j0.0466 & 0.0546 + j0.0033 \\
0.0604 + j0.0760 & 0.0530 - j0.0034 & 0.0528 - j0.0075 & 0.0546 + j0.0033 & 0.0604 + j0.0760 \\
0.0530 - j0.0034 & 0.0562 + j0.0447 & 0.0539 + j0.0004 & 0.0524 - j0.0036 & 0.0530 - j0.0034 \\
0.0528 - j0.0075 & 0.0539 + j0.0004 & 0.0634 + j0.1488 & 0.0526 - j0.0058 & 0.0528 - j0.0075 \\
0.0546 + j0.0033 & 0.0524 - j0.0036 & 0.0526 - j0.0058 & 0.0583 + j0.0566 & 0.0546 + j0.0033 \\
0.0604 + j0.0760 & 0.0530 - j0.0034 & 0.0528 - j0.0075 & 0.0546 + j0.0033 & 0.0604 + j0.0860
\end{matrix}$$

The equivalent bus impedance matrix is extracted from the complete bus impedance matrix from the entries belonging to the four generator internal buses. That portion of the matrix is highlighted by the thick border, and is reproduced below:

$$\mathbf{Z}_{\text{eqBUS(pre-fault)}} = \begin{bmatrix} 0.0562 + j0.0447 & 0.0539 + j0.0004 & 0.0524 - j0.0036 & 0.0530 - j0.0034 \\ 0.0539 + j0.0004 & 0.0634 + j0.1488 & 0.0526 - j0.0058 & 0.0528 - j0.0075 \\ 0.0524 - j0.0036 & 0.0526 - j0.0058 & 0.0583 + j0.0566 & 0.0546 + j0.0033 \\ 0.0530 - j0.0034 & 0.0528 - j0.0075 & 0.0546 + j0.0033 & 0.0604 + j0.0860 \end{bmatrix}$$

The equivalent bus impedance matrix is inverted to produce the equivalent bus admittance matrix. This equivalent represents the entire network including the loads as seen from the generator internal buses.

$$\mathbf{Y}_{\text{pre}} = \begin{bmatrix} 3.4888 - j13.7442 & 0.6965 + j2.8056 & 1.3505 + j5.3907 & 0.9469 + j3.5914 \\ 0.6965 + j2.8056 & 0.6362 - j5.8241 & 0.4675 + j1.5462 & 0.3126 + j0.8810 \\ 1.3505 + j5.3907 & 0.4675 + j1.5462 & 2.5152 - j12.4058 & 0.7491 + j4.0634 \\ 0.9469 + j3.5914 & 0.3126 + j0.8810 & 0.7491 + j4.0634 & 1.3011 - j9.3975 \end{bmatrix}$$

In-Fault Equivalent Admittance Matrix

When a fault occurs at bus 6, it coincides with ground, and is removed from the network. All lines which formerly connected to bus 6 are now grounded, and the resulting bus admittance matrix has no axis corresponding to bus 6. The in-fault bus admittance matrix is thus 16 × 16:

$\mathbf{Y}_{\text{BUS (in-fault)}} = 10^2 \times$ [columns 1–8]

$$\begin{bmatrix}
0.0952 - j1.153 & -j0.020 + j0.200 & 0 & -j0.015 + j0.250 & -j0.020 + j0.300 & 0 & 0 & 0 \\
-j0.020 + j0.200 & 0.1002 - j0.7093 & -j0.040 + j0.500 & 0 & 0 & 0 & 0 & 0 \\
0 & -j0.040 + j0.500 & 0.0668 - j0.7874 & 0 & 0 & -j0.015 + j0.200 & 0 & 0 \\
-j0.015 + j0.250 & 0 & 0 & 0.0674 - j0.553 & 0 & 0 & 0 & 0 \\
-j0.020 + j0.300 & 0 & 0 & 0 & 0.1404 - j1.342 & -j0.040 + j0.450 & 0 & 0 \\
0 & 0 & -j0.015 + j0.200 & 0 & -j0.040 + j0.450 & 0.1418 - j1.865 & -0.0150 + j0.200 & 0 \\
0 & 0 & 0 & 0 & 0 & -j0.015 + j0.200 & 0.1195 - j1.305 & -j0.030 + j0.400 \\
0 & 0 & 0 & 0 & -j0.030 + j0.300 & 0 & 0 & -j0.030 + j0.400 \\
-j0.030 + j0.400 & 0 & 0 & 0 & 0 & 0 & 0 & 0 \\
0 & 0 & -j0.005 + j0.100 & 0 & 0 & 0 & 0 & -j0.030 + j0.300 \\
0 & 0 & 0 & 0 & 0 & 0 & 0 & 0 \\
0 & 0 & 0 & 0 & 0 & 0 & 0 & 0 \\
0 & 0 & 0 & 0 & 0 & 0 & 0 & 0 \\
0 & 0 & 0 & 0 & 0 & 0 & 0 & 0 \\
0 & 0 & 0 & 0 & 0 & 0 & 0 & 0 \\
0 & 0 & 0 & 0 & 0 & 0 & 0 & 0
\end{bmatrix}$$

[columns 9–16]

$$\begin{bmatrix}
0 & -j0.030 + j0.400 & 0 & 0 & 0 & 0 & 0 & 0 \\
0 & 0 & 0 & 0 & 0 & 0 & 0 & 0 \\
0 & 0 & -j0.005 + j0.100 & 0 & 0 & 0 & 0 & 0 \\
-j0.030 + j0.300 & 0 & 0 & 0 & 0 & 0 & 0 & 0 \\
0 & 0 & 0 & 0 & 0 & 0 & 0 & 0 \\
0 & 0 & 0 & 0 & 0 & 0 & 0 & 0 \\
0 & 0 & 0 & -j0.030 + j0.300 & 0 & 0 & 0 & 0 \\
0.1758 - j1.441 & 0 & 0 & 0 & -j0.010 + j0.150 & 0 & 0 & 0 \\
0 & 0.0300 - j2.399 & 0 & 0 & 0 & 0 + j2.000 & 0 & 0 \\
0 & 0 & 0.0050 - j0.500 & 0 & 0 & 0 & 0 + j0.40 & 0 \\
-j0.010 + j0.150 & 0 & 0 & 0.0300 - j1.299 & 0 & 0 & 0 & 0 + j1.00 \\
0 & 0 + j2.0000 & 0 & 0 & 0.0100 - j1.150 & 0 & 0 & 0 \\
0 & 0 & 0 & 0 & 0 & 0 - j2.0000 & 0 & 0 \\
0 & 0 & 0 + j0.400 & 0 & 0 & 0 & 0 - j0.400 & 0 \\
0 & 0 & 0 & 0 + j1.00 & 0 & 0 & 0 & 0 - j1.000 \\
0 & 0 & 0 & 0 & 0 + j1.00 & 0 & 0 & 0 - j1.000
\end{bmatrix}$$

The process of inverting the admittance matrix, extracting the equivalent bus impedance matrix, and then inverting it to obtain the equivalent bus admittance matrix under faulted conditions is repeated as before. The result is:

$$\mathbf{Y}_{\text{fault}} = \begin{bmatrix} 1.8824 - j18.9138 & 0.1499 + j1.3326 & 0.0410 + j0.5822 & 0.0622 + j0.7034 \\ 0.1499 + j1.3326 & 0.4513 - j6.2672 & 0.0153 + j0.1535 & 0.0160 + j0.0818 \\ 0.0410 + j0.5822 & 0.0153 + j0.1535 & 1.4876 - j18.1657 & 0.0153 + j0.9066 \\ 0.0622 + j0.7034 & 0.0160 + j0.0818 & 0.0153 + j0.9066 & 0.7904 - j11.5149 \end{bmatrix}$$

Post-Fault Equivalent Admittance Matrix

The fault is cleared by removing line 6-8 from the network. The network once again has 17 buses. The resulting bus admittance matrix is

[Large 17×17 bus admittance matrix shown split across two halves of the page]

Once again, the impedance matrix is calculated, equivalent bus impedance matrix is extracted, and its inverse calculated. The result for the post-fault equivalent admittance matrix $\mathbf{Y}_{\text{post}}$ is given below:

$$\mathbf{Y}_{\text{post}} = \begin{bmatrix} 3.5734 - j13.5020 & 0.7270 + j2.8796 & 1.3122 + j5.0025 & 0.9676 + j3.6307 \\ 0.7270 + j2.8796 & 0.6470 - j5.8016 & 0.4488 + j1.4260 & 0.3197 + j0.8928 \\ 0.3122 + j5.0025 & 0.4488 + j1.4260 & 2.4329 - j11.8181 & 0.73253 + j3.9979 \\ 0.9676 + j3.6307 & 0.3197 + j0.8928 & 0.7325 + j3.9979 & 1.3057 - j9.3913 \end{bmatrix}$$

Electric Power Engineering   55

1.6.5 Mechanical Power Input to Generators

As explained earlier, in a stability program using classical models for generators the internal voltage magnitude of each generator is assumed to be constant and the angle is assumed to move according to the equations of motion. The mechanical power input to each generator is assumed to be equal to the electrical power output in the pre-fault steady state. In general, the electrical output power of a generator with its internal voltage E_{GEN} is calculated by first calculating the generator current $\mathbf{I}_{GEN} = \mathbf{Y}_{eq}\mathbf{E}_{GEN}$, and then taking real part of the product $\mathbf{P}_{GEN} = \text{Real}(\mathbf{E}_{GEN}\mathbf{I}^*_{GEN})$. The admittance matrix $\mathbf{Y}_{eq}$ is the equivalent of the network as seen from the generator internal buses. These matrices for the three states of the network have been calculated in Section 1.6.4.

Performing the indicated calculation in the pre-fault state,

$$\mathbf{I}_{GEN0} = \begin{bmatrix} 3.4888 - j13.7442 & 0.6965 + j2.8056 & 1.3505 + j5.3907 & 0.9469 + j3.5914 \\ 0.6965 + j2.8056 & 0.6362 - j5.8241 & 0.4675 + j1.5462 & 0.3126 + j0.8810 \\ 1.3505 + j5.3907 & 0.4675 + j1.5462 & 2.5152 - j12.4058 & 0.7491 + j4.0634 \\ 0.9469 + j3.5914 & 0.3126 + j0.8810 & 0.7492 + j4.0634 & 1.3011 - j9.3975 \end{bmatrix} \begin{bmatrix} 1.0381 + j0.2193 \\ 1.0154 + j0.2412 \\ 1.0534 + j0.2958 \\ 0.9997 + j0.0131 \end{bmatrix}$$

$$\mathbf{I}_{GEN0} = \begin{bmatrix} 7.3935 - j0.8046 \\ 2.4945 - j0.0436 \\ 7.3365 - j0.6771 \\ 1.3111 + j0.0301 \end{bmatrix}$$

which is the same current found earlier in Section 1.6.4. The four generator electrical output powers in the pre-fault state are calculated, and set equal to their mechanical input powers:

$$\mathbf{P}_{GEN0} = \mathbf{P}_m = \text{Real} \begin{bmatrix} 1.0381 + j0.2193 \\ 1.0154 + j0.2412 \\ 1.0534 + j0.2958 \\ 0.9997 + j0.0131 \end{bmatrix} \times \begin{bmatrix} 7.3935 + j0.8046 \\ 2.4945 + j0.0436 \\ 7.3365 + j0.6771 \\ 1.3111 - j0.0301 \end{bmatrix} = \begin{bmatrix} 7.4986 \\ 2.5225 \\ 7.5279 \\ 1.3111 \end{bmatrix}$$

The product of the two column vectors above implies term-by-term multiplication to produce the four generator powers. These generator input mechanical powers are assumed to be constant throughout the stability calculation in the classical model.

1.6.6 Solving the Equations of Motion

The four generator rotors are governed by the following equations of motion:

$$\frac{d}{dt}\begin{bmatrix}\delta_{14}\\\delta_{15}\\\delta_{16}\\\delta_{17}\\\omega_{14}\\\omega_{15}\\\omega_{16}\\\omega_{17}\end{bmatrix}=\begin{bmatrix}\omega_{14}\\\omega_{15}\\\omega_{16}\\\omega_{17}\\\frac{\omega_s}{2H_{14}}(7.4986-P_{e-14})\\\frac{\omega_s}{2H_{15}}(2.5225-P_{e-15})\\\frac{\omega_s}{2H_{16}}(7.5279-P_{e-16})\\\frac{\omega_s}{2H_{17}}(1.3111-P_{e-17})\end{bmatrix}$$

The electrical power output of each generator is calculated as explained in the previous section, using appropriate equivalent admittance matrix at each instant. For example, during the pre-fault period, the electrical output powers equal the mechanical input powers and, the initial velocities being zero, the rotor angles will continue in their steady state. When the fault occurs, the in-fault bus admittance matrix Y_{pre} is used. And finally, when the fault is removed, the post-fault bus admittance matrix Y_{post} is used.

The equations of motion can be solved by any of the standard numerical integration techniques. The most commonly used method is the fourth-order Runge–Kutta method. In the present case, we will use this technique as supplied in the MATLAB software package. Before the advent of computer techniques, the above equations were solved by hand, and some very clever approximations to Euler's method of integration were developed in order to obtain excellent results.

Let the inertia constants of the four machines be [∞, 50, 50, 25] on the system base respectively. (Recall that the H constant on a system base that is different from the machine base is obtained by multiplying the machine constant on its own base with the ratio of the machine base to system base. Thus, if the machine rating is 1000 MVA, and the system base is 100 MVA, the H constant on the system base will be 50 s.) The inertia of ∞ implies an infinite bus, such as a connection to a very large system. In this case, the generator will not change its angle of its speed, and the resulting oscillations of all other generators will be judged for instability against its rotor. Later on we will consider the effect of having all inertias of finite size.

Further, let us assume that the fault occurs at 0.1 s after the start of the simulation, and line 6-8 is removed at 0.2 s. The entire simulation is to be run for 3 seconds. Thus, the use of the admittance matrices in calculating the electrical power output of the generators is to be scheduled as follows:

Time	Bus Admittance Matrix
$0 < t < 0.1$	Y_{pre}
$0.1 < t < 0.2$	Y_{fault}
$0.2 < t < 3.0$	Y_{post}

The result of the numerical integration for the above case is shown in Fig. 1.11. It should be noted that if the fault is cleared in 0.1 s after its occurrence, a stable oscillation of machine rotors results, as indicated by the fact that all swing curves seem to be oscillatory, and in practice with damping taken into account will settle at a new steady state value. On the other hand, if the fault lasts for 0.3 s (i.e., it is cleared at $t = 0.4$ s), the resulting oscillations are as shown in Fig. 1.12. Here, clearly, the rotor of machine no. 15 is going out of step, and will be tripped from the system.

Now consider the case where there is no infinite bus in the system, so that all machine rotors have a finite inertia. Let the inertia of the machine at bus no. 14 be 50. The resulting oscillations for a fault clearing time of 0.4 s are given in Fig. 1.13.

In this case, the loads do not consume as much power in the post-fault case as they did in the pre-fault case, since the network has changed.

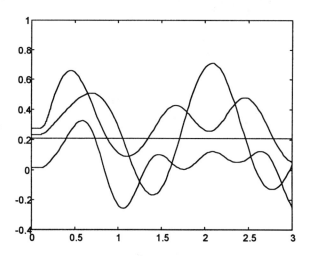

Figure 1.11 Oscillations of machine rotors when fault occurs at 0.1 sec after the start of the study and is removed at 0.2 sec. The system oscillations are stable.

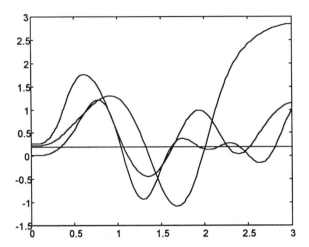

Figure 1.12 Oscillations of machine rotors when fault occurs at 0.1 sec after the start of the study and is removed at 0.4 sec. The system oscillations are unstable.

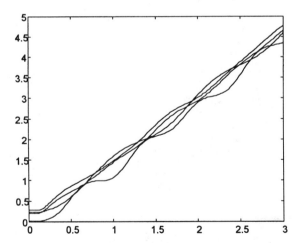

Figure 1.13 Oscillations for the same case as in Fig. 1.10, but with finite inertias for all the generators. This also is a stable case, although the system frequency is higher than normal.

Consequently, in the post-fault case, the aggregate of the generator input mechanical powers is slightly greater than their combined output power. The result of this mismatch is that the generators operate at a speed slightly greater than the synchronous speed, although their relative phase angle differences remain bounded. Thus, the system is stable. In practice, if the generators speed up (or slow down), the automatic generation control will take over and alter the input mechanical powers in order to keep the machines operating at normal speed. Since such controls are not represented in this study, we do not see this effect. Nevertheless, it is clear that as far as the transient stability oscillation is concerned, this also is a stable condition.

2
Electromagnetics

Edmund K. Miller*
Consultant, Santa Fe, New Mexico

2.1 INTRODUCTION

Electromagnetics is the study of electric charges and the fields that they produce. It is usual to identify two separate, but related, topics within the overall discipline:

1. statics and quasi-statics, where the observation time is short with respect to relevant time scales, so that phenomena of interest are exactly or approximately constant; and
2. dynamics, where significant changes occur over observation times of interest. Both areas are described by the Maxwell equations (MEs), with the latter the primary topic to be considered.

Electric charges are the source of all electromagnetic fields. Two kinds of charges have been discovered, conventionally called positive (any atom stripped of one or more electrons has a positive charge) and negative (the charge assigned to an electron) respectively. Magnetic charges are also postulated, but for our purposes can be considered to be simulated by electric

---

* Retired from Los Alamos National Laboratory, Los Alamos, New Mexico.

charges in circular motion, thus giving rise to a magnetic dipole. Magnetic monopoles have been sought, but are rare, if indeed they exist at all.

Charges at rest produce a static electric field (denoted by **E**), which has the properties of both magnitude and direction, i.e., it is a vector field. By convention, the electric field is considered to originate at a positive charge and terminate on a negative charge. Charges in constant motion relative to an observer comprise a current that produces in addition to a static electric field, a static magnetic field relative to the current (denoted by **H**). Both kinds of static field fall off with distance, r, in three-dimensional (3D) space, at least as fast a $1/r^2$. Charges whose motion varies in time, i.e. charges which are accelerated, in either speed or direction, produce a radiating electromagnetic field whose propagation involves a time-varying combination of electric and magnetic field components which are orthogonal to each other and transverse to the direction of propagation in the 'far' field, where they each fall off as $1/r$. Since the power flow is proportional to the vector cross product of **E** and **H**, as $\mathbf{S} = \mathbf{E} \times \mathbf{H}$, with **S** called the Poynting vector, the far-field power flow falls as $1/r^2$ in a lossless medium, a fundamental property of electromagnetic fields. Thus, the total power flow crossing a spherical surface of radius $R \gg$ wavelength (denoted by λ) centered on the current, becomes a constant in the far field.

Solution methods derived from MEs can be described as first-principles formulations (FPFs) when there are no intrinsic approximations in the analytical formulations on which they are based. On the other hand, when the formulation itself is based on certain approximations, such as the wavelength approaching infiniy or zero (the frequency approaching zero or infinity, respectively), the method is called asymptotic. Computational models based on high-frequency asymptotic methods are usually referred to as optics-based, since in the limit ray-like descriptions result. Models based on low-frequency asymptotic methods are referred to as quasi-static, implying their nearly static nature. Both FPFs and asymptotic approximations are used extensively to develop numerical or computer models of EM behavior.

In electromagnetics, as in many other like disciplines, FPFs can be based on integral equation (IE) or differential equation (DE) descriptions. Maxwell's equations in their original form are DEs, but these have an equivalent integral form, where for either case the unknowns are spatial fields and sources, to be determined by boundary or initial values. Integral equations, on the other hand, employ a Green's function, i.e., the field due to a point source obtained from a solution to the DEs, which is integrated over an unknown source distribution on a boundary (usually) to match known boundary, or other, fields. In either approach, there is required to be some known excitation, e.g. a plane wave or a local voltage,

Electromagnetics

usually referred to as the "source" term, to quantify the constants of integration that arise in solving differential equations.

These FPFs can also be formulated in either the frequency domain (FD) or the time domain (TD). For the former, a time-harmonic variation is assumed, where all of the sources and fields vary as $e^{j\omega t}$, whereas in the latter case the time variation is arbitrary to the extent permitted by the constraints imposed on the numerical model. Thus, FD solutions are valid at the single frequency for which they have been obtained, whereas TD solutions can provide information over a band of frequencies. All FPFs EM models can be identified as using one of these four approaches or their equivalents.

2.2 ANALYTICAL AND NUMERICAL BACKGROUND

2.2.1 Modeling as a Transfer Function

A common feature of these FPFs and approximation methods for solving MEs is their incorporation of an analytical relationship between a cause (forcing function, input or source) and its effect (the response or output). This relationship may be viewed as a generalized transfer function (see Fig. 2.1) [1] in which two basic problem types become apparent. For the analysis or direct problem, the input is known and the transfer function is derivable from the problem description, with the output or response to be determined. The solution process, in other words, proceeds in the direction of the arrows in the figure. For the case of the synthesis or inverse problem, two subclasses may be identified for which the solution process "reverses" direction in one or more respects. The "easier" synthesis problem involves finding the input given the output and transfer function. An example of this easier synthesis problem is that of determining the driving voltages or induced

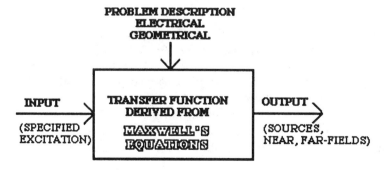

Figure 2.1 Transfer function model for electromagnetics.

currents needed to produce an observed pattern from a known antenna geometry.

The "harder" synthesis problem itself separates into two yet-different problems. One is that of finding the transfer function, given the input and output, an example of which is that of finding *a* source distribution that produces a given far field. This problem does not usually have a unique solution, as there can be many equivalent source distributions enclosed by a given surface that produce the same far field to some specified precision. The other, and still more difficult problem, is that of finding *the* object geometry that produces an observed far field from a known exciting field. This problem does have a unique solution, as there is only one geometrical configuration on which the required boundary conditions will be satisfied that radiates the observed far field. The latter problem is the most difficult of the three synthesis problems to solve because it is intrinsically transcendental and non-linear. All of these problems, whether direct or inverse, involve relationships between sources, an example of which would be unknown currents on physical boundaries, and the fields that they produce. This source–field relationship takes the form of a field "propagator," so called because the fields produced by the sources carry or propagate information about those sources throughout the solution space.

Thus, an important distinguishing property of the various approaches to solving MEs is the kind of field propagator they use, classified accordingly as:

Solution Approach	Propagator
Integral equations	Green's function
Modal representations	Green's function developed in special coordinate systems
Differential equations	Maxwell curl equations
Optics approximations	Ray tubes, diffraction and refraction coefficients

The field propagator can be thought of as a cause–effect relationship, as the "cause" mentioned above normally involves a source whose fields produce the "effect." It is therefore valid to conclude that the essence of electromagnetics is the study and determination of field propagators, and it follows that computational electro magnetics (CEM) inevitably does also. This conclusion, while perhaps appearing transparent, is actually an extremely fundamental one, as it provides a focus for what CEM is all about, and provides a basis for classification of model types as we now discuss.

Electromagnetics

2.2.2 Developing Analytical Solutions

Prior to the widespread availability of the digital computer, electromagneticists were limited to obtaining numerical results for relatively simple problems, since even when analytical, "closed form" solutions could be developed, their numerical evaluation was often very difficult. These usually involved various special functions, e.g., Bessel functions and Legendre polynomials, which arise in separation-of-variables solutions. Consequently, much attention was devoted to approximation techniques that might lead to analytical expressions whose numerical evaluation was practicable on mechanical calculators. These included forerunners of the numerical techniques now widely used, variational methods being one example. Some examples of analytical solutions available for classical problems are included in the problem section below. However, because so much of electromagnetics problem solving now depends on numerical models, a brief outline is given of computational EM in the following sections.

2.2.3 Developing Numerical Solutions

Some Issues Involved in Developing a Computer Model

In order to establish an appropriate perspective for subsequent discussion, we briefly consider here a classification of model types, the steps involved in developing a computer model, the desirable attributes of a computer model, and finally the role of approximation throughout the modeling process.

Classification of Model Types

It is convenient to classify solution techniques for electromagnetic modeling in terms of the field propagator that might be used, the anticipated application, and the problem type for which the model is intended to be used, as is outlined in Table 2.1. It may be seen that the most appropriate choice for a particular problem involves trading off the attributes among these categories.

Development of a Computer Model

Development of a computer model in electromagnetics or literally any other disciplinary activity can be decomposed into a small number of basic, generic steps. These steps might be described by different names, but would include at a minimum: (1) conceptualization; (2) formulation; (3) numerical implementation; (4) computation and (5) validation. Note that by its nature, validation is an open-ended process which cumulatively can absorb more effort than all the other steps together.

Table 2.1 Field Propagator Used, Intended Application, and Problem Type Determine Various Kinds of Modeling Approaches

Field Propagator	Description Based on
Integral operator	Green's function for infinite medium or special boundaries
Differential operator	Maxwell curl equations or their integral counterparts
Modal expansions	Solutions of Maxwell's Equations in particular coordinate system and expansion
Optical description	Rays and diffraction coefficients
Application	**Requires**
Radiation	Determining the originating sources of a field
Propagation	Obtaining the fields distant from a known source
Scattering	Determining the perturbing effects of medium inhomogeneties
Problem Type	**Characterized by**
Solution domain	Time or frequency
Solution space	Configuration or wavenumber
Dimensionality	1D, 2D, 3D
Electrical properties of dielectric	Lossy; perfectly conducting; anisotropic; medium and/or boundary inhomogeneous; nonlinear
Boundary geometry	Linear; curved; segmented; compound; arbitrary

Desirable Attributes of a Computer Model

A computer model must have some minimum set of basic properties to be useful. From the long list of attributes that might be desired, we consider: (1) accuracy, (2) efficiency, and (3) utility the three most important. Accuracy is put foremost, since results of insufficient or unknown accuracy have uncertain value and may even be harmful. On the other hand, a program that produces accurate results but at unacceptable cost will have hardly any more value. Finally, a program's applicability in terms of the depth and breadth of the problems for which it can be used determines its utility.

The Role of Approximation

Approximation is an intrinsic part of each step involved in developing a computer model, ranging from conceptualization to the actual computation. The approximations made can affect the two sources of error that may cause a computed result to differ from a "true" result for the problem being modeled. These are the physical modeling error (PME) and the numerical modeling error (NME). The PME arises because the numerical model used is normally an idealized mathematical representation of the actual physical problem of interest, while the NME arises because the numerical results obtained are only approximate solutions to that idealized representation.

2.3 ANALYTICAL ISSUES IN DEVELOPING A COMPUTER MODEL

As mentioned above, selection of a field propagator is a first step in developing an electromagnetic computer model. Further attention here is limited to propagators which employ either the Maxwell curl equations or source integrals which employ a Green's function. We consider briefly first selection of the solution domain and then selection of the field propagator, with discussion here limited to integral and differential equation models.

2.3.1 Selection of Solution Domain

Either the IE or DE propagator can be formulated in the time domain, where time is treated as an independent variable, or in the frequency domain where the harmonic time variation $\exp(j\omega t)$ is assumed. Whatever propagator and domain are chosen, the analytically formal solution can be numerically quantified via use of the method of moments (MoM) [2], leading ultimately to a linear system of equations as a result of developing a discretized and sampled approximation to the continuous (generally) physical reality being modeled. Developing the approach that may be best suited to a particular problem involves making trade-offs among a variety of choices throughout the analytical formulation and numerical implementation. In the following discussion, we consider some aspects of these choices and their influence on the utility of the computer model which eventually results.

2.3.2 Selection of Field Propagator

We briefly discuss and compare below the characteristics of IE- and DE-based models in terms of their development and applicability.

Integral Equation (IE) Model

The basic starting point for developing an IE model in electromagnetics is the selection of a Green's function appropriate for the problem class of interest. While there are a variety of Green's functions from which to choose, a typical starting point for most IE MoM models is that for a point current source in an infinite medium. Although the formulation might be accomplished in various ways, one of the more straightforward is based on the scalar Green's function and Green's theorem. This leads to the Kirchhoff integrals [3], from which the fields in a given contiguous volume of space can be written in terms of integrals over the surfaces which bound it and volume integrals over those sources located within it.

Analytical manipulation of a source integral which incorporates the selected Green's function as part of its kernel function then follows, with the specific details depending on the particular formulation being employed. Perhaps the simplest is that of boundary condition matching, wherein the behavior required of the electric and/or magnetic fields at specified surfaces which define the problem geometry is explicitly imposed. Alternative formulations, for example, the Rayleigh–Ritz variational method and Rumsey's reaction concept might be used instead, but as pointed out by Harrington [2], from the viewpoint of a numerical implementation any of these approaches leads to formally equivalent models.

This analytical formulation leads to an integral operator, whose kernel can include differential operators as well, which acts on the unknown source or field. Although it would be more accurate to refer to this as an integro-differential equation, it is usually called simply an integral equation. Two kinds of integral equation are obtained, one known as a Fredholm integral equation of the first kind in which the unknown appears only under the integral, and the other a second-kind equation in which the unknown also appears outside the integral.

Differential Equation (DE) Model

A DE MoM model, being based on the defining Maxwell's Equations, requires intrinsically less analytical manipulation than does derivation of an IE model. Numerical implementation of a DE model, however, can differ significantly from that used for an IE formulation in a number of ways for several reasons:

1. The differential operator is a local rather than global one, in contrast to the Green's function upon which the integral operator is based. This means that the spatial variation of the fields must be developed from sampling in as many dimensions as possessed by the

problem, rather than one fewer as the IE model permits if an appropriate Green's function is available.
2. The integral operator includes an explicit radiation condition.
3. The differential operator includes a capability to treat medium inhomogeneities, non-linearities, and time variations in a more straightforward manner than does the integral operator.

2.4 NUMERICAL ISSUES IN DEVELOPING A COMPUTER MODEL

2.4.1 Sampling Functions

At the core of numerical analysis is the idea of polynomial approximation, an observation made by Arden and Astill [4] in facetiously using the subtitle "Numerical Analysis or 1001 Applications of Taylor's Series." The basic idea is to approximate quantities of interest in terms of sampling functions, often polynomials, that are then substituted for these quantities in various analytical operations. Thus, integral operators are replaced by finite sums and differential operators are similarly replaced by finite differences. For example, use of a first-order difference to approximate a derivative of the function $F(x)$ in terms of samples $F(x_+)$ and $F(x_-)$ leads to

$$\frac{dF(x)}{dx} \approx \frac{F(x_+) - F(x_-)}{h}, \quad x_- \leqslant x \leqslant x_+ \tag{2.1a}$$

and implies a linear variation for $F(x)$ between x_+ and x_-, as does use of the trapezoidal rule

$$\int_{x_-}^{x_+} F(x)dx \approx \frac{h}{2}[F(x_+) + F(x_-)] \tag{2.1b}$$

to approximate the integral of $F(x)$, where $h = x_+ - x_-$. The central difference approximation for the second derivative,

$$\frac{d^2 F(x)}{dx^2} \approx \frac{1}{h^2}[F(x_+) - 2F(x_0) + F(x_-)] \tag{2.1c}$$

similarly implies a quadratic variation for $F(x)$ around $x_0 = x_+ - h/2 = x_- + h/2$, as does use of Simpson's rule

$$\int_{x_-}^{x_+} F(x)dx \approx \frac{h}{6}[F(x_+) + 4F(x_0) + F(x_-)] \tag{2.1d}$$

to approximate the integral. Other kinds of polynomials and function sampling can be employed, as discussed in a large volume of literature, some examples of which are Abramowitz and Stegun [5], Acton [6], and Press *et al*. [7]. It is interesting to see that numerical differentiation and integration can be accomplished using the same set of function samples and spacings, differing only in the signs and values of some of the associated weights. Note also that when the function samples can be unevenly spaced, as in Gaussian quadrature, the result will always be more accurate (for well-behaved functions) for a given number of samples. This suggests the benefits that might be derived from using unequal sample sizes in MoM modeling should a systematic way of determining the best sampling scheme be developed.

2.4.2 The Method of Moments (MoM)

Numerical implementation of the moment method is a relatively straightforward, and intuitively logical extension, of these basic elements of numerical analysis, as described in the well-known book by Harrington [2] and discussed and used extensively in CEM (see for example Mittra [8, 9], Strait [10], Strait and Adams [11], Harrington *et al*. [12], Perini and Buchanan [13], Ney [14], Itoh [15], Poggio and Miller [16], and Miller and Burke [17]. The MoM inevitably leads to a linear system of equations or matrix approximation of the original integral or differential operators.

When using IE techniques, the coefficient matrix in the linear system of equations which results is most often referred to as an impedance matrix, because in the case of the E-field form, its multiplication of the vector of unknown currents equals a vector of electric fields or voltages. The inverse matrix similarly is often called an admittance matrix because its multiplication of the electric field or voltage vector yields the unknown current vector. In this discussion we instead use the terms "direct matrix" and "solution matrix," since they are more generic descriptions whatever the forms of the originating integral or differential equations. As illustrated below, development of the direct matrix and solution matrix dominates both the computer time and storage requirements of numerical modeling.

In the particular case of an IE model, the coefficients of the direct or original matrix are the mutual impedances of the multi-port representation which approximates the problem being modeled, and the coefficients of its solution matrix (or equivalent thereof) are the mutual admittances. Depending on whether a sub-domain or entire-domain basis has been used (see following section), these impedances and admittances represent either spatial or modal interactions among the N ports of the numerical model. In either case, these coefficients possess a physical relatability to the problem being modeled, and ultimately provide all the information available concerning any electromagnetic observables that are subsequently obtained.

Similar observations might also be made regarding the coefficients of the DE models, but whose multi-port representations describe local rather

Electromagnetics

than global interactions. Because the DE model almost always leads to a larger, albeit less dense, direct matrix, its inverse (or equivalent) is rarely computed. It is worth noting that there are two widely used approaches for DE modeling: finite difference (FD) and finite element (FE) methods [4, 18, 19]. They differ primarily in how the differential operators are approximated and the differential equations are satisfied, although the FE method commonly starts from a variational viewpoint while the FD approach begins from the defining differential equations. The FE method is generally better suited for modeling problems with complicated boundaries to which it provides a piecewise linear approximation as opposed to the cruder stairstep approximation of FD.

Factors Involved in Choosing Basis and Weight Functions

Basis and weight function selection plays a critical role in determining the accuracy and efficiency of the resulting computer model. One goal of the basis and weight function selection is to minimize computer time while maximizing the accuracy for the problem set to which the model is to be applied. Another, possibly conflicting, goal might be that of maximizing the collection of problem sets to which the model is applicable. A third might be to replicate the problem's physical behavior with as few samples as possible.

Basis Function Selection

We note that there are two classes of bases used in MoM modeling: sub-domain and entire-domain functions. The former involves the use of bases which are applied in a repetitive fashion over sub-domains or sections (segments for wires, patches for surfaces, cells for volumes) of the object being modeled. The simplest example of a sub-domain basis is the single-term basis given by the "pulse" or stairstep function, which leads to a single, unknown constant for each sub-domain. Multi-term bases involving two or more functions on each sub-domain and an equivalent number of unknowns are more often used for sub-domain expansions.

The entire-domain basis, on the other hand, uses multi-term expansions extending over the entire object, for example a circular harmonic expansion in azimuth for a body of revolution. As for sub-domain expansions, an unknown is associated with each term in the expansion. Examples of hybrid bases can also be found, where sub-domain and entire-domain bases are used on different parts of an object [20].

Although sub-domain bases are probably more flexible in terms of their applicability, they have a disadvantage generally not exhibited by the entire-domain form, which is the discontinuity that occurs at the domain boundaries. This discontinuity arises because an n_s-term sub-domain function can provide at most $(n_s - 1)$th continuity to an adjacent basis of the unknown it represents, assuming one of the n_s constants is reserved for the unknown itself. For example, the three-term or sinusoidal sub-domain basis $a_i + b_i \sin(ks) + c_i \cos(ks)$ used for wire modeling can represent a current continuous at most up to its first derivative. This provides continuous charge density, but produces a discontinuous first derivative in charge equivalent to a tri-pole charge at each junction.

As additional terms are used to develop a sub-domain basis, higher-order continuity can be achieved in the unknown that the basis represents, assuming still that one constant is reserved for the unknown. In the general case of the n_s-term sub-domain basis, up to $n_s - 1$ constants can be determined from continuity conditions, with the remainder reserved for the unknown. The kind of basis function employed ultimately determines the degree of fit that the numerical result can provide to the true behavior of the unknown for a given order of matrix. An important factor that should influence basis function selection then is how closely a candidate function might resemble the physical behavior of the unknown it represents.

Weight Function Selection

The simplest weight that might be used is a delta function which leads to a point-sampled system of equations. But point sampling of the field operators can reveal any numerical anomalies that might arise as a result of basis function discontinuities. Distributed multi-term weight functions can also be employed on either a sub-domain or an entire domain basis to provide a further smoothing of the final equations to be solved. One example of this is the special case where the same functions are used for both the bases and weights, a procedure known as Galerkin's method. The kind of testing function employed ultimately determines the degree to which the equations can be matched for a given basis function and number of unknowns.

Computing the System Matrix

We observe that obtaining the coefficients of the system matrix (also called the direct or formulation matrix) in IE modeling is generally a two-step process. The first step is that of integrating the defining integral equation in which the unknown is replaced by the basis functions selected. The second step involves integration of this result multiplied by the weight function selected. When using delta function weights this second step is numerically trivial. But when using non-delta weights, such as the case in a Galerkin approach where the same function is used for both basis and weights, this second step can be analytically and numerically challenging.

Among the factors affecting the choice of the basis and weight functions, therefore, one of the most important is that of reducing the computational effort needed to obtain the coefficients of the system matrix. This is one of the reasons, aside from their physical appeal, why sinusoidal bases are often used for wire problems. In this case, where piecewise linear filamentary current sources are most often used in connection with the thin-wire approximation, field expressions are available in easily evaluated analytical expressions [21, 22]. This is the case as well where Galerkin's method (where the weight and basis functions are the same or conjugates) is employed [23].

Aside from such special cases, however, numerical evaluation of the system matrix coefficients will involve the equivalent of point sampling of whatever order is needed to achieve the desired accuracy as illustrated below. Using a wire-like one-dimensional problem to illustrate this point, we observe that at its most elementary level evaluation of the ijth matrix coefficient then involves evaluating integrals of the form

$$Z_{ij} = \int w_i(s) \int [b_j(s')K(s,s')]ds \approx \sum_m \sum_n p_m q_n w_i(s_n) b_j(s'_m) K(s_n, s'_m)$$

$$= \sum_m \sum_n p_m q_n z(i,j,n,m); \qquad (2.2)$$

$$m = 1, \ldots, M(i,j),$$

$$n = 1, \ldots, N(i,j), \, i,j = 1, \ldots, N$$

where $K(s,s')$ is the IE kernel function, and s_n and s'_m are the nth and mth locations of the observation and source integration samples. Thus, the final, system matrix coefficients can be seen to be "constructed" from sums of the more elementary coefficients $z(i,j,n,m)$ weighted by the quadrature coefficients p_m and q_n used in the numerical integration, which will be the case whenever analytical expressions are not available for the $Z_{i,j}$. These elementary coefficients, given by $w_i(s_n)b_j(s'_m)K(s_n, s'_n)$, can in turn be seen to be simply products of samples of the integral equation kernel or operator and sampled basis and testing functions. It should be apparent from this expanded expression for the system matrix coefficients that interchanging the basis and weight functions leaves the final problem description unchanged, although the added observation that two different integral equations can yield identical matrices when using equivalent numerical treatments is less obvious [24].

Computing the Solution Matrix

Once the system matrix has been computed, the solution can be obtained numerically using various approaches. These range from inversion of the system matrix to developing a solution via iteration. A precautionary comment is in order with respect to the accuracy with which the solution matrix might be obtained. As computer speed and storage have increased, the number of unknowns employed in modeling has also increased, from a few tens in the early years to thousands now when using IE techniques. The increasing number of operations involved in solving these larger matrices increases the sensitivity of the results to round-off errors. This is especially the case when the system matrix is not well conditioned, i.e., the solution process "amplifies" the effects of errors in the system matrix or the right-hand side. It is therefore advisable to perform some sensitivity analyses to determine the system-matrix condition number and to ascertain the possible need for performing some of the computations in double precision.

Obtaining the Solution

When a solution matrix has been developed from the system matrix using inversion or factorization, subsequently obtaining the solution is computationally straightforward, involving multiplication of the right-hand side (RHS) source vector by the solution matrix. When an iterative approach is used, a solution matrix is not computed but the solution is instead developed from RHS-dependent manipulation of the system matrix. Motivation for the latter comes from the possibility of reducing the N_x^3 dependency of the direct procedure. As the problem size increases, the computation cost will be increasingly dominated by the solution time.

2.5 PROBLEMS HAVING PRIMARILY ANALYTICAL SOLUTIONS

Although most EM analysis is done using numerical methods, the analytical origin of these computational models remains important. Thus, we examine several of the "classical" problems whose solutions may be regarded as being primarily analytical, even though obtaining numerical results from them can require extensive computation, before presenting some of their computational counterparts.

2.5.1 Problem PA1: Fields and Radiation Produced by a Hertzian Dipole

Finding the near and far fields of a specified current is one of the oldest problems in electromagnetics. Before computers became widely available, the variety of problems whose solutions were obtained in "closed form" was understandably quite limited. Furthermore, these were primarily limited to the far fields, where only the $1/r$ components are present, as the near fields require a much more careful treatment to include the higher-order field terms.

Fortunately, for some simple filamentary currents, e.g. where the current flows on the z-axis of a coordinate system with a special variation in the z-direction, the complete fields, both near and far, can be found in closed form. The simplest of these special current sources is a time-harmonic Hertzian dipole, which consists of a current constant in the z-direction and of length $L \ll \lambda$, with λ the wavelength, which for simplicity we take to have its center at $z = 0$.

First note that the vector potential of a general, volumetric current in a medium of permeability μ is given by Balanis [25]:

$$\mathbf{A}(\mathbf{r}) = \frac{\mu}{4\pi} \int\int\int_V \mathbf{J}(\mathbf{r}') \frac{e^{-jkR}}{R} dv' \qquad (2.3)$$

with $\mathbf{r}$ and $\mathbf{r}'$ the observation and source coordinates, respectively, $R = |\mathbf{r} - \mathbf{r}'|$, and V is the volume to which the current density, $\mathbf{J}(\mathbf{r}')$ (amps/m^2), is confined. If the current is confined to the surface of a perfect electric conductor (PEC) defined by S, then the vector potential can be written

$$\mathbf{A}(\mathbf{r}) = \frac{\mu}{4\pi} \int\int_S \mathbf{J}_s(\mathbf{r}') \frac{e^{-jkR}}{R} ds' \qquad (2.4)$$

with $\mathbf{J}_s(\mathbf{r}')$ (amps/m) the surface current density. Finally, if the current is confined to a filament of radius $\ll \lambda$, described by C, the corresponding vector potential can be simplified further to a line integral given by

$$\mathbf{A}(\mathbf{r}) = \frac{\mu}{4\pi} \int_C \mathbf{I}(\mathbf{r}') \frac{e^{-jkR}}{R} dl' \qquad (2.5)$$

with $\mathbf{I}(\mathbf{r}')$ (amps) the line current at point $\mathbf{r}'$.

With the Hertzian dipole of interest defined by

$$\mathbf{I}(\mathbf{r}') = \mathbf{a}_z I_0 (0, 0, z'), -L/2 \leqslant z' \leqslant L/2 \qquad (2.6)$$

where the dipole length $L \ll \lambda$ and $\mathbf{a}_z$ is the z-directed unit vector, the vector potential can be evaluated to be

$$\mathbf{A}(\mathbf{r}) = \mathbf{a}_z \frac{\mu I_0}{4\pi r} e^{-jkr} \int_{-L/2}^{L/2} dz' = \mathbf{a}_z \frac{\mu I_0 L}{4\pi r} e^{-jkr} \tag{2.7}$$

with $r = (x^2 + y^2 + z^2)^{1/2}$. Observe that the vector potential has only a z-component and that the phase is referred to the center of the dipole, thereby ignoring any phase change over its length. It's usually more convenient to work in spherical coordinates, especially in the far field, so now write $\mathbf{A}$ in terms of (r, θ, φ) using the rectangular-to-spherical transformation

$$\begin{bmatrix} A_r \\ A_\theta \\ A_\varphi \end{bmatrix} = \begin{bmatrix} \sin\theta\cos\varphi & \sin\theta\sin\varphi & \cos\theta \\ \cos\theta\cos\varphi & \cos\theta\sin\varphi & -\sin\theta \\ -\sin\varphi & \cos\varphi & 0 \end{bmatrix} \begin{bmatrix} A_x \\ A_y \\ A_z \end{bmatrix} \tag{2.8}$$

which, since $A_x = A_y = 0$, simplifies to

$$A_r = A_z \cos\theta = \frac{\mu I_0 L}{4\pi r} e^{-jkr} \cos\theta,$$

$$A_\theta = -A_z \sin\theta = \frac{\mu I_0 L}{4\pi r} e^{-jkr} \sin\theta,$$

$$A_\varphi = 0. \tag{2.9}$$

Having obtained the vector potential, the electric and magnetic fields can be obtained from the defining equations, i.e.

$$\mathbf{H} = \frac{1}{\mu} \nabla \times \mathbf{A} \tag{2.10}$$

and

$$\mathbf{E} = -j\omega \mathbf{A} - j\frac{1}{\omega\mu\varepsilon} \nabla(\nabla \cdot \mathbf{A}) = \frac{1}{j\omega\varepsilon} \nabla \times \mathbf{H} \tag{2.11}$$

so that in cylindrical coordinates we find

$$H_r = H_\theta = 0 \text{ and } H_\varphi = j\frac{kI_0 L \sin\theta}{4\pi r}\left[1 + \frac{1}{jkr}\right] e^{-jkr} \tag{2.12}$$

and

$$E_r = \eta \frac{I_0 L \cos\theta}{2\pi r^2}\left[1 + \frac{1}{jkr}\right] e^{-jkr},$$

$$E_\theta = \eta \frac{kI_0 L \sin\theta}{4\pi r}\left[1 + \frac{1}{jkr} - \frac{1}{(kr)^2}\right] e^{-jkr}, \tag{2.13}$$

$$E_\varphi = 0.$$

Equations (2.12) and (2.13) are "exact" to the degree that $kL \ll 1$, and are valid in the near field (except at $r = 0$) and the far field. Note that the magnetic field has only a φ component, a result to be expected from our knowledge of statics, where a straight filament of current was found to have only an azimuthal magnetic field. But in addition to the static (i.e., frequency-independent) $1/r^2$ term that is predicted by the Biot–Savart law, there is a $1/r$ term whose magnitude is proportional to frequency via $k = \omega/c$, with c the speed of light. The electric field similarly has not only the radial component that can be anticipated from a stationary point charge, but a θ component as well, whose magnitude is also proportional to frequency. These k-dependent components obviously arise from the oscillatory nature of the Hertzian dipole current, and can be deduced to be associated with electromagnetic radiation, as each is proportional to $1/r$.

Also observe that the field expressions in Eqs. (2.12) and (2.13) serve as the Green's function for an electric filamentary current of length $dz = L$ and amplitude I_0. Thus, the field of a relatively arbitrary current filament described by a contour C in space could be developed from the above expressions and the integral of Eq. (2.5). If used that way, it would usually be more convenient to work in a rectangular coordinate field, which requires two steps. Although there are various ways to accomplish this, the first would involve expressing the fields above in rectangular form, transforming them from their spherical components using the inverse of Eq. (2.8). Second, a differential current element of length $\mathbf{dl}$ would also be expressed in rectangular coordinates, with the fields of each separate component then written in x–y–z form. This is left as an exercise for the reader.

It is straightforward to determine the Poynting vector (which can be interpreted as the pointwise power flow) produced by the Hertzian dipole from

$$\mathbf{S} = \frac{1}{2}\mathbf{E} \times \mathbf{H}^* = \frac{1}{2}(\mathbf{a}_r E_r + \mathbf{a}\theta E_\theta) \times \mathbf{a}_\varphi H_\varphi = \frac{1}{2}(\mathbf{a}_r E_\theta H_\varphi^* - \mathbf{a}_\theta E_r H_\varphi^*) \quad (2.14)$$

so that the radial and theta-dependent components become

$$S_r = \frac{\eta}{8}\left[\frac{|I_0|L\sin\theta}{\lambda r}\right]^2\left[1 - j\frac{1}{(kr)^3}\right],$$

and

$$S_\theta = j\eta\frac{k|I_0 L|^2 \cos\theta\sin\theta}{16\pi^2 r^3}\left[1 + \frac{1}{(kr)^2}\right]. \quad (2.15)$$

Observe that only the radial component of the Poynting vector has a $1/r^2$ term, whose integral over a far-field sphere centered on the dipole is thus a constant. The other radial component and all of the theta components vary as $1/r^n$, where $n \geq 3$, and therefore do not represent radiation; rather, they account for the near-field power flow associated with energy storage in the space near the dipole.

For the radial power flow we obtain

$$P = \int_0^{2\pi}\int_{-\pi/2}^{\pi/2} S_r \, ds = \int_0^{2\pi}\int_{-\pi/2}^{\pi/2} S_r r^2 \sin\theta \, d\theta \, d\varphi = \eta\frac{\pi}{3}\left[\frac{|I_0|L}{\lambda}\right]^2\left[1 - j\frac{1}{(kr)^3}\right] \tag{2.16}$$

of which the radiation term is the first one in the bracket and the second represents imaginary (reactive or quadrature) power flow. The radiation resistance, R_{rad} of the Hertzian dipole can now be determined from

$$P_{\text{rad}} = \eta\frac{\pi}{3}\left[\frac{|I_0|L}{\lambda}\right]^2 = \frac{1}{2}|I_0|^2 R_{\text{rad}} \tag{2.17}$$

to give

$$R_{\text{rad}} = \eta\left(\frac{2\pi}{3}\right)\left(\frac{L}{\lambda}\right)^2. \tag{2.18}$$

While this has been derived for the special case of a Hertzian dipole, it is found that for all linear antennas for which $L \ll \lambda$, $R_{\text{rad}} \propto (L/\lambda)^2$.

The Hertzian dipole is useful to determine the asymptotic behavior of current sources short compared with the wavelength. But most practical antennas are such that $L \sim \lambda$, whose radiation properties need more detailed analysis. This is now usually achieved using numerical models such as are discussed in the next section, but there are two specific current filaments whose characteristics provide a bound on the expected behavior of wire antennas. These are the sinusoidal current filament (SCF) and the constant current filament (CCF), which are given by

$$I(z) = I_0 \sin[\omega t - k(L/2 - |z|)], \ |z|^2 L/2 \tag{2.19a}$$

and

$$I(z) = I_0, \ |z|^2 L/2 \tag{2.19b}$$

respectively. The far electric field of a current filament of length L lying on the z-axis with its center at the origin can be expressed as [3]

$$E_\theta(\theta) = j\eta \frac{ke^{-jkr}}{4\pi r} \sin(\theta) \int_{-L/2}^{L/2} I(z')e^{jkz'\cos(\theta)}dz' \tag{2.20}$$

Electromagnetics

which yields

$$E_\theta(\theta) = j\eta \frac{I_0 e^{-jkr}}{2\pi r} \left(\frac{\cos\left[\frac{kL}{2}\cos(\theta)\right] - \cos\left[\frac{kL}{2}\right]}{\sin(\theta)} \right) \quad (2.21a)$$

for the SCF and

$$E_\theta(\theta) = j\eta \frac{I_0 e^{-jkr}}{2\pi r} \frac{\cos\left[\frac{kL}{2}\cos(\theta)\right]}{\cos(\theta)} \quad (2.21b)$$

for the CCF, while the magnetic field for both is given by

$$H_\varphi(\theta) = \frac{E_\theta(\theta)}{\eta}. \quad (2.22)$$

The radially directed radiated power is then given by $E_\theta^2(\theta)/\eta$. Integration of these expressions provides the total radiated power, which is plotted in Fig. 2.2 as a function of L for $I_0 = 1$ amp, together with Eq. (2.18) for the Hertzian dipole, the latter over the range $L/\lambda < 1$. It can be

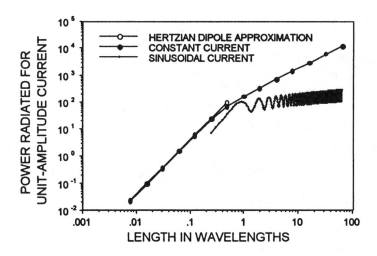

Figure 2.2 Power radiated by three unit-amplitude current filaments. Both the constant and sinusoidal filaments exhibit the L^2 dependence for $L < \lambda/2$ or so, in agreement with the analytical behavior of the Hertzian dipole approximation. For longer filaments, however, while the constant filament radiates a power proportional to its length, the sinusoidal filament grows at a $\log(kL)$ rate, due to their different radiation mechanisms.

seen that the power radiated by both the SCF and CCF vary as L^2 as $L \to 0$, in agreement with the behavior predicted by analysis of the Hertzian dipole, but where there is an offset between them because the moment (integral of $I dz$) of the SCF is about half that of the CCF. The behavior with increasing L is very different as a result of the differing radiation mechanisms for the two current filaments, a phenomenon not covered further here.

2.5.2 Problem PA2: Fields in Simple Waveguides

One of the simplest problems to solve analytically using a separation-of-variables (SoV) solution is that of determining the fields in a rectangular waveguide. Two distinct kinds of fields can exist in such a waveguide, denoted as TE (transverse electric) and TM (transverse magnetic) waves, depending on whether they have no z-component of the electric or magnetic field, respectively. The most general case would contain a combination of both, but since it can always be decomposed into TE and TM components, no generality is lost in analyzing them separately. The relevant equation is thus

$$\frac{\partial^2 \psi}{\partial x^2} + \frac{\partial^2 \psi}{\partial y^2} + \frac{\partial^2 \psi}{\partial z^2} + k^2 \psi = 0 \qquad (2.23)$$

where ψ is E_z for the TM case and H_z for the TE case and is sometimes called the generating potential, as all of the other field components can be determined from it.

Although the most general case has $\psi = F(x, y, z)$, because the rectangular guide can be described by constant-coordinate surfaces in the independent variables it is possible to write

$$\psi = X(x) Y(y) Z(z) \qquad (2.24)$$

which vastly decreases the difficulty of developing a solution and where ψ is now taken to represent E_z. Upon substituting the SoV representation in the equation for ψ there is obtained

$$YZ \frac{\partial^2 X}{\partial x^2} + XZ \frac{\partial^2 Y}{\partial y^2} + XY \frac{\partial^2 Z}{\partial z^2} + k^2 XYZ$$
$$= 0 = X''YZ + XY''Z + XYZ'' + k^2 XYZ \qquad (2.26)$$

and upon dividing through by XYZ it follows that

$$\frac{X''}{X} + \frac{Y''}{Y} + \frac{Z''}{Z} + k^2 = 0 \qquad (2.27)$$

Electromagnetics

It can be deduced that since the variables are independent, the only way that this equation can be satisfied is if its separate components are constants, or if

$$-k_x^2 - k_y^2 + k_z^2 + k^2 = 0 \tag{2.28a}$$

so that

$$X'' + k_x^2 X = 0, \ Y'' + k_y^2 Y = 0, \ \text{and} \ Z'' - k_z^2 Z = 0 \tag{2.28b}$$

where the signs of the x, y, and z propagation constants are chosen in anticipation of the kinds of solutions expected for a wave propagating in the z-direction and consisting of standing waves in the x- and y-directions. The resulting solutions can be written as

$$\begin{aligned} X(x) &= e_{x1} \cos(k_x x) + e_{x2} \sin(k_x x), \\ Y(y) &= e_{y1} \cos(k_y y) + e_{y2} \sin(k_y y), \\ Z(z) &= e_{z+} e^{k_z z} + e_{z-} e^{-k_z z}. \end{aligned} \tag{2.29}$$

If we are interested in a wave propagating in the $+z$ direction, then the c_{z+} term must be dropped to ensure a finite-valued field as $z \to \infty$ and E_z for the TM case is given by

$$E_z(x, y, z) = [e_{x1} \cos(k_x x) + e_{x2} \sin(k_x x)][e_{y1} \cos(k_y y) + e_{y2} \sin(k_y y)]e_{z-} e^{-k_z z}. \tag{2.30a}$$

An exactly analogous expression can be derived for H_z in the TE case

$$H_z(x, y, z) = [h_{x1} \cos(k_x x) + h_{x2} \sin(k_x x)][h_{y1} \cos(k_y y) + h_{y2} \sin(k_y y)]h_{z-} e^{-k_z z}. \tag{2.30b}$$

The other field components can then be found from the Maxwell curl equations to obtain

$$E_x = -\frac{k_z}{k_T^2} \frac{\partial E_z}{\partial x} - \frac{j\omega\mu}{k_T^2} \frac{\partial H_z}{\partial y}, \ E_y = -\frac{k_z}{k_T^2} \frac{\partial E_z}{\partial y} + \frac{j\omega\mu}{k_T^2} \frac{\partial H_z}{\partial x} \tag{2.31a}$$

and

$$H_x = \frac{j\omega\varepsilon}{k_T^2} \frac{\partial E_z}{\partial y} - \frac{k_z}{k_T^2} \frac{\partial H_z}{\partial x}, \ H_y = -\frac{j\omega\varepsilon}{k_T^2} \frac{\partial E_z}{\partial x} - \frac{k_z}{k_T^2} \frac{\partial H_z}{\partial y} \tag{2.31b}$$

where $k_T^2 = k_x^2 + k_y^2$.

So far, the only boundary condition used to develop a solution has been that of requiring the fields to be finite as $z \to \infty$. To complete the solution, we now impose the condition that the tangential electric fields vanish on the inside walls of the waveguide, considering the TM and TE cases in order.

The TM Case

In this case, $H_z = 0$, so that the applicable boundary condition is that

$$E_z = 0 \text{ at } x = 0, a \text{ and at } y = 0, b \tag{2.32}$$

where the waveguide occupies the space $0 \le x \le a, 0 \le y \le b, z \ge 0$. Consequently, the cosine terms in the field solutions drop out and E_z becomes

$$E_z = E_0 \sin(k_x x) \sin(k_y y) e^{-k_z z} \tag{2.33}$$

where $E_0 = c_{x2} c_{y2}$, so that the boundary condition requires that

$$\sin(k_x a) = 0 \text{ and } \sin(k_y b) = 0$$

which leads to

$$k_x a = m\pi \text{ and } k_y b = n\pi$$

where $m, n = 1, 2, 3, \ldots$ or

$$k_x = \frac{m\pi}{a} \text{ and } k_y = \frac{n\pi}{b}.$$

The final form of E_z thus becomes

$$E_z = E_0 \sin\left(\frac{m\pi x}{a}\right) \sin\left(\frac{n\pi y}{b}\right) e^{-k_z z} \tag{2.34}$$

with the other field components derivable from E_z. A complete solution for a multi-moded waveguide requires summing over all possible values of mode numbers m and n, but it is normally desirable to operate a waveguide in a single mode, i.e., where m and n have only a single value. The propagation constants are then given by

$$k_T^2 = \left(\frac{m\pi}{a}\right)^2 + \left(\frac{n\pi}{b}\right)^2 \text{ and } k_z = \sqrt{\left(\frac{m\pi}{a}\right)^2 + \left(\frac{n\pi}{b}\right)^2 - k^2} = \alpha + j\beta \tag{2.35a}$$

where α and β are the real and imaginary parts of k_z.

It can be seen that k_z is either pure real or pure imaginary. When the former is true, the fields are evanescent, varying with z as $e^{-\alpha z}$, whereas for the latter the fields are propagating and vary as $e^{-j\beta z}$. The frequency where the fields change from evanescent to propagating is called the "cutoff" frequency, f_c, and it clearly depends on m and n, being determined from

$$k_c^2 = \omega_c^2 \mu \varepsilon = \left(\frac{m\pi}{a}\right)^2 + \left(\frac{n\pi}{b}\right)^2 \tag{2.35b}$$

Electromagnetics

so that

$$f_c = \frac{\omega_c}{2\pi} = \frac{1}{2\pi\sqrt{\mu\varepsilon}}\sqrt{\left(\frac{m\pi}{a}\right)^2 + \left(\frac{n\pi}{b}\right)^2} = \frac{v}{2}\sqrt{\left(\frac{m}{a}\right)^2 + \left(\frac{n}{b}\right)^2} \qquad (2.36)$$

where $f > f_c$ is needed for propagation to occur and v is the speed of light in an infinite medium having electrical parameters, μ, ε of the medium that fills the guide. It can be seen that the lowest order TM mode, $TM_{m,n} = TM_{1,1}$ has the lowest cutoff frequency of all the modes.

The propagation constant can be written in terms of f_c as

$$\beta = \frac{\omega}{v}\sqrt{1 - \left(\frac{f_c}{f}\right)^2} \qquad (2.37)$$

while the intrinsic wave impedance of any mode is given by

$$\eta_{TM} = \frac{E_x}{H_y} = -\frac{E_y}{H_x} = \frac{\beta}{\omega\varepsilon} = \eta\sqrt{1 - \left(\frac{f_c}{f}\right)^2} \qquad (2.38)$$

with η the wave impedance of an infinite medium having the waveguide's electrical parameters, i.e. $\eta = (\mu\varepsilon)^{1/2}$.

The TE Case

For the TE case, the boundary conditions on the electric field result in requiring the x and y spatial derivatives of H_z to be zero, so that

$$H_z = H_0 \cos\left(\frac{m\pi x}{a}\right)\cos\left(\frac{n\pi y}{a}\right)e^{-k_z z} \qquad (2.39)$$

where it may be seen that either m or n, but not both at the same time, may be zero. Thus, the lowest order TE modes are $T_{m,n} = T_{1,0}$ or $T_{0,1}$. Expressions for the cutoff frequency and propagation constant for the TE mode are the same as for the TM mode, but the wave impedance is given by

$$\eta_{TE} = \frac{\eta}{\sqrt{1 - (f_c/f)^2}} \qquad (2.40)$$

It is interesting to observe that

$$\eta_{TE}\eta_{TM} = \frac{\eta}{\sqrt{1 - (f_c/f)^2}} \eta\sqrt{1 - \left(\frac{f_c}{f}\right)^2} = \eta^2. \qquad (2.41)$$

A plot of the wave impedances as a function of frequency is shown in Fig. 2.3. Note that the TE mode impedance approaches an open circuit, and that of the TM mode a short circuit, as f approaches f_c from above.

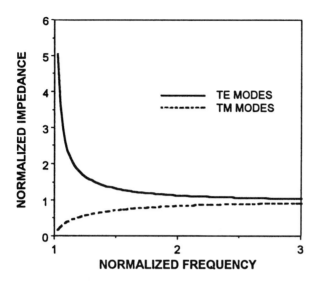

Figure 2.3 The effective wave impedances of TE and TM modes propagating in a rectangular waveguide are reciprocals of each other with their product being η^2, with η the wave impedance of the material that fills the waveguide. The results here are normalized by η with the normalized frequency being f/f_c. The impedance approaches that of the material filling the waveguide with increasing frequency as the propagating modes become more like plane waves.

2.5.3 Problem PA3: Scattering From an Infinite Circular Cylinder

The problem of an infinite, circular, PEC illuminated by a plane wave is one of the more straightforward exterior boundary value problems [26]. Its solution is developed from the scalar Helmholtz equation in cylindrical coordinates, as a complete solution for this problem can be developed from its separate TE and TM components. These in turn are obtained from the z-components of the magnetic and electric fields, respectively, each of which is found from solutions of the scalar Helmholtz equation.

Thus, we begin with

$$\frac{1}{\rho}\frac{\partial}{\partial \rho}\left(\rho \frac{\partial \psi}{\partial \rho}\right) + \frac{1}{\rho^2}\frac{\partial^2 \psi}{\partial \varphi^2} + \frac{\partial^2 \psi}{\partial z^2} + k^2 \psi = 0 \qquad (2.42)$$

where ψ is E_z for the TM case and H_z for the TE case. For an SoV solution to hold, ψ in Eq. (2.42) must have the form

Electromagnetics

$$\psi(\rho, \varphi, z) = R(\rho)\Phi(\varphi)Z(z) \tag{2.43}$$

i.e., a solution in terms of the three variables of the cylindrical coordinate system must be able to be written as a product of three solutions, one for each of these variables. Upon substituting Eq. (2.43) in (2.42) and dividing by ψ, we obtain

$$\frac{1}{\rho R}\frac{d}{d\rho}\left(\rho\frac{\partial R}{d\rho}\right) + \frac{1}{\rho^2 \Phi}\frac{d^2\Phi}{d\varphi^2} + \frac{1}{Z}\frac{d^2 Z}{dz^2} + k^2 = 0 \tag{2.44}$$

where it can be seen that the third term involves Z and z alone. This means that it must also be independent of z if Eq. (2.44) is to be true independent of the values of (ρ, φ, z). Thus

$$\frac{1}{Z}\frac{d^2 Z}{dz^2} = -k_z^2 \tag{2.45}$$

with k_z a constant (the z separation variable). Then Eq. (2.44) can be rewritten as

$$\frac{\rho}{R}\frac{d}{d\rho}\left(\rho\frac{dR}{d\rho}\right) + \frac{1}{\Phi}\frac{d^2\Phi}{d\varphi^2} + (k^2 - k_z^2)\rho^2 = 0 \tag{2.46}$$

after multiplying by ρ^2.

The second term in Eq. (2.46) involves only φ while the others do not, so that it must be true that

$$\frac{1}{\Phi}\frac{d^2\Phi}{d\varphi^2} = -n^2 \tag{2.47}$$

with the constant n the φ separation variable. Then Eq. (2.46) can be written as

$$\frac{\rho}{R}\frac{d}{d\rho}\left(\rho\frac{dR}{d\rho}\right) - n^2 + (k^2 - k_z^2) = 0 \tag{2.48}$$

which now involves only the ρ variable. Thus, the original partial differential equation, (2.42), has been separated into three ordinary differential equations in one variable each: (2.45), (2.47), and (2.48).

Now define

$$k_\rho^2 + k_z^2 = k^2 \tag{2.49}$$

so that the final separated equations become

$$\frac{\rho}{R}\frac{d}{d\rho}\left(\rho\frac{dR}{d\rho}\right) + [(k_\rho\rho)^2 - n^2]R = 0 \tag{2.50a}$$

$$\frac{d^2\Phi}{d\varphi^2} + n^2\Phi = 0 \tag{2.50b}$$

and

$$\frac{d^2Z}{dz^2} + k_z^2 Z = 0 \tag{2.50c}$$

Equation (2.50a) is a well-studied equation known as Bessel's equation of order n, while the two other equations have solutions in terms of exponential functions. That for Eq. (2.50b) can be written

$$\Phi_n = e^{\pm jn} \tag{2.51a}$$

where n must be an integer for Φ_n to be a single-valued function of φ, while the z-variable solution is given by

$$Z_{k_z} = e^{\pm jk_z z} \tag{2.51b}$$

The radial solution can only be written symbolically, since closed form expressions of the kind found for the Φ and Z solutions are not available. A common notation for the radial solution is

$$B_n(k_\rho\rho) \sim J_n(k_\rho\rho), Y_n(k_\rho\rho), H_n^{(1)}(k_\rho\rho), H_n^{(2)}(k_\rho\rho) \tag{2.51c}$$

where the B_n stands for one of the solutions to Bessel's equation, and J_n, Y_n, and H_n are the Bessel, Neumann, and Hankel functions, respectively, with the superscript on the latter denoting it to be a first kind or second kind as given by

$$H_n^{(1)}(x) = J_n(x) + jY_n(x) \text{ and } H_n^{(2)}(x) = J_n(x) - jY_n(x). \tag{2.51d}$$

A solution of the Helmholtz equation in cylindrical coordinates thus can be expressed as

$$\psi_{k_\rho, n, k_z}(\rho, \varphi, z) = B_n(k_\rho\rho)\Phi_n(\varphi)Z(k_z z) \tag{2.52}$$

for a particular set of separation variable values. The specific range of values that n, and k_z (or k_ρ, since they are related to k and each other as shown in Eq. (2.49)) are determined by the excitation. While n is restricted to discrete values, k_z may not be, and the general solution of the Helmholtz equation can be written as a double sum or a sum–integral combination as

$$\psi = \sum_n a \sum_{k_z} A_{n,k_z} \psi_{k_\rho, n, k_z} \text{ or } \psi = \sum_n \int_{k_z} A_n(k_z)\psi_{k_\rho, n, k_z} dk_z \tag{2.53}$$

where A_n is to be determined from the boundary condition required of the field on the cylinder's surface.

Electromagnetics

For the problem of interest, let us now consider a plane wave incident normal to the axis of the cylinder from the $-x$-axis with its electric field parallel, and it magnetic field perpendicular, to the cylinder axis, which is known as TM (transverse magnetic) excitation. Then the incident electric field is

$$E_z^i = E_0 e^{-jkx} = E_0 e^{-jk\rho\cos\varphi}. \tag{2.54}$$

In order to complete the problem's description, this excitation needs to be written in terms of the functions used for the Helmholtz equation solution. It is straightforward to show that

$$e^{-jkx} = e^{-jk\rho\cos\varphi} = \sum_{n=-\infty}^{\infty} j^{-n} J_n(k\rho) e^{jn} \tag{2.55}$$

which gives the incident field when multiplied by E_0. For this particular incident field, $k_z = 0$, so that the general solution requires a summation over only n in Eq. (2.53).

We must now choose that solution for that scattered field from the results above that represents an outward propagating wave (i.e. one whose phase varies as $-jk_\rho\rho$ as $\rho \gg \lambda$). From the Hankel functions in Eq. (2.51d), it can be determined that the second kind has this property in the same way tht $e^{-jkx} = \cos(kx) - j\sin(kx)$ has for a plane wave in rectangular coordinates. Also, the scattered field must have the same variation as the incident field, so it has $k_z = 0$, or no z variation as well. Thus, from the boundary condition the electric field must vanish on the surface of the PEC cylinder, or

$$E_z^i + E_z^s = E_z$$

so that we can write

$$E_z = E_0 \sum_{n=-\infty}^{\infty} j^{-n}[J_n(k_\rho\rho) + A_n H_n^{(2)}(k_\rho\rho)] e^{jn\varphi} = 0 \tag{2.56}$$

from which it follows that

$$A_n = \frac{-J_n(k_\rho a)}{H_n^{(2)}(k_\rho a)} \tag{2.57}$$

as the solution for the TM polarization, where we note that for normal incidence $k_\rho = k$. The scattered electric field can then be written

$$E_z^s = -E_0 \sum_{n=-\infty}^{\infty} j^{-n} \frac{J_n(k_\rho a)}{H_n^{(2)}(k_\rho a)} H_n^{(2)}(k_\rho\rho) e^{jn} \tag{2.58}$$

which can be verified to satisfy $E_z = 0$ when $\rho = a$.

The current on the circular cylinder can be shown to be given by

$$J_z = H_\varphi|_{\rho=a} = \frac{1}{j\omega\mu}\frac{\partial E_z}{\partial \rho}|_{\rho=a} = \frac{-2E_0}{\omega\mu\pi a}\sum_{n=-\infty}^{\infty}\frac{j^{-n}e^{jn\varphi}}{H_n^{(2)}(ka)} \quad (2.59)$$

from using the "Wronskian" relationship

$$J_n(x)Y_n'(x) - J_n'(x)Y_n(x) = \frac{2}{\pi x} \quad (2.60)$$

where the prime indicates differentiation with respect to the argument. For a thin wire, defined to be one where $ka \ll 1$, only the $n = 0$ term remains non-negligible, so that the total current on the wire becomes

$$I = \int_0^{2\pi} J_z a \, d\varphi \approx \frac{2\pi E_0}{j\omega\mu \log(ka)} \quad (2.61)$$

which can be seen to approach 0 as $1/\log(ka)$, being in phase quadrature with the incident field.

The case of a TE incident wave follows in a similar fashion to the above, with the incident and scattered z-directed magnetic fields given by the same expression as used for the electric field, i.e.

$$H_z = H_0 \sum_{n=-\infty}^{\infty} j^{-n}[J_n(k_\rho\rho) + B_n H_n^{(2)}(k_\rho\rho)]e^{jn} \quad (2.62)$$

but the applicable boundary condition is now

$$E_\varphi = E_\varphi^i + E_\varphi^s = 0. \quad (2.63)$$

The φ-component of the electric field is obtained as

$$E_\varphi = \frac{1}{j\omega\varepsilon}\mathbf{a}_\varphi \cdot \nabla \times \mathbf{H} = \frac{jk}{\omega\varepsilon}H_0\sum_{n=-\infty}^{\infty} j^{-n}[J_n'(k_\rho\rho) + B_n H_n^{'(2)}(k_\rho\rho)] \quad (2.64)$$

where a prime denotes differentiation with respect to the argument and the B_n is the modal amplitude of the scattered z-directed magnetic field. Thus

$$B_n = -\frac{J_n'(ka)}{H_n^{'(2)}(ka)} \quad (2.65)$$

and the current is then

$$J_\varphi = H_z(a) = j\frac{2H_0}{\pi ka}\sum_{n=-\infty}^{\infty}\frac{j^{-n}e^{jn\varphi}}{H_n^{'(2)}(ka)} \quad (2.66)$$

Electromagnetics

which is formally identical to Eq. (2.59) except that there is a derivative of the Hankel function in the denominator of the series. A plot of the current induced on the front side of a circular PEC cylinder by a normally incident TE plane wave is shown in Fig. 2.4.

2.5.4 Problem PA4: Radiation From an Infinite Circular Cylinder

Not surprisingly, whether excited as a scatterer, as just discussed, or as an antenna, a solution for an infinite cylinder takes the same general form. However, for the plane wave scattering problem the individual solutions whose sum yields the final result is a "discrete" spectrum in the z-direction, so called because k_z must have the same value as the incident plane wave. For the antenna problem to be discussed now, the solution involves a "continuous" spectrum in z, as will be seen below.

Because it provides a starting point for the finite length wire antenna, the problem considered here has the infinite cylinder excited by a z-directed electric field applied across a circumferential gap of width δ centered at $z = 0$. For simplicity we assume that this exciting field is constant across the gap and independent of φ, being given by $E_z^i(z) = -V/\delta$, and is zero elsewhere. This behavior is consistent with a linear variation of the potential

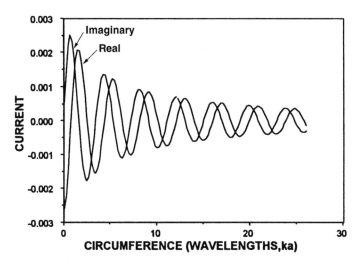

Figure 2.4 The current induced on the front side of a circular PEC cylinder illuminated by a normally incident plane wave as obtained from a separation-of-variables solution.

across the gap. We already know from the discussion of Problem PA3 that the general solution has the form

$$\psi = \sum_n \sum_{k_z} A_{n,k_z} \psi_{k_\rho,n,k_z} \quad \text{or} \quad \psi = \sum_n \int_{k_z} A_n(k_z)\psi_{k_\rho,n,k_z} dk_z$$

with our task to determine which of these applies to the antenna case. The answer comes from a consideration of the excitation, which is needed in a form compatible with this general solution. First note that E_z^i is independent of φ and ρ, being a function of z only. Second, its z variation must be expressed in the form of the z-variable solution, Eq. (2.51b). However, as specified above, E_z^i does not appear related to $\exp(\pm jk_z z)$, since the latter has a spatial description whereas the solution has a spectral form in terms of the z separation variable k_z. The needed form of E_z^i is obtained by representing it in terms of a weighted integral of the z solutions, which will be recognized to be a Fourier transform. We thus use the transform pair

$$E_z^i(z) = \frac{1}{2\pi} \int_{-\infty}^{\infty} e^{jk_z z} e_z^i(k_z) dk_z \qquad (2.67a)$$

and

$$e_z^i(k_z) = \int_{-\infty}^{\infty} e^{-jk_z z} E_z^i(z) dz \qquad (2.67b)$$

where Eq. (2.67b) provides the spectral form of the excitation needed for our separation-of-variables solution, and since there is no φ-dependence in effect only the $n = 0$ term is used. Upon using the form assumed for $E_z^i(z)$, Eq. (2.67b) yields [27]

$$e_z^i(k_z) = -2\pi V \frac{\sin(k_z \delta/2)}{k_z \delta/2} \qquad (2.68)$$

i.e., the spectrum of the excitation in terms of the spatial parameter δ and the spectral wavenumber k_z. Using the boundary condition that the total electric field is zero, i.e., that

$$E_z^i(z) + E_z^r(z) = 0 = \frac{1}{2\pi} \int_{-\infty}^{\infty} e^{jk_z z} e_z^i(k_z) dk_z + \int_{-\infty}^{\infty} A_0(k_z) e^{jk_z z} H_0^{(2)}(k_\rho a) dk_z \qquad (2.69)$$

where E^r is the radiated field. Since cancellation of the two terms in the integrand of Eq. (2.69) ensures that E_z will be zero, it follows that

$$-2\pi V \frac{\sin(k_z \delta/2)}{k_z \delta/2} + A_0(k_z) H_0^{(2)}(k_\rho a) = 0 \qquad (2.70)$$

so that $A_0(k_z)$ is found to be

$$A_0(k_z) = 2\pi V \frac{\sin(k_z\delta/2)}{k_z\delta/2} \frac{1}{H_0^{(2)}(k_\rho a)}. \tag{2.71}$$

Thus, the radiated electric field is given by

$$E_z^r(\rho, z) = \int_{-\infty}^{\infty} 2\pi V \frac{\sin(k_z\delta/2)}{k_z\delta/2} \frac{H_0^{(2)}(k_\rho \rho)}{H_0^{(2)}(k_\rho a)} e^{jk_z z} dk_z \tag{2.72}$$

The quantity of primary interest for the antenna is its axial current, which can be written

$$I_z(z) = H_\varphi(a, z) = \int_0^{2\pi} \frac{1}{j\omega\mu} \frac{\partial E_z(a, z)}{\partial \rho} d\varphi$$

$$= 2ja\omega\varepsilon V \int_0^\infty \cos(k_z z) \frac{\sin(k_z\delta/2)}{(k_z\delta/2)} \frac{H_0'^{(2)}(k_\rho a)}{k_\rho H_0^{(2)}(k_\rho a)} dk_z \tag{2.73}$$

$$= -2ja\omega\varepsilon V \int_0^\infty \cos(k_z z) \frac{\sin(k_z\delta/2)}{(k_z\delta/2)} \frac{H_1^{(2)}(k_\rho a)}{k_\rho H_0^{(2)}(k_\rho a)} dk_z$$

Evaluation of the integral (2.73) can be difficult, either numerically or analytically, although asymptotic answers for the latter approach can be obtained. The current at $z = \delta/2$ is then

$$I_z(\delta/2) = -2ja\omega\varepsilon V \int_0^\infty \cos(k_z\delta/2) \frac{\sin(k_z\delta/2)}{(k_z\delta/2)} \frac{H_1^{(2)}(k_\rho a)}{k_\rho H_0^{(2)}(k_\rho a)} dk_z \tag{2.74}$$

and the input admittance is obtained as

$$Y_0 \equiv \frac{I_z(\delta/2)}{V} = G_0 + jB_0 \tag{2.75}$$

where G_0 is the conductance and B_0 is the susceptance. Evaluation of Y_0 from Eq. (2.74) can be attempted analytically and numerically. Note that G_0 is determined from $0 \le k_z \le k$, since for $k_z > k$ the Hankel function ratio becomes pure imaginary, thus canceling out the j arising from k_ρ and resulting in a pure real integrand. Furthermore, with $k\delta/2 \ll 1$, the integrand does not depend on δ over $0 \le k_z \le k$, so that G_0 does not as well, i.e. the conductance is independent of the exciting gap width.

The susceptance determination is not as straightforward, however, because of non-integrable singularities in the integrand near $k = k_z$. From a numerical viewpoint, the

$$\int_0^\infty dk_z/k_z$$

infinite integration range presents a further complication. In the limit that $\delta \to 0$, the integrand varies as $1/k_z$ for $k_z \gg k$, resulting in Y_0 becoming infinite since

$$\int_0^\infty dk_z/k_z$$

is not convergent. That this is a reasonable result can be seen by observing that the capacitance of a parallel plate capacitor of area A is proportional to A/δ, and its admittance is given by $B = j\omega C$; thus $B \to \infty$ as $\delta \to 0$, i.e. for zero gap width.

Analytical approximations have been developed [27, 28] for Eq. (2.74), to obtain

$$G_{0,\text{app}} = -\sqrt{\frac{\mu}{\varepsilon}} \frac{\pi}{\ln(\Gamma K/2)} \tag{2.76a}$$

and

$$B_{0,\text{app}} = -2a\omega\varepsilon \left(\frac{0.9}{2K \ln[\Gamma K/2]} \right) + \ln(k/2) + \ln(\delta) \tag{2.76b}$$

where the susceptance is seen to have a log dependence on the exciting gap width.

A comparison of these approximations for the infinite antenna admittance with numerical evaluation of Eq. (2.75) is shown in Fig. 2.5. The conductance values agree to better than a 1% difference, but the susceptance values are about 10% apart, evidently due to differences in how B is defined.

2.5.5 Problem PA5: Radiation From a Hertzian Dipole Near an Infinite Plane

An important canonical problem in analyzing EM radiation is that of an antenna located near an interface between two homogeneous half spaces, a situation that arises whenever an antenna is operated near the Earth's surface. One starting point for modeling the behavior of wire antennas in this environment is to formulate the problem of a horizontal or vertical Hertzian current source, or dipole, located at the planar interface thus formed. A solution for these two current sources then permits representation of a straight, arbitrarily oriented current source by decomposing it into horizontal and vertical components. For simplicity, we describe only the solution for the vertical current source.

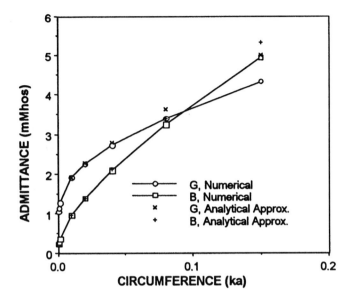

Figure 2.5 The input admittance (open symbols) as obtained from a numerical integration of the analytical separation-of-variables solution of a PEC circular cylinder of radius $a = 1$ cm and feed gap width $\delta = 0.1$ cm compared with an analytical approximation ($\times$ and $+$ symbols) [27].

Vertical Hertzian Current Source at the Interface

Let the axis of the vertical electric dipole (VED) source lie at $z = 0$ on the z-axis of a rectangular coordinate system, where the plane $z = 0$ represents the Earth's surface, so that points for which $z < 0$ are beneath the air–ground interface and where the propagation constant is

$$k_- = \omega(\mu_0\mu_-\varepsilon_0\varepsilon_-)^{1/2} = \omega(\mu_-\varepsilon_-)^{1/2}/c = \omega[\mu_-(\varepsilon_{-r} + \sigma_-/j\omega\varepsilon_0)]^{1/2}/c$$

while it is $k_+ = \omega(\mu_0\mu_+\varepsilon_0\varepsilon_+)^{1/2} = \omega(\mu_+\varepsilon_+)^{1/2}/c$ in the air above, where $z > 0$, with μ_0 and ε_0 are the permeability and permittivity of free space, c is the speed of light and the subscripts "$+$" and "$-$" denote the relative permeability and permittivity in the upper and lower half spaces, respectively. In the cylindrical coordinate system r, φ, z, distance from the origin is given by $R = \sqrt{(r^2 + z^2)^{1/2}}$, and the fields are independent of φ because of symmetry. The fields of a VED in an infinite medium can be represented by a Hertzian vector in the same direction that varies as $1/R$ near the origin, so a general solution for the VED near the interface can be written as [3]

$$\Pi_-(\rho, z) = \frac{1}{R}e^{jk_-R} + F_-(k_-; \rho, z)$$
$$\Pi_+(\rho, z) = \frac{1}{R}e^{jk_+R} + F_+(k_+; \rho, z) \tag{2.77}$$

where the first term accounts for the singularity and F_+ and F_- are finite everywhere, satisfy the wave equation and the radiation condition at infinity, and are to be determined by imposing continuity of the tangential E and H fields at the interface. Also, a multiplying term given by $IdL/(4\pi j\omega\varepsilon_0\varepsilon_+)$ is omitted for simplicity, where ε_+ is the relative permittivity of the upper medium. The first terms can also be regarded as the primary or incident fields, and the F terms as the secondary or scattered fields. The total fields are then given by

$$\mathbf{E} = \nabla\nabla\cdot\mathbf{\Pi}, \mathbf{H} = \frac{k^2}{j\omega\mu}\nabla\times\mathbf{\Pi} \tag{2.78}$$

so that the tangential (i.e. ρ, φ) components are obtained as

$$E_\rho = \frac{\partial^2}{\partial\rho\partial z}, H_\varphi = \frac{k^2}{j\omega\mu}\frac{\partial\Pi}{\partial\rho} \tag{2.79}$$

Thus, at any point on the interface, the functions (2.77) must satisfy

$$\frac{\partial^2\Pi_+}{\partial\rho\partial z} = \frac{\partial^2\Pi_-}{\partial\rho\partial z}, k_+^2\frac{\partial\Pi_+}{\partial\rho} = k_-^2\frac{\partial\Pi_-}{\partial\rho}, \tag{2.80}$$

and since they can be integrated with respect to r and the functions and their derivatives must vanish as $r \to \infty$, the boundary conditions simplify to

$$\frac{\partial\Pi_+}{\partial z} = \frac{\partial\Pi_-}{\partial z}, k_+^2\Pi_+ = k_-^2\Pi_- \text{ at } z = 0. \tag{2.81}$$

The next step in developing an analytical solution for the fields of the VED in both half spaces is to expand both the incident and scattered fields in cylindrical wavefunctions of the type

$$\psi(\rho, \varphi, z) = e^{jn\varphi}J_n(\lambda\rho)e^{jhz} \tag{2.82}$$

where h, a parameter, is the effective wavenumber in the z-direction and $\lambda = (k^2 - h^2)^{1/2}$ is the wavenumber in the radial direction, i.e. along the interface. Note that since the fields are angle-independent, only the $n = 0$ term appears in the final solution.

To complete the solution, we can multiply Eq. (2.82) by a function $f(\lambda)$, and integrate from 0 to ∞ to obtain

$$\Pi_+ = \frac{e^{jk_+R}}{R} + \int_0^\infty f_+(\lambda)J_0(\lambda\rho)\exp(-z\sqrt{\lambda^2-k_+^2})d\lambda,$$
$$\Pi_- = \frac{e^{jk_+R}}{R} + \int_0^\infty f_-(\lambda)J_0(\lambda\rho)\exp(+z\sqrt{\lambda^2-k_+^2})d\lambda \quad (2.83)$$

where we have used $jh = \pm(\lambda^2 - k^2)^{1/2}$, whose sign must always be chosen such that the fields vanish as $z \to \pm\infty$, which is done by making the real part of $(\lambda^2 - k^2)^{1/2}$ always positive.

The primary fields are also required in the same expansion so that the boundary conditions can be expressed in terms of their spectral components. The Fourier–Bessel theorem [3] permits us to write

$$f(\rho) = \int_0^\infty \lambda d\lambda \int_0^\infty r dr f(r) J_0(\lambda r) J_0(\lambda\rho) \quad (2.84)$$

if $f(\rho)$ vanishes as $\rho \to \infty$, so that the integral of $f(\rho)\rho^{1/2}$ converges absolutely. Now we let $f(\rho) = \exp(jk\rho)/\rho$ so that

$$\frac{e^{jk\rho}}{\rho} = \int_0^\infty \lambda d\lambda J_0(\lambda\rho) \int_0^\infty dr e^{jkr} J_0(\lambda r) \quad (2.85)$$

and observe that integration over r can be achieved if the Bessel function $J_0(\lambda r)$ is replaced by its integral representation [3], or

$$\int_0^\infty e^{jkr} J_0(\lambda r) dr = \frac{1}{2\pi} \int_{-\pi}^{+\pi} d\beta \int_0^\infty dr e^{j(k+\lambda\cos\beta)r} = \frac{1}{2\pi j} \int_{-\pi}^{+\pi} \frac{d\beta}{k+\lambda\cos\beta} \quad (2.86)$$

where it is assumed that k has a positive imaginary part so that the integrand becomes zero at $r = \infty$.

The latter integral in Eq. (2.86) turns out to be equivalent to integration about a closed contour, so that upon using

$$u = e^{j\beta}, \, du = ju d\beta$$

$$\frac{2}{2\pi j}\int_{-\pi}^{+\pi}\frac{d\beta}{k+\lambda\cos\beta} = \frac{1}{\pi}\oint\frac{du}{2ku+\lambda(1+u^2)} = \frac{1}{\pi\lambda}\oint\frac{du}{(u-u_1)(u-u_2)}, \quad (2.87)$$

with the integration path a unit circle centered at the origin of the complex u plane. The quantities u_1 and u_2 are the roots of

$$u^2 + \frac{2ku}{\lambda} + 1 = 0$$

and since their product is unity, one root (u_1) lies inside, and one (u_2) outside, of the unit circle. Thus, Eq. (2.87) can be written, using Cauchy's theorem, as

$$\frac{1}{\pi\lambda}\oint\frac{du}{(u-u_1)(u-u_2)} = \frac{1}{\pi\lambda(u_1-u_2)}\oint\frac{du}{u-u_1} = \frac{2j}{\lambda(u_1-u_2)} = \frac{1}{\sqrt{\lambda^2-k^2}} \tag{2.88}$$

A cylindrical wave expansion of the primary field Hertz vector can then be expressed as

$$\frac{e^{jk\rho}}{\rho} = \int_0^\infty \frac{J_0(\lambda\rho)}{\sqrt{\lambda^2-k^2}}\lambda d\lambda, \text{ at } z=0 \tag{2.89}$$

and upon using Eq. (2.82), a representation for $z \neq 0$ follows as

$$\frac{e^{jk_+R}}{R} = \int_0^\infty \frac{J_0(\lambda\rho)\exp\left(-z\sqrt{\lambda^2-k_+^2}\right)}{\sqrt{\lambda^2-k_+^2}}\lambda d\lambda,$$

$$\frac{e^{jk_-R}}{R} = \int_0^\infty \frac{J_0(\lambda\rho)\exp\left(+z\sqrt{\lambda^2-k_-^2}\right)}{\sqrt{\lambda^2-k_-^2}}\lambda d\lambda \tag{2.90}$$

so that the Hertz vectors of the total fields can be written

$$\Pi_+ = \int_0^\infty \left(\frac{\lambda}{\sqrt{\lambda^2-k_+^2}} + f_+(\lambda)\right)J_0(\lambda\rho)\exp\left(-z\sqrt{\lambda^2-k_+^2}\right)d\lambda,$$

$$\Pi_- = \int_0^\infty \left(\frac{\lambda}{\sqrt{\lambda^2-k_-^2}} + f_-(\lambda)\right)J_0(\lambda\rho)\exp\left(+z\sqrt{\lambda^2-k_-^2}\right)d\lambda. \tag{2.91}$$

Finally then the boundary condition that the tangential fields are continuous at $z=0$, Eq. (2.81) provides a solution for f_+ and f_- as

$$f_+(\lambda) = \frac{\lambda}{\sqrt{\lambda^2-k_+^2}}\frac{k_-^2[\lambda^2-k_+^2]^{1/2} - k_+^2[\lambda^2-k_-^2]^{1/2}}{k_-^2[\lambda^2-k_+^2]^{1/2} + k_+^2[\lambda^2-k_-^2]^{1/2}},$$

$$f_-(\lambda) = \frac{\lambda}{\sqrt{\lambda^2-k_-^2}}\frac{k_+^2[\lambda^2-k_-^2]^{1/2} - k_-^2[\lambda^2-k_+^2]^{1/2}}{k_+^2[\lambda^2-k_-^2]^{1/2} + k_-^2[\lambda^2-k_+^2]^{1/2}}. \tag{2.92}$$

The Hertz vectors can be considerably simplified upon substituting Eq. (2.92) into Eq. (2.91) and collecting terms to obtain

$$\Pi_+ = 2k_-^2 \int_0^\infty \frac{J_0(\lambda\rho)}{N} \exp\left(-z\sqrt{\lambda^2 - k_+^2}\right) \lambda \, d\lambda,$$

$$\Pi_- = 2k_+^2 \int_0^\infty \frac{J_0(\lambda\rho)}{N} \exp\left(-z\sqrt{\lambda^2 - k_-^2}\right) \lambda \, d\lambda,$$

where

$$N = k_+^2 \sqrt{\lambda^2 - k_-^2} + k_-^2 \sqrt{\lambda^2 - k_+^2}. \tag{2.93}$$

Vertical Electric Dipole Above Interface

Extending the previous development to the case where the VED is in the upper medium at a height z_+ is straightforward, with the Hertzian potentials then becoming [29]

$$\Pi_+ = \int_0^\infty \left[\exp\left(-\sqrt{\lambda^2 - k_+^2}|z - z_+|\right) + \frac{k_-^2\sqrt{\lambda^2 - k_+^2} - k_+^2\sqrt{\lambda^2 - k_-^2}}{k_-^2\sqrt{\lambda^2 - k_+^2} + k_+^2\sqrt{\lambda^2 - k_-^2}} \right.$$
$$\left. \exp\left(-\sqrt{\lambda^2 - k_+^2}|z + z_+|\right) \right] \frac{J_0(\lambda\rho)}{\sqrt{\lambda^2 - k_+^2}} \lambda \, d\lambda,$$

$$\Pi_- = 2\frac{k_+^2}{k_-^2} \int_0^\infty \exp\left(z\sqrt{\lambda^2 - k_-^2}\right) \frac{k_-^2}{k_-^2\sqrt{\lambda^2 - k_+^2} + k_+^2\sqrt{\lambda^2 - k_-^2}} J_0(\lambda\rho) \lambda \, d\lambda$$

$$\tag{2.94}$$

If $|k_+\alpha N_1| \gg 1$, then it can be shown [29] that the change in the impedance of a Hertzian dipole in the upper medium can be written

$$\frac{\Delta Z}{R_+} \approx \frac{3j e^{-jk_+\alpha}}{(k_+\alpha)^3}\left(\frac{N_1 - 1}{N_1 + 1}\right)\left\{\left(1 + \frac{4}{N_1}\right) + j\left(k_+\alpha + \frac{6}{k_+\alpha N_1}\left(1 - \frac{2N_1 - 3}{N_1^2}\right)\right)\right\} \tag{2.95a}$$

where

$$N_1^2 = \frac{k^2}{k_+^2}, \quad R_+ = 20(k_+ dL)^2, \quad \alpha = 2k_+ z_+ \tag{2.95b}$$

and R_+ is the radiation resistance of the same dipole located in free space. Some results for ΔZ as a function of z_0 for various ground parameters are shown in Fig. 2.6.

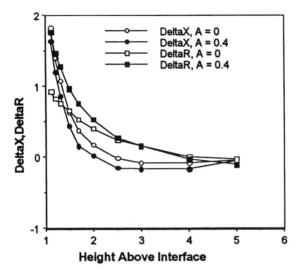

Figure 2.6 Change in impedance, divided by R_0, of a Hertzian dipole as a function of height in units of $2k_+z_0$ above a half space [29]. The parameter $A = \pi/4 - \beta$ where $\beta = 0.5\tan^{-1}(\varepsilon_-\omega/\sigma_-)$ and $|N_1|^2 = [\varepsilon_-^2 + (\sigma_-/\omega\varepsilon_0)^2] = 10$. The input impedance of an above-ground antenna on the subsurface increases upon approaching the interface, an effect that is sometimes compensated for by using a ground screen under the antenna. This sensitivity to the ground can also be used for subsurface exploration.

2.6 PROBLEMS REQUIRING COMPUTER SOLUTIONS

2.6.1 Problem PC1: Frequency Domain Modeling Using a Wire Integral Equation

An IE for a wire antenna in an infinite medium can take various forms, one of which is [16]

$$\hat{\mathbf{s}} \cdot \mathbf{E}^i(s) = \frac{1}{4\pi j\omega\varepsilon_\infty} \int_{C(\mathbf{r})} I(s') G_\infty(s, s') ds, \text{ with } G_\infty(s, s') \\ = [k_\infty^2 \hat{\mathbf{s}} \cdot \hat{\mathbf{s}}' + (\hat{\mathbf{s}} \cdot \nabla)(\hat{\mathbf{s}}' \cdot \nabla)] g_\infty(R) \quad (2.96)$$

In the above equations:

- s and s' denote the axial coordinates at the observation and source points, respectively;
- $\mathbf{r}$ and $\mathbf{r}'$ are the vector coordinates at the observation and source points;

Electromagnetics

- $\hat{\mathbf{s}} = \dfrac{\nabla C(\mathbf{r})}{|C(\mathbf{r})|}$ and $\hat{\mathbf{s}}' = \dfrac{\nabla C(\mathbf{r}')}{|C(\mathbf{r}')|}$ are unit tangent vectors at $\mathbf{r}$ and $\mathbf{r}'$;
- $C(\mathbf{r})$ is the range of integration over the wire;
- $a(\mathbf{r})$ is the wire radius at $\mathbf{r}$
- $s \in C(\mathbf{r}) + a(\mathbf{r})$ so that $R \geq a(\mathbf{r})$ as required by the thin-wire approximation;
- $g_\infty(R) = \dfrac{e^{-jk_\infty R}}{R}$;
- $R = [a(\mathbf{r}')^2 + |\mathbf{r} - \mathbf{r}'|^2]^{1/2}$;
- $k_\infty = \omega/(\mu_\infty \varepsilon_\infty)^{1/2}$ is the wavenumber of the infinite medium in which the wire is located;
- the superscript "i" denotes an incident field quantity;
- and the "^" is used to denote a unit length vector.

A generic form of the IE above can be written as [2]

$$L(s, s')f(s') = g(s) \tag{2.97}$$

where $L(s, s')$ is the integral operator, $f(s')$ is the unknown current and $g(s)$ is the known source or forcing function, which for our application are sampled values of a specified tangential electric field. The method of moments (MoM) is an intuitively logical way of solving this operator equation, whereby the unknown is expressed (or sampled) in terms of a set of basis or expansion functions $f_i(s')$ and unknown coefficients a_i

$$f(s') = \sum_{i=1}^{N_s} a_i f_i(s') \tag{2.98}$$

so that the operator equation can then be written

$$L(s, s') \sum_{i=1}^{N_s} a_i f_i(s') = \sum_{i=1}^{N_s} a_i L(s, s') f_i(s') = g(s) \tag{2.99}$$

where N_s is the number of spatial unknowns. Upon then sampling the operator equation (2.98) using a set of testing or weight functions

$$\{w_j(s)\}, m = 1, \ldots, M \tag{2.100}$$

there is obtained

$$\sum_{i=1}^{N_s} a_i \langle w_j(s), L(s, s'f_i(s')) \rangle = \langle w_j(s), g(s) \rangle; \; m = 1, \ldots, M \tag{2.101}$$

where the $\langle \, , \, \rangle$ signify what is called an "inner product" (the inner product of two functions $p(\mathbf{r})$ and $q(\mathbf{r})$ over a surface S is defined as $\int_S p(\mathbf{r}) q(\mathbf{r}) d^2 s$).

The original operator equation can now be written in a discretized, sampled approximation as

$$\sum_{i=1}^{N_s} Z_{ji}a_i = b_j; j = 1, \ldots, M \quad (2.102)$$

where

$$Z_{ji} = \langle w_i(s), L(s, s')f_i(s') \rangle \text{ and } b_i = \langle w_i(s), g(s) \rangle \quad (2.103)$$

Finally, the coefficients that quantify the numerical solution for the current distribution are obtained as

$$a_i = \sum_{j=1}^{M} Y_{ij}b_j, i = 1, \ldots, N_s \quad (2.104)$$

where Z_{ij} is known as the impedance matrix and Y_{ij} is the admittance matrix, being its inverse. Equations (2.102) and (2.104) can be written more compactly in symbolic form as

$$\vec{Z} \cdot \mathbf{A} = \mathbf{B} \to \mathbf{A} = \vec{Y} \cdot \mathbf{B}$$

The admittance matrix is a numerical representation of the complete EM properties of the object being modeled, within the approximations made in its computation, and as such it has some interesting properties. First, it provides a solution for arbitrary excitation. Second, it can be stored for subsequent reuse. Perhaps most interesting, it possesses in a very real sense the properties of a hologram, as can be seen by writing the field radiated by the object as

$$E_r = \sum_{i=1}^{N_s} E_i^o \sum_{j=1}^{N_s} Y_{ij} E_j^i \quad (2.105)$$

where the first sum yields a solution for the current distribution and the second sums the field at a specified observation point due to that current. Also

$$E_i^o = \hat{\mathbf{s}}_i \cdot \mathbf{E}^o(s_i) \text{ and } E_j^i = \hat{\mathbf{s}}_j \cdot \mathbf{E}^i(s_j) \quad (2.106)$$

where the exciting field $\mathbf{E}^i$ and the "observation" field $\mathbf{E}^o$ are tangential projections onto the object of an arbitrary incident field and from a point source located at the observation point.

Much work has been done to identify the "best" expansion and testing functions. A variety of combinations is described by Poggio and Miller [30]. Those in numerical electromagnetics code (NEC) use what is called "subsectional collocation" wherein the wire is divided into N_s "segments" with

Electromagnetics

$$f_i(s') = A_i + B_i \sin[k(s' - s_i)] + C_i\{\cos[k(s' - s_i)] - 1\},$$
$$s_i - \Delta_i/2 \leq s' \leq s_i + \Delta_i/2, i = 1, \ldots, N_s \qquad (2.107a)$$

and

$$w_j(s) = \delta(s - s_j), j = 1, \ldots, N_s \qquad (2.107b)$$

where Δ_i is the length of segment i and with $\delta(s - s_j)$ a delta function, thus producing point sampling of the tangential electric field. Of the $3N_s$ unknowns in Eq. (2.107a), $2N_s$ are eliminated by enforcing current and current–slope (charge) continuity at the junctions between segments, so that the final number is N_s.

The treatment just outlined can be used for developing a numerical solution for both IEs and DEs, making it possible to obtain results for a wide variety of problems for which analytical solutions are unavailable. An example is presented here as obtained from the thin wire, electric field IE (Eq. 2.96), using NEC. The input admittance of a wire antenna excited at its center by an idealized voltage source is compared with the result obtained when the same antenna is driven by a two-wire transmission line, as illustrated in Fig. 2.7, is shown in Fig. 2.8. The increased capacitive effect of the transmission line is seen to shift the resonance pattern of the admittance upward in frequency.

2.6.2 Problem PC2: Frequency Domain Scattering from a PEC Cylinder

Starting with the Helmholtz equation,

$$(\nabla^2 + k^2)F(\rho, \varphi) = 0, \qquad (2.108)$$

where $F(\rho, \varphi)$ is the field quantity whose solution is sought, we obtain the discretized form in some solution space [31]:

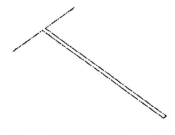

Figure 2.7 Model for a dipole antenna fed from a two-wire transmission line.

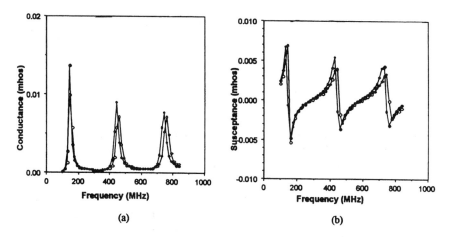

Figure 2.8 (a) Input conductance and (b) susceptance for a transmission line-fed (open circles) (G. J. Burke, personal communication) and gap-excited dipole antenna (closed circles) as obtained from the NEC model. An upward shift in the resonance structure occurs because of the increased capacitance of the transmission line feed.

$$[Z]X] = Y], \quad (2.109)$$

where [Z] is the DE system matrix which arises from discretizing $(\nabla^2 + k^2)$, and X] denotes the vector of unknown field samples. We use the terms "scattered field formulation" (SFF) and "total field formulation" (TFF), depending on whether X] represents the scattered field or the total field. The vector Y] is a known quantity and would usually represent an incident field evaluated either over the surface of the object being modeled or over some closure boundary at which the mesh of field samples is terminated. For simplicity, we restrict our attention to a normally incident plane wave on a perfectly conducting, infinite circular cylinder.

The Helmholtz equation in polar coordinates is

$$[\partial^2/\partial \rho^2 + \rho^{-2}\partial^2/\partial \varphi^2 + \rho^{-1}\partial/\partial \rho + k^2]F(\rho, \varphi) = 0, \quad (2.110)$$

for which the computational molecule centered at discretized coordinates r, a with sampling intervals $\Delta \rho$ and $\Delta \varphi$ in a polar mesh can be written in terms of samples $x_{r,a} = F(\rho_r, \varphi_a)$, $a = 1, \ldots, A$ and $r = 1, \ldots, R$, as

$$(x_{r+1,a} + x_{r-1,a} - 2x_{r,a})/(\Delta\rho)^2 + (x_{r,a+1} + x_{r,a-1} - 2x_{r,a})/(\rho_r\Delta\varphi)^2$$
$$+ (x_{r,a} - x_{r-1,a})/(\rho_r\Delta\rho) + k^2 x_{r,a} = 0.$$

$$(2.111)$$

Upon collecting terms, Eq. (2.111) becomes

$$x_{r+1,a}(\Delta\rho)^{-2} + x_{r,a}[k^2 + (\rho_r\Delta\rho)^{-1} - 2(\Delta\rho)^{-2} - 2(\rho_r\Delta\varphi)^{-2} \\ + x_{r-1,a}[(\Delta\rho)^{-2} - (\rho_r\Delta\rho)^{-1}] + (x_{r,a+1} + x_{r,a-1})/(\rho_r\Delta\varphi)^2 = 0. \quad (2.112)$$

The $x_{r,a}$ are the z-directed electric and magnetic fields for the transverse magnetic (TM) and transverse electric (TE) polarizations, respectively, which serve as the potentials from which the other field components can be obtained using finite difference approximations to the curl equations, assuming an $e^{j\omega t}$ time dependence, as

$$(H_\varphi)_{r,a} = (-j/k\eta)\partial(E_z)_{r,a}/\partial\rho \approx (-j/k\eta)(x_{r+1,a} - x_{r,a})/\Delta\rho, \quad (2.113a)$$

$$(H_\rho)_{r,a} = (j/k\eta\rho)\partial(E_z)_{r,a}/\partial\varphi \approx (j/k\eta\rho)(x_{r,a+1} - x_{r,a})/\Delta\varphi, \quad (2.113b)$$

$$(E_\varphi)_{r,a} = (-j\eta/k)\partial(H_z)_{r,a}/\partial\rho \approx (-j\eta/k)(x_{r+1,a} - x_{r,a})/\Delta\rho, \quad (2.114a)$$

and

$$(E_\rho)_{r,a} = (j\eta/k\rho)(\partial H_z)_{r,a}/\partial\varphi \approx (j\eta/k\rho)(x_{r,a+1} - x_{r,a})/\Delta\varphi. \quad (2.114b)$$

The overall system matrix that results from such straightforward discretization of the Helmholtz equation can thus be expressed as

$$|Z|_{AR \times AR} = \begin{bmatrix} [c] & [d] & [0] & \cdots & & [0] \\ [d] & [c] & [d] & [0] & \cdots & [0] \\ [0] & [d] & [c] & [d] & [0] & \vdots \\ [0] & [0] & [d] & [c] & [d] & \vdots \\ \vdots & \vdots & \vdots & \vdots & \vdots & \vdots \\ [0] & \cdots & [0] & [d] & [c] & [b] \\ [b] & \cdots & [0] & [0] & [b] & [c] \end{bmatrix} \quad (2.115a)$$

where the various blocks have the general form

$$|c|_{A \times A} = \begin{bmatrix} X & X & 0 & \cdots & 0 & 0 & X \\ X & X & X & \cdots & 0 & 0 & 0 \\ 0 & X & X & \cdots & 0 & 0 & 0 \\ \vdots & \vdots & \vdots & \vdots & \vdots & \vdots & \vdots \\ 0 & 0 & 0 & \cdots & X & X & 0 \\ 0 & 0 & 0 & \cdots & X & X & X \\ X & 0 & 0 & \cdots & 0 & X & X \end{bmatrix} \quad (2.115b)$$

and

$$|\mathbf{d}|_{A \times A} = \begin{bmatrix} X & 0 & 0 & \cdots & 0 & 0 & 0 \\ 0 & X & 0 & \cdots & 0 & 0 & 0 \\ 0 & 0 & X & \cdots & 0 & 0 & 0 \\ \vdots & \vdots & \vdots & \vdots & \vdots & \vdots & \vdots \\ 0 & 0 & 0 & \cdots & X & 0 & 0 \\ 0 & 0 & 0 & \cdots & 0 & X & 0 \\ 0 & 0 & 0 & \cdots & 0 & 0 & X \end{bmatrix} \qquad (2.115c)$$

This specific form of the submatrix [c] results from the simplest kind of central differencing to approximate the wave equation, which is equivalent to using a quadratic basis and point testing. Use of higher-order bases and/or testing functions, as are encountered in what are called "finite element" as opposed to finite difference models, would change the form of Eq. (2.115) principally by adding further blocks of similar form away from the main diagonal because the spatial sampling would then extend two (or more) mesh points in angle or radius away from the testing or observation point that defines the wave equation. Similarly, irregular meshes could produce varying patterns of non-zero coefficients and thus complicate the "bookkeping" required of the procedure described below, but the overall result remains a sparse system matrix having the general nature demonstrated here.

The "X"s represent general numerical coefficients determined by the kinds of basis and testing functions employed in the discretization and sampling operation, and the submatrices [b], not all of which would be present at the same time, are determined by the closure condition used to terminate the DE mesh, i.e., to keep it from extending to infinity for exterior problems. The need for a closure condition to permit truncation of the spatial mesh can be seen from Eq. (2.112), where computing $x_{r,a}$ requires values at $r+1$ and $r-1$. But the values at $r+1$ themselves require those at $r+2$, etc. A closure condition provides a way of ending this "leapfrog" effect, but it is a topic that is beyond the scope of the present discussion, about which further details and references can be found in [31]. The results shown in Figs. 2.9 and 2.10 were obtained using a modal closure condition and a stepwise solution of Eq. (2.112)

2.6.3 Problem PC3: Transient Radiation from Wire Antenna

Equations analogous to those available in the frequency domain can be derived for many situations directly in the time domain. For a wire in free space, an IE comparable to the Pocklington IE can be developed as [32, 33]

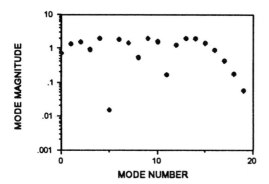

Figure 2.9 Comparison of results for mode magnitudes obtained from a numerical differential equation model (circles) with an analytical separation-of-variables solution (crosses) for TM plane-wave scattering from perfectly conducting circular cylinder of five wavelengths diameter shows the good agreement that a numerical model can provide, with the results being graphically indistinguishable [31].

$$\hat{\mathbf{s}} \cdot \mathbf{E}^i(s, t) = \frac{\mu_\infty}{4\pi} \int_{C(\mathbf{r})} \left(\frac{\hat{\mathbf{s}} \cdot \hat{\mathbf{s}}'}{R} \frac{\partial I(s', t')}{\partial t'} + \left[\frac{c_\infty \mathbf{s} \cdot \mathbf{R}}{R^2} \right] \left[\frac{\partial I(s', t')}{\partial s'} - \frac{c_\infty}{R} Q(s', t') \right] \right) ds',$$

$$Q(s', t') = -\int_{-\infty}^{t'} \frac{\partial I(s', t')}{\partial s'} dt'$$

(2.116)

where $Q(s, t)$ is the charge density at space location s and time t, and with the "retarded time" $t' = R/c_\infty$, where c_∞ is the speed of light in the medium. A time domain IE like that in Eq. (2.116) can be readily solved using "time stepping," whereby the solution is developed as a function of space and time as briefly outlined here.

First, write the IE in operator notation as

$$O(s, t; s', t') F(s', t') = G(s, t), \quad (2.117)$$

where t' is used here to denote the retarded time, i.e. $t' = t - R/c_\infty$. Proceeding in a fashion analogous to that used for solving the frequency domain IE, we might then represent the space and time dependence of $F(s', t')$ as

$$F(s', t') = \sum_{i=1}^{N_s} \sum_{j=1}^{N_t} A_{ij} P_{ij}(s', t') \quad (2.118)$$

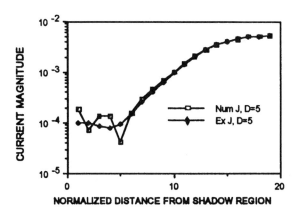

Figure 2.10 Magnitude of surface current as obtained using a differential equation numerical model (squares) and from analytical solution (diamonds) for TM incidence and a five-wavelength diameter cylinder. The numerical results are in agreement to about 1% relative to the peak.

where the A_{ij} are space–time samples of the unknown (i being the space index and j the time index) of which there are N_s and N_t space and time samples respectively, so that Eq. (2.118) becomes

$$\sum_{i=1}^{N_s}\sum_{j=1}^{N_t} A_{ij} O(s,t;s',t') P_{ij}(s',t') = G(s,t) \tag{2.119}$$

Upon sampling the field using the testing functions

$$\{W_{km}(s,t)\} = \{w_k(s)v_m(t)\}, k = 1, \ldots, N_s, m = 1, \ldots, N_t \tag{2.120}$$

the operator equation can be written in a discretized, sampled approximation as

$$\sum_{i=1}^{N_s}\sum_{j=1}^{N_t} A_{ij} \langle v_m(t), \langle w_k(s), O(s,t';s',t') P_{ij}(s',t') \rangle \rangle = \langle v_m(t), \langle w_k(s), G(s,t) \rangle \rangle \tag{2.121}$$

where the inner product now involves not only a space integration, but a time integration as well.

Employing subsectional collocation in both space and time, the unknown current can be written as

$$P_{ij}(s',t') = A_{1,ij} + A_{2,ij}(s'-s_i) + A_{3,ij}(s'-s_i)^2 + A_{4,ij}(t'-t_j)$$
$$+ A_{5,ij}(t'-t_j)^2 + A_{6,ij}(s'-s_i)(t'-t_j) + A_{7,ij}(s'-s_i)^2(t'-t_j)$$
$$+ A_{8,ij}(s'-s_i)(t'-t_j)^2 + A_{9,ij}(s'-s_i)^2(t'-t_j)^2$$
$$s_i - \Delta_i/2 \leq s' \leq s_i + \Delta_i/2, i = 1, \ldots, N_s$$
$$t_j - \delta/2 \leq t' \leq t_j + \delta/2, j = 1, \ldots, N_t$$

(2.122a)

with the constant time step being δ, while the testing function is given by

$$W_{km}(s,t) = \delta(s-s_j)\delta(t-t_j) \quad (2.122b)$$

so that the tangential field is point sampled in both space and time and the time solution is developed from time stepping. There are nine unknown coefficients associated with each space–time step, eight of which are eliminated in a fashion similar to that described for NEC, by matching the current to its neighbors in the adjacent eight space–time samples [32] at $i \pm 1, j; i, j \pm 1; i \pm 1, j \pm 1$. An alternate approach would be to write the basis function as $P_{ij}(s',t') = R_i(s')S_j(t')$, which would then involve five unknowns (by setting the constant term in $S_j(t')$ to unity) with four being eliminated by matching at $i \pm 1, j; i, j \pm 1$. Finally, the space–time current samples can be written in the form

$$I_{ij} = I(s_i, t_j) = Y_{ik}E_{kj}^t = Y_{ik}(E_{kj}^i + E_{kj}^s) = Y_{ik}\left(E_{kj}^t + \sum_{i'=1}^{N_s} Z_{ii'}I_{i',m-f(i,i')}\right)$$

(2.123)

where E_{ij}^t = total E_{ij}, E_{ij}^s = scattered E_{ij}.

For a straight wire, $f(i, I') = |i - I'|$, but it is generally a more complicated function of object geometry. Also, Y_{ik} represents the inverse of the time-independent matrix that accounts for field–current interactions in the time step j and $Z_{ii'}$ accounts for the fields produced over the object from currents that occurred at earlier times. Developed in this way, the time-stepping model is called "implicit," while if no interactions are allowed within time step j the model is "explicit."

This kind of approach has been used to develop a general-purpose TD model for wires called TWTD (thin-wire time domain) [33]. The current at the feedpoint of a wire excited by a Gaussian voltage pulse, i.e. $V = V_0 \exp[-t^2 a^2]$, and the broadside radiated field that it produces are illustrated in Fig. 2.11. Because there are no end reflections during the time history shown in this figure, the wire is infinitely long in effect, so that the behavior of an infinite cylindrical antenna can be modeled. The negative undershoot of the current is due to a small, continuous reflection of

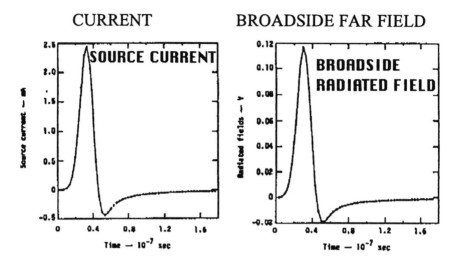

Figure 2.11 The source current and broadside-radiated field of an infinite PEC cylinder excited by a Gaussian voltage pulse [33]. The negative undershoot of the source current indicates a continuous reflection of the outward-propagating current pulse leaving the source region. This reflection is accompanied by a charge acceleration that produces a negative radiated field compared with the positive pulse caused by the initial charge acceleration due to initiation of the current.

the outward-propagating current and charge pulse, which produces a corresponding negative undershoot in the radiated field following the original positive pulse due to the charge acceleration that accompanies initiation of the current.

In Fig. 2.12 are shown space snapshots of the current and charge on a finite length center-fed dipole, where more physics of the radiation process is revealed by the slowly diminishing amplitudes of the current and charge pulses as they propagate down the wire, and the more noticeable decreases that occur upon end reflection. Space–time contour plots of the far field are presented in Fig. 2.13, which further illustrate the radiation process by the circular (since the wire is rotationally symmetric they are cuts from a sphere) fields centered at the wire's feedpoint and its ends. Further radiation circles would be seen were the plot to include more time history and hence more end reflections.

Electromagnetics

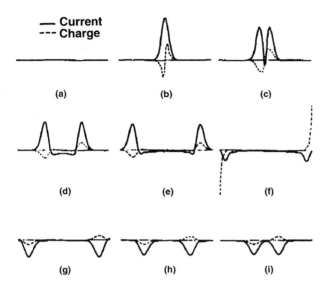

Figure 2.12 Time snapshots of the current (solid line) and charge (dashed line) pulses produced by a Gaussian voltage pulse applied at the center of a wire dipole. The slight decay of the pulses as they propagate towards the wire ends is caused by reflection of some of the current and charge back towards the source, while the greater decay upon end reflection is due to the larger radiation that occurs there.

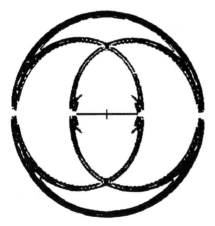

Figure 2.13 Space–time contour plots of the far radiated field from the dipole whose current and charge are illustrated in Fig. 2.10. The dominant radiation sources are associated with the initial charge acceleration when the exciting voltage pulse is applied and the subsequent acceleration that occurs when the pulse reflect from the wire ends.

2.6.4 Problem PC4: Determining Transfer Functions

An ubiquitous need in electromagnetics, indeed in any field where wavelike behavior is of concern, is to obtain the response of a system or component as a function of frequency. This is normally done by: (1) sampling a response at discrete frequencies across a band of interest; and (2) plotting the results to evaluate the behavior. If the response is found to not be well-enough resolved, it is necessary to continue repeating steps (1) and (2) until the response appears to be adequately sampled and its variation with frequency can be approximated by straight-line interpolation between the available samples.

This approach can be more costly than necessary, since, understandably, the spectrum is often over-sampled to avoid missing any important features of the response. On the other hand, it can also be error-prone if the response contains fine details such as high-Q resonances that are unanticipated and therefore easily missed. What is needed is a way to minimize the number of samples needed to develop an approximate version of the spectrum to a specified degree of uncertainty. This is the kind of problem to which an adaptive sampling strategy would be well suited if one can be developed, which is the subject of this section.

Observe that the essential problem here is to find a "reduced order" or fitting model (FM) that can approximate the sampled data provided from what we will call a generating model (GM) (e.g., in EM a model based directly on Maxwell's equations) and to devise a way of testing how well this FM represents that sampled data. Fortunately, frequency responses in EM, acoustics, structural analysis, and similar fields are very well approximated by a pole series, i.e. [34, 35],

$$F(X) \approx F_p(X) + F_{np}(X) = \sum R_\alpha/(X - s_\alpha) + F_{np}(X); \alpha = 1, \ldots, P \quad (2.124)$$

where $F(X)$ is the fitting model. The subscript "p" refers to that part of $F(X)$ that can be represented by a pole series, while the "np" designates the remaining, or non-pole, part of $F(X)$. This is necessary, since a pole series alone may not provide the needed accuracy.

Assuming the availability of samples of $F(X)$ denoted by $F(X_i) = F_i$, we equate the FM to these sampled values, i.e.,

$$F_i = F(X_i) = \sum R_\alpha/(X_i - s_\alpha) + F_{np}(X_i); \alpha = 1, \ldots, P \quad (2.125)$$

where X_i represents the sampling frequency. Usually, sampling would be done along the $j\omega$-axis so that $X_i = j\omega_i$, with $\omega_i = 2\pi f_i$.

Note that a simple pole series can be developed into a particular rational function where the denominator order exceeds that of the numerator by one. A general rational function has no specified connection between

Electromagnetics

the orders of the polynomials which comprise it, however. The capability of a pole series to model resonances can be retained while changing the numerator polynomial order relative to that of the denominator. The possibility of using various numerator and denominator polynomial orders provides a way to approximate the effect of the non-pole term by simply increasing the order of the numerator polynomial. For example, an increase of one in the numerator order has the effect of representing F_{np} by a constant, which, when absorbed into the rational function, results in equal numerator and denominator orders. If F_{np} is represented by a constant and a term linear in X, this has the effect of making the numerator order one greater than the denominator. Thus, by varying the relative orders of the polynomials which comprise the FM, various models of the non-pole contribution are implicitly included.

Therefore, in general we use a FM given by

$$F(X) = N(X)/D(X), \qquad (2.126a)$$

$$N(X) = N_0 + N_1 X + N_2 X^2 + \cdots + N_n X^n, \qquad (2.126b)$$

and

$$D(X) = D_0 + D_1 X + D_2 X^2 \cdots + D_d X^d. \qquad (2.126c)$$

The coefficients of the FM are obtained from sampled values of Eqs. (2.126) as given by

$$F_i D_i = N_i, \, i = 0, \ldots, D-1 \qquad (2.127a)$$

where

$$F_i = F(X_i), \qquad (2.127b)$$

$$D_i = D(X_i) = D_0 + D_1 X_i + D_2 (X_i)^2 + \cdots + D_d (X_i)^d, \qquad (2.127c)$$

and

$$N_i = N(X_i) = N_0 + N_1 X_i + N_2 (X_i)^2 + \cdots + N_n (X_i)^n. \qquad (2.127d)$$

where it may be seen that there are $d + n + 2$ unknown coefficients in the two polynomials $D(X)$ and $N(X)$. A constraint or additional condition is needed to make the sampled equations inhomogeneous, and while there is no unique choice for this constraint, if we set $D_d = 1$ then the following equations result:

$$F_0 D_0 + X_0 F_0 D_1 + \ldots + F_0(X_0)^{d-1} D_{d-1} \\ - N_0 - X_0 N_1 - \ldots - (X_0)^n N_n = -(X_0)^d F_0 \\ F_1 D_0 + X_1 F_1 D_1 + \ldots + F_1(X_1)^{d-1} D_{d-1} \\ - N_0 - X_1 N_1 - \ldots - (X_1)^n N_n = -(X_1)^d F_1 \\ \vdots \\ F_{D-1} D_0 + X_{D-1} F_{D-1} D_1 + \ldots + F_{D-1}(X_{D-1})^{d-1} D_{d-1} \\ - N_0 - X_{D-1} N_1 - \ldots - (X_{D-1})^n N_n = -(X_{D-1})^d F_{D-1}$$

(2.128)

where $D \geq n + d + 1$ is then required.

Equation (2.128) provides a solution for the FM coefficients from samples of $F(X)$. Aside from providing a continuous estimate of $F(X)$ from its sampled values, it's also possible to further reduce the number of samples needed by locating them adaptively in frequency. This possibility is illustrated conceptually in Fig. 2.14. The basic idea is to employ three or more overlapping FMs that cover the frequency range of interest, where

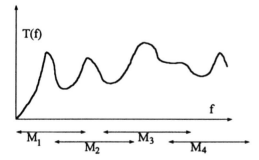

Figure 2.14 The possibility of developing an FM representation of a transfer function over a wide frequency interval by employing a numer of subinterval lower-order fitting models, M_i, is illustrated here. Each FM represents the response over a frequency interval $F_i = F_{i,u} - F_{i,l}$, where the adjacent FMs share some frequency samples and the difference between them $\Delta MM_{i,j}(f) = [|M_i(f) - M_j(f)|]/[|M_i(f)| + |M_j(f)|]$ provides a normalized, mutual-error estimate between FMs i and j. By evaluating $\Delta MM_{i,j}(f)$ as a function of frequency, new samples, either function or derivative, can be placed where $\Delta MM_{i,j}(f)$ exhibits values that exceed a specified amount and where, therefore, the corresponding FM results are most uncertain.

Electromagnetics

samples are successively added at the frequency where the maximum difference between the overlapping FMs occurs until a specified error criterion is satisfied across the entire bandwidth.

The result of using such an adaptive-sampling approach is demonstrated in Fig. 2.15 for the problem of the current induced on a circular, PEC cylinder by a normally incident plane wave, as previously illustrated in Problem PA3. It can be seen that, beginning with a relatively sparse sampling of the current's frequency response, computing a small number of additional samples based on the FM differences yields an analytical approximation of the exact response that is accurate to the pre-specified uncertainty of 0.1%.

2.6.5 Problem PC5: Useful Numerical Methods

Shanks' Method

Numerical computations very often involve dealing with sequences of numbers as some parameter is increased, with the goal that when it becomes large enough a converged result accurate to some specified level will be obtained. In many cases, for example the summing of infinite series or evaluating infinite range integrals, the parameter is the number of terms actually summed or the range over which the integral is computed.

Numerical methods having this goal exhibit a signal-processing nature. As a matter of fact, Shanks [36], who developed a method for accelerating

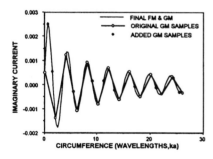 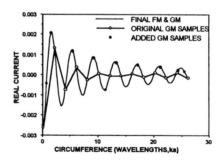

Figure 2.15 The frequency dependencies of the real and imaginary currents on the front side of a PEC circular cylinder due to a normally incident TE plane wave are shown as initially sampled (the open circles), with the additional samples automatically placed by adaptive sampling (the solid circles), and with the continuous estimate based on a rational function fitting model (the solid line). The FM estimate matches the exact result between the samples used to within 0.1%.

the convergence of sequences of partial sums, referred to the set of samples that such an operation produces as a "mathematical transient" for which two applications arise. These are to estimate the "limit" or "anti-limit" of the sequence as a parameter goes to infinity or zero. Shanks' method is a very general procedure, of which the lowest-order form is given by

$$S_{ij} = \frac{S_{i-1,j-1}S_{i-1,j+1} - S_{i-1,j}^2}{S_{i-1,j-1} + S_{i-1,j+1} - 2S_{i-1,j}} \qquad (2.129)$$

where $j = 1 - 1, \ldots, N - i + 1$, and $i = 1, \ldots, \text{Int}[(N + 1)/2,]$ with $N + 1$ the number of terms in the original sequence, the values of which are given by $S_{1,0}$ to $S_{1,N}$, and it should be noted that Eq. (2.129) is nonlinear in the $S_{i,j}$. Shanks' Method works in much the same way as a difference table, involving computation of successive sequences obtained from values of the preceding sequence.

Application of Shanks' Method leads to a triangular array of numbers where convergence occurs, although not necessarily monotonically, for increasing values of both i and j, as

$$\begin{bmatrix} S_{1,0} \\ S_{1,1} & S_{2,1} \\ S_{1,2} & S_{2,2} & \cdots \\ S_{1,3} & S_{2,2} & \vdots & S_{N/2,N/2} \\ & & \vdots & \cdots \\ \vdots & S_{2,N-1} \\ S_{1,N} \end{bmatrix} N \text{ even,}$$

$$\begin{bmatrix} S_{1,0} \\ S_{1,1} & S_{2,1} \\ S_{1,2} & S_{2,2} & \cdots \\ S_{1,3} & S_{2,2} & \vdots & S_{(N+1)/2,(N+1)/2-1} \\ & & & S_{(N+1)/2,(N+1)/2} \\ \vdots & \vdots & \cdots \\ & S_{2,N-1} \\ S_{1,N} \end{bmatrix} N \text{ odd} \qquad (2.130)$$

A simple example is provided by the Leibniz series for π, given by $\pi = 4(1 - 1/3 + 1/5 - 1/7 + 1/9 - 1/11 + \cdots)$, which gives an accuracy of about x places using approximately 10^x terms (!) of the series. By contrast, using Shanks' Method on the Leibniz series gives an accuracy of x places after using approximately only x terms, a result that is demonstrated below

Electromagnetics

using $N = 10$. This kind of dramatic result cannot always be expected of Shanks' Method, but it does demonstrate that 'information' about π's value is implicit, although not obviously available, in the original series.

4.000000000
2.666666666 3.166666666
3.466666666 3.133333333 3.142105263
2.895238095 3.145238095 3.141450216 3.141599357
3.339682539 3.139682539 3.141643324 3.141590860 3.141592714
2.976046176 3.142712843 3.141571290 3.141593231 3.141592637 3.141592654
3.283738484 3.140881349 3.141602742 3.141592438 3.141592659
3.017071817 3.142071817 3.141587321 3.141592743
3.252365935 3.141254824 3.141595655
3.041839619 3.141839619
3.232315809

where the left-most column is the original sequence of answers, $S_{1,0}$ to $S_{1,10}$ and the final estimate is $S_{6,6} = 3.141592654$, which can be compared with an 11-digit answer for π given by 3.1415926536.

Another practical computational application made of Shanks' Method is in computing the Sommerfeld integrals that arise when analyzing the fields of current sources located near infinite planar interfaces. An example of using Shanks' Method for evaluating a simpler integral, which is also computed using Romberg's method in the next section [37], is given below. The integral tested here is

$$\int dx/x, 1 \leq x \leq 3 = 1.098\,612\,288\,668\,109\,691\,395\,245\,236\,922\,525\,7 \quad (2.131)$$

whose Shanks' table, expressed in digits of agreement with the above, is

0.6703
1.2080   2.5127
1.7843   3.4890   4.9803
2.3782   4.5867   6.4480   8.3950
2.9781   5.7527   8.0764 10.3459 12.7087
3.5796   6.9458   9.8116 12.5034 15.2093 18.1376
4.1815   8.1470 11.5957 14.8029 17.9113
4.7835   9.3504 13.3959 17.1749
5.3856 10.5543 15.2006
5.9877 11.7584
6.5897

from which the most accurate answer is 1.098 612 288 668 109 690 59. Thus, nearly 12 orders of magnitude improvement is achieved from the trapezoidal

Richardson Extrapolation and Romberg Quadrature

Another technique that increases accuracy for some sequences and which is used in Romberg's Method for numerical quadrature (integration), is Richardson extrapolation [37, 38]. Richardson extrapolation is based on the idea that the error in a numerical process which depends on a sampling interval h is proportional to h raised to an integer power. Suppose for example that

$$f(h) = f_{\text{true}} + E(h) \tag{2.132}$$

where $f(h)$ is a computed result based on sampling the process of interest at an interval spacing of h, f_{true} is the true result, and $E(h)$ is an error term with the property $E(h) \to 0$ when $h \to 0$. If the error is simply linearly dependent on h, then Eq. (2.132) can be written

$$f(h) = f_{\text{true}} + a_1 h. \tag{2.133}$$

Observe that in Eq. (2.133) f is a computable quantity once h is specified, and f_{true} and a_1 are unknowns. Thus, if f_1 is computed for $h = h_1$ and f_2 for $h = h_2$, there then result two equations in two unknowns, from which f_{true} can be obtained as

$$f_{\text{true}} = (f_2 h_1 - f_1 h_2)/(h_1 - h_2) \tag{2.134}$$

In the special case $h_1 = 2h_2$, then

$$f_{\text{true}} = 2f_2 - f_1 \tag{2.135}$$

This procedure can be easily adapted to the more general case where $E(h)$ is given by

$$E(h) = a_1 h + a_2 h^2 + a_3 h^3 + \cdots + a_n h^n \tag{2.136}$$

to require $n+1$ evaluations $f_i = f(h_i)$ using $n+1$ different values of h_i to obtain the error term coefficients and f_{true}. As remarked by Ralston, this approach cannot be guaranteed to converge, since it depends on the behavior of the higher-order derivatives of f_{true}, but the basic idea does lead to an elegant extension of the trapezoidal rule for numerical quadrature called Romberg integration, which is now developed.

We begin by writing $E(h)$ as

$$E(h) = \sum_{j=K}^{\infty} a_j h^j \tag{2.137}$$

where the lowest-order term is h^K and all higher-order terms in h are included for completeness. Since, on a normalized basis, it is usually the case that $h \ll 1$, the $j = K$ term is the largest contributor to the error. Retaining only the $j = K$ and $j = K + 1$ terms in the error (the reason for which is seen immediately below) and proceeding as above, we have

$$f_1 = f_{\text{true}} + a_K h_1^K + a_{K+1} h_1^{K+1} \text{ and } f_2 = a_0 + a_K h_2^K + a_{K+1} h_2^{K+1} \quad (2.138)$$

so that upon solving for f_{true} we obtain

$$f_{\text{true}} = \frac{1}{h_1^K - h_2^K}(h_1^K f_2 - h_2^K f_1) + a_{K+1}\frac{h_1^K h_2^{K+1} - h_2^K h_1^{K+1}}{h_1^K - h_2^K} \quad (2.139)$$

which for the special case $h_1 = 2h_2 = 2h$ becomes

$$f_{\text{true}} = \frac{1}{2^K - 1}(2^K f_2 - f_1) + a_{K+1}\frac{2^K - 2^{K+1}}{2^K - 1} h^{K+1} \quad (2.140)$$

Thus, the explicit effect of the Richardson extrapolation is to increase the power of the smallest power of h in the error from K to $K + 1$, explaining the power of this technique. Note that any computation based on a fixed spacing is a candidate for Richardson extrapolation. The use of $h_1 = 2h_2$ is especially efficient since f_2 can be computed reusing the set of samples needed for f_1 plus about the same number of additional samples located midway between them, rather than twice that number of new samples that would be needed otherwise.

Romberg quadrature exploits this idea to obtain improved accuracy from a trapezoidal rule, for which the lowest-order error term is proportional to h^2, in evaluating an integral using numerical quadrature and leads to

$$T_{i,j} = \frac{1}{4^i - 1}(4^i T_{i-1,j+1} - T_{i-1,j}) \quad (2.141)$$

where $i = 0, 1, \ldots, k, j = k, k - 1, \ldots, 0, i + j = k$ and where

$$T_{0,k} = \frac{b - a}{2^k}\left(\frac{1}{2}f_0 + f_1 + f_2 + \cdots + f_{2^k-1} + \frac{1}{2}f_{2^k}\right) \quad (2.142)$$

is the trapezoidal rule approximation for $2^k + 1$ samples (2^k subintervals) over the integration range $b - a$.

Note that $T_{i,j}$ can be arranged in a triangular array of results with the leftmost vertical column comprised of the trapezoidal rule answers with convergence towards an improved answer taking place down the diagonal and along each row of this arry with increasing k as

$$\begin{bmatrix} T_{0,0} & & & & & \\ T_{0,1} & T_{1,0} & & & & \\ T_{0,2} & T_{1,1} & T_{2,0} & & & \\ \vdots & \vdots & \vdots & \cdots & & \\ \vdots & \vdots & \vdots & \cdots & \cdots & \\ T_{0,n} & T_{1,n-1} & T_{2,n-2} & \cdots & \cdots & T_{n,0} \end{bmatrix} \qquad (2.143)$$

where the array for the integration example used for Shanks' Method is given by

```
0.6703
1.2080   1.9440
1.7843   2.8985   3.2300
2.3782   3.9875   4.5924   4.7793
2.9781   5.1513   6.1754   6.5645   6.6809
3.5796   6.3439   7.8963   8.6386   8.9072   8.9848
4.1815   7.5450   9.6767  10.9096  11.4750  11.6691  11.7236
4.7835   8.7483  11.4761  13.2742  14.2917  14.7353  14.8805  14.9203
5.3856   9.9523  13.2805  15.6705  17.2372  18.0952  18.4507  18.5624  18.5924
5.9877  11.1563  15.0862  18.0757  20.2295  21.6189  22.3525  22.6422  22.7302  22.7536
6.5897  12.3604  16.8923  20.4832  23.2351  25.2059  26.4521  27.0860  27.3256  27.3963  27.4148
```

in terms of the digits of agreement with a 36-digit answer. The best result from the Romberg integration is

1.098 612 288 668 109 691 395 245 237 345 193 11.

It is clear the Romberg integration performs better for this example than does Shanks' Method, increasing the accuracy of the trapezoidal rule by more than 20 orders of magnitude. How this can happen may be illustrated by the array below, which depicts how the lowest-order error term varies with Romberg matrix entry.

$$\begin{bmatrix} h_0^K & & & & & \\ h_1^K & h_1^{K+1} & & & & \\ h_2^K & h_2^{K+1} & h_2^{K+2} & & & \\ h_3^K & h_3^{K+1} & h_3^{K+2} & \cdots & & \\ h_4^K & h_4^{K+1} & h_4^{K+2} & \cdots & & \\ \vdots & \vdots & \vdots & \vdots & h_{n-1}^{K+n-1} & \\ h_n^K & h_n^{K+1} & h_n^{K+2} & \cdots & h_n^{K+n-1} & h_n^{K+n} \end{bmatrix} \qquad (2.144)$$

where $h_0 = b - a = 2$ and $h_i = h_0/2^i$. Each successive appliction of Richardson extrapolation is seen to increase the order of the error term by one to achieve a total increase of h^n when the highest-order trapezoidal rule answer employs $2^{n-1} + 1$ samples over the range $b - a$. In this case, $n = 10$ and with $K = 2$, we see that an accuracy improvement of up to $(2/2^{10})^2$ might be expected from increasing the number of samples used

Electromagnetics

for the trapezoidal rule from 2 to $2^n + 1$, or in this example about 4×10^{-6}, a result close to the difference between the first and last entries in the left column of the matrix above. The resultant accuracy improvement to be expected from generating the Romberg array would similarly be of order $(2^9)^{(2+10)}$, or about 3×10^{32}, not too far from the improvement actually achieved of about 10^{27}. Note that the actual error depends not only on h_n^{K+n} but on its product with a derivative of that same order of the function being integrated with h [37].

Romberg quadrature can be implemented by increasing the order of the trapezoidal rule until $|T_{n,0} - T_{n-1,1}|$ is less than some error criterion. This approach can, however, lead to many more function samples over the interval $b - a$ than are necessary if a sharp peak occurs within it. As an alternative, an adaptive sampling procedure can then be used where only a limited number of samples are permitted per subinterval, to avoid sampling with equal density everywhere over the integration range [39]. It is worth noting that although Romberg integration has been found to perform better than Shanks' Method for the example above, the latter is applicable to a wider variety of numerical "signals," in contrast to the former which is designed specifically for operations based on subdivision.

Functions From Noise

It is well known that many classical functions have recurrence relations, which provide a convenient way to obtain values of the function for a fixed argument and variable order. A typical recurrence relation has the form

$$f_n(z) = a_1 f_{n-1}(z) + a_2 f_{n-2}(z) \qquad (2.145)$$

where the recursion coefficients can depend on the argument z (possibly complex) and order n (possibly non-integer). For cylindrical Bessel functions, $a_1 = 2n/z$ and $a_2 = -1$. A recurrence relation needs only two starting values among the three f_n, f_{n-1}, and f_{n-2}, from which the other f_m, for $m = n-3, n-4, \ldots$, or $m = n+1, n+2, \ldots$, can be obtained, in princple at least. However, round-off errors can accumulate if the recursion is in the direction of diminishing function value, which for Bessel functions occurs for increasing n, since f_n decreases approximately as $(z/n)^n$ when $n > z$.

While recursion to smaller order will preserve accuracy, there remains the problem of determining the starting values for f. This can be accomplished by beginning the recursion at some value of $n > z + D$, with D the number of decimal places accuracy desired in f_m for $m \leq z$, using arbitrary (e.g., 0 and 1) values for the starting values for f [5]. The results are then normalized using a series relationship such as

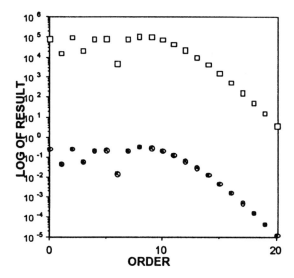

Figure 2.16 An example of using backward recursion to compute Bessel functions $J_n(10)$ of variable order n. The recursion began using values of 0 and 1 for $J_{30}(10)$ and $J_{29}(10)$, respectively, producing the sequence of results (magnitudes) shown by the open squares. Upon using their sum to normalize their values, the results shown by the open circles are obtained, with the crosses indicating directly computed correct values, with an agreement of 10 digits or more obtained between them.

$$1 = f_0 + 2f_2 + 2f_4 + \cdots$$

which applies to Bessel functions. Thus, by beginning with "noise" for the starting values or input, the recurrence relationship acts in the manner of a digital filter used to produce speech from a white noise input, producing in this case an accurate numerical output. An example of using this approach to compute Bessel functions is demonstrated in Fig. 2.16.

REFERENCES

1. E. K. Miller (1988). A Selective Survey of Computational Electromagnetics, invited tutorial-review paper, *IEEE Trans. Antennas and Propagat.*, September, 1281–1305.
2. R. F. Harrington (1968). *Field Computation by Moment Methods*. Macmillan, New York.
3. J. A. Stratton (1941). *Electromagnetic Theory*. McGraw-Hill, New York.

4. B. W. Arden and K. N. Astill (1970). *Numerical Algorithms: Origins and Applications*. Addison-Wesley, Reading MA.
5. M. Abramowitz and I. A. Stegun (1964). *Handbook of Mathematical Functions*, Applied Mathematics Series, Vol. 55. Washington, National Bureau of Standards.
6. F. S. Acton (1970). *Numerical Methods that Work*. Harper & Row, New York.
7. W. H. Press, B. R. Flannery, S. A. Teukolsky, and W. T. Vettering (1986). *Numerical Recipes*. Cambridge University Press, London.
8. R. Mittra (ed.) (1973). *Computer Techniques for Electromagnetics*. Pergamon Press, New York.
9. R. Mittra (ed.) (1975). *Numerical and Asymptotic Techniques in Electromagnetics*. Springer-Verlag, New York.
10. B. J. Strait (ed.) (1980). *Applications of the Method of Moments to Electromagnetic Fields*. SCEEE Press, St Cloud FL.
11. B. J. Strait and A. T. Adams (1980). On contributions at Syracuse University to the moment method, *IEEE Trans. Electromagn. Comp.*, **EMC-22**, 228–237.
12. R. F. Harrington, D. R. Wilton, C. M. Butler, R. Mittra, and C. L. Bennett (1981). *Computational Methods in Electromagnetics*. SCEEE Press, St Cloud FL.
13. J. Perine and D. J. Buchanan (1982). Assessment of MoM techniques for shipboard applications, *IEEE Trans. Electromagn. Comp.*, **EMC-24**, 32–39.
14. M. M. Ney (1985). Method of moments as applied to electromagnetic problems, *IEEE Trans. Microwave Theory Tech.*, **MTT-33**, 972–980.
15. T. Itoh (1986). An overview on numerical techniques for modeling miniaturized passive components, *Ann. Telecommun.*, **41**, 449–462.
16. A. J. Poggio and E. K. Miller (1988). Low-frequency analytical and numerical methods for antennas, in *Antenna Handbook* (Y. T. Lo and S. W. Lee, eds). Van Nostrand Reinhold, New York. pp. 3-1 to 3-98.
17. E. K. Miller and G. J. Burke (1992). Low-frequency computational electromagnetics for antenna analysis, invited article for special issue on antennas of *Proc. IEEE*, **80**(1), 24–43.
18. R. P. Silvester and R. L. Ferrari (1983). *Finite Elements for Electrical Engineers*. Cambridge University Press, London.
19. J. L. Mason and W. J. Anderson (1985). Finite element solution for electromagnetic scattering from two-dimensional bodies. *Int. J. Numer. Meth. Eng.*, **21**, 909–928.
20. J. M. Bornholdt and L. N. Medgyesi-Mitschang (1986). *Mixed domain Galerkin expansions in electromagnetic problems*, National Radio Science Meeting, Boulder CO.
21. J. H. Richmond (1974). Radiation and scattering by thin-wire structures in a homogeneous, conducting medium, *IEEE Trans. Antennas and Propagat.*, **AP-22**, 365.
22. E. K. Miller and F. J. Deadrick (1975). Some computational aspects of thin-wire modeling, in *Numerical and Asymptotic Techniques in Electromagnetics*, Springer-Verlag, New York. pp. 89–127.

23. J. H. Richmond (1966). A wire-grid model for scattering by conducting bodies, *IEEE Trans. Antennas and Propagat.*, **AP-14**, 782–786.
24. D. R. Wilton and C. M. Butler (1981). Effective methods for solving integral and integro-differential equations, *Electromagnetics*, **1**.
25. C. A. Balanis (1982). *Antenna Theory, Analysis and Design*, Harper & Row, New York.
26. R. F. Harrington (1961). *Time-Harmonic Electromagnetic Fields*, McGraw-Hill, New York.
27. E. K. Miller (1967). Admittance Dependence of the Infinite Cylindrical Antenna Upon Exciting Gap Thickness, *Radio Science*, **2**, 1431–1435.
28. R. L. Fante (1966). On the admittance of the infinite cylindrical antenna, *Radio Science*, **1**(9), 181–187; errata: *Radio Science*, **1**(10), 1234.
29. J. R. Wait (1969). Characteristics of antennas over lossy earth, in *Antenna Theory*, Part 2 (R. E. Collin and F. J. Zucker, eds.). McGraw-Hill, New York. Chapter 23.
30. A. J. Poggio and E. K. Miller (1973). Integral-equation solutions of three-dimensional scattering problems, in *Computer Techniques for Electromagnetics*. Pergamon Press, New York. pp. 159–264.
31. E. K. Miller and M. A. Gilbert (1991). Solving the Helmholtz equation using multiply-propagated fields, *International Journal of Numerical Modeling in Engineering*, **4**, 123–138.
32. E. K. Miller, A. J. Poggio, and G. J. Burke (1973). An integro-differential equation technique for the time-domain analysis of thin-wire structures. Part I: the numerical method, *Journal of Computational Physics*, **12**, 24–48.
33. E. K. Miller and J. A. Landt (1980). Direc time-domain techniques for transient radiation and scattering from wires, invited paper in *Proc. IEEE*, **68**, 1396–1423.
34. G. J. Burke, E. K. Miller, S. Chakrabarti, and K. R. Demarest (1989). Using model-based parameter estimation to increase the efficiency of computing electromagnetic transfer functions, *IEEE Trans. Magnetics*, **25**(4), 2807–2809.
35. E. K. Miller and G. J. Burke (1991). Using model-based parameter estimation to increase the physical interpretability and numerical efficiency of computational electromagnetics, *Computer Phys. Commun.*, **68**, 43–75.
36. D. Shanks (1955). Nonlinear transformations of divergent and slowly convergent sequences. *J. Mathematics and Physics*, The Technology Press, Massachusetts Institute of Technology, Cambridge MA. pp. 1–42.
37. A. Ralston (1965). *A First Course in Numerical Analysis*. McGraw-Hill, New York. pp. 119–121.
38. R. C. Booton, Jr. (1992). *Computational Methods for Electromagnetics and Microwaves*. Wiley Interscience, New York.
39. E. K. Miller (1970). A variable interval width quadrature technique based on Romberg's method, *Journal of Computational Physics*, **5**, 265–279.

# 3
Algorithms Used in Signal Analysis

Hugh F. VanLandingham
Virginia Polytechnic Institute and State University, Blacksburg, Virginia

3.1 INTRODUCTION

This chapter is concerned with signal representation and manipulation. Since most signal operations are performed in computers, the focus of these calculations is on discrete arrays, representing samples of continuous-time signals. There are nine sections in this chapter. Section 3.1 is the introduction; Sect. 3.2 – signal representation – discusses the vector notation used for signal manipulation; Sect. 3.3 illustrates the ideas of creating new signals by system transformation and the convolution of two signals; Sect. 3.4 represents signals in the frequency domain; Sect. 3.5 illustrates the efficiency of using an FFT algorithm; Sect. 3.6 presents methods of obtaining frequency selective digital systems (filters); and Sects. 3.7 and 3.8 discuss eigensystem decomposition and singular value decomposition, respectively, with discussion of solving sets of linear algebraic equations. Finally, Sect. 3.9 relates to the generation of specific types of (pseudo)random numbers, namely those having time correlation and those having non-standard distributions.

3.2 SIGNAL REPRESENTATION

In signal processing it is often necessary to include specific signals as test signals, or simply as an integral part of a simulation. Signals can be elementary sinusoids and exponential functions or complex time-correlated stochastic signals. In this first section a brief overview of simple signal representation and processing is given. We will follow MATLAB notation as a means of describing specific operations.

3.2.1 Digital Signal Processing

Modern signal processing is predominantly performed in digital computers. As a result we must consider the problems of signal conversion and the computer operations which are used as equivalent analog operations. Figure 3.1 depicts a generic DSP (digital signal processing) system which is interfaced to the analog world. The actual DSP function, shown as the central block, is quite versatile, limited only to the imagination of the programmer. The interface blocks are usually implemented in hardware and typically consist of filters in addition to the basic analog-to-digital (A/D) and digital-to-analog (D/A) operations. A simple, but effective model which incorporates these interface blocks along with the CT system dynamics is given in Chapter 5, Section 2.

To begin, we must be able to describe *signals*. A deterministic signal can be thought of as a function of time. Therefore, a time base can be established to represent the time axis. With discrete time (DT) signals the time base is a set of points, usually with equal increments; e.g. we might define:

$$t = [0 : 0.1 : 10] \tag{3.1}$$

which is MATLAB notation for an array of points called t, starting with 0 and incrementing by 0.1 up to a value of 10. Now, if the utility program (MATLAB in this case) has a built-in function, say $\sin(x)$, we can establish a function of that form. MATLAB is built around array manipulation. Consequently, the expression

$$u(t) = A^* \sin(w^* t) \tag{3.2}$$

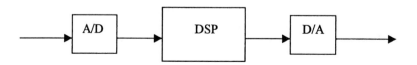

Figure 3.1 Digital signal processing.

Algorithms Used in Signal Analysis

creates a sinusoid of amplitude A and (radian) frequency w. If the signal $u(t)$ is plotted versus the array t, the result is, as expected, a sine function that extends from $t = 0$ to $t = 10$. A similar process can be used for any other built-in function, like $\exp(x)$, or combinations, like $A^* \exp(-a^*t).^* \sin(w^*t + B)$. In this last function the two arrays "exp" and "sin" are multiplied together on a term-by-term basis (corresponding to the ".*" operation). Small "functions" can be programmed that offer more "design" parameters in constructing signals. For example, one might want to control not only the frequency and time duration of the signal, but also its phase and sampling representation.

Example 3.1 (Sinusoid Representation)

Figure 3.2 illustates a cosine signal with a unit amplitude and a 45° phase. As can be seen from the figure, the frequency is 10 Hz, the duration is 0.2 s, and the sampling frequency is 120 Hz. A simple MATLAB code for generating such functions follows.

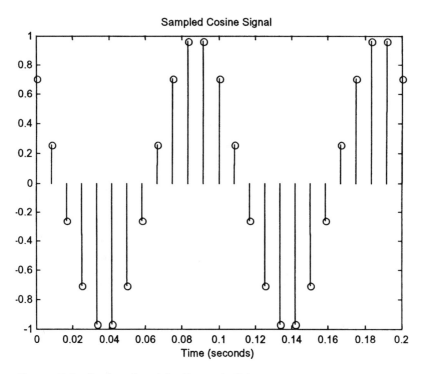

Figure 3.2 Cosine signal for Example 3.1.

Computer Code
```
function y = cosine(Af,T,fs)
a = real(A); b = imag(A); t = 0:1/fs:T;
c = cos(2*pi*f*t); s = sin(2*pi*f*t); y = a*c-b*s;
```

Phasors

In the case of sinusoidal functions at a specific frequency a complex number A can contain the combined information of the amplitude and the phase of the signal. Such representations are called *phasors*, and can be used to add single-frequency sinusoidal signals of different amplitudes and phases, i.e. by simply adding the complex *phasors*, representing the individual signals, and subsequently converting back to a sinusoidal representation. The cosine will be used as the reference for the phasor representation. To illustrate, let us assume the following example.

Example 3.2 (Sum of Sinusoidal Signals)

$$3\cos(2\pi f_0 t) + 2\sin(2\pi f_0 t + 60°) \Rightarrow (3\angle 0°) + (2\angle 60°) = (4.3589\angle 23.41°)$$

The first step was to convert the sinusoids into phasors, as shown above. After adding the complex phasors to obtain the above result, the final sum can be expressed as

$$\cos(2\pi f_0 t + 23.41°)$$

Periodic and Almost Periodic Signals

A signal $x(t)$ is said to be *periodic* if $x(t) = x(t + T)$, where the smallest such T is called the *period* of the signal. The process of summing sinusoidal signals can be extended to the creation of more general periodic signals as a single function operation by placing the (complex) amplitudes in a vector A and corresponding frequencies in another vector F as shown in the following code, which calls the function of Example 3.1 for each sinusoid. If the individual sinusoids have frequencies which are rationally related, then the sum will be periodic; otherwise, for example if one frequency is the square root of another, the sum will be *almost periodic*.

Computer Code
```
function y = sumsig(A,F,T,fs)
Y = [ ]; M = size(A,1);
for k = 1:M, y = cosine(A(k),F(k),T,fs); Y = [Y; y]; end
y = zeros(1,size(Y,2));
for k = 1:size(Y,1), y = y + Y(k,:); end
```

Algorithms Used in Signal Analysis

Random Signals

Another convenient mechanism present in most utility programs is the ability to generate *pseudo-random* numbers. A digital computer accomplishes this by a series of numerical operations in which the remaining least significant digits are "unpredictable," i.e. "random." Even though not strictly random in the probabilistic sense, these numbers are very useful in generating signals that *look like* random signals. Thus (still using the array t), the expression

$$v(t) = a + b^*\text{rand}(t) \tag{3.3}$$

represents a random signal $v(t)$ with erratic amplitude variations between a and $a + b$, as a result of "rand" being pseudo-random numbers between 0 and 1. The expression in (3.3) could, for example, be used to model a disturbance signal in a larger system model.

It should be clear that combinations of signals may be formed. For example, writing

$$w(t) = u(t) + v(t) \tag{3.4}$$

specifies a signal $w(t)$ which is a sinusoidal variation, from Eq. (3.2), superimposed on the random signal from Eq. (3.3).

3.3 SYSTEMS AND SIGNAL CONVOLUTION

Now that we have the rudiments of signal representation, let us consider "transforming" one signal into another. Such an operation is accomplished using a "system." If we are to use signals effectively, we must understand systems also. This chapter is therefore closely coupled with Chapter 5, since a system is simply an operation on one signal to create another. Chapter 5 utilizes state space system models because of their generality and amenability to algorithmic computation. Briefly, a DT state space model consists of 4-matrices which we shall denote A, B, C, and D, related to the following equations:

$$x(t + 1) = Ax(t) + Bu(t) \tag{3.5}$$

$$y(t) = Cx(t) + Du(t) \tag{3.6}$$

The first equation involves an $n \times n$ matrix A and an $n \times m$ matrix B which multiply the signals $x(t)$, called the state vector, and $u(t)$, the input vector. The second equation involves a $p \times n$ matrix C and a $p \times m$ matrix D which also multiply the signals $x(t)$ and $u(t)$ to obtain the output vector $y(t)$. It is convenient to visualize these equations as representing an operation which converts the m-dimensional signal vector $u(t)$ to the p-dimensional output vector $y(t)$, using the n-dimensional vector $x(t)$ as an internal vector which carries the information of the instantaneous "state" of the system; e.g. $x(0)$ would represent the initial conditions associated with energy stored in the system at $t = 0$. Such a visualization is illustrated in Figure 3.3, where the system is represented by a unit-response (matrix) function, $h(t)$, which can be calculated from the state space model.

There are several means of describing a system operation. The state space model of Eqs. (3.5) and (3.6) is one. Another is the *unit pulse response function* $h(t)$, which is the characteristic response of the system to an elementary "unit pulse" signal. Our notation for the response of a (linear, single-input, single-output, time-invariant) system to a unit pulse, denoted $\delta(t - n)$, at time n is $h(t - n)$. Considering that any single-sided "positive time" input (DT) signal $x(t)$ can be expanded (term by term) into

$$x(t) = \sum_{n=0}^{\infty} x(n)\delta(t - n) \tag{3.7}$$

we can say that the output signal $y(t)$ will be equal to

$$y(t) = L[x(t)] = \sum_{n=0}^{\infty} x(n)L[\delta(t - n)] \tag{3.8}$$

where $L[\cdot]$ represents the symbolic linear operator. It follows that in response to an arbitrary input signal $x(t)$, the output signal $y(t)$ equals the convolution summation:

$$y(t) = L[x(t)] = \sum_{n=0}^{\infty} x(n)h(t - n) \tag{3.9}$$

Conceptually, Eq. (3.9) can be visualized in terms of each sequence in the summation being typed on its own strip of paper, $x(t)$ forward and $h(t)$ in reverse order, as illustrated in Fig. 3.4. Referring to Fig. 3.4, which is

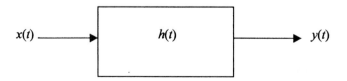

Figure 3.3 A system transformation.

Algorithms Used in Signal Analysis

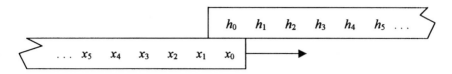

Figure 3.4 Visualizing DT convolution.

lined up for $t = 0$, $y(0)$ is, as required by (3.9), the overlapping product summed together, i.e. $x(0)\,h(0)$. For the next time step one strip is slid one value relative to the other strip so that $x(0)$ and $h(1)$ now line up. Consequently, $y(1) = x(0)h(1) + x(1)h(0)$. This procedure may be continued as necessary to calculate the next terms of $y(t)$. We will complete this section with an example that uses a system to "smooth" a random signal.

Example 3.3

Figure 3.5 shows both the input signal and the smoothed output signal. The raw signal is a unit amplitude cosine signal with added random noise which varies between -0.5 and $+0.5$. The smoothed signal is effectively a 10 point sliding average of the original signal (with the phase delay removed). Although the effect of the added random noise has not been completely removed, it is clear that the system operation did diminish it.

3.4 THE FREQUENCY DOMAIN

The frequency domain is based on the Fourier transform, which is a mathematical process to convert the information of a signal from the time domain to the "frequency" domain. Since DT signals are numerical sequences, separated uniformly in time by a "sampling interval" T, the Fourier transform of a general signal $x(t)$ is given by

$$X(f) = \mathcal{F}\{x(t)\} = \sum_{n=0}^{\infty} x(n)\exp(-j2\pi nTf) \qquad (3.10)$$

Equation (3.10) follows from Eq. (3.7) and the fact that the Fourier transform of a unit pulse delayed by the T seconds is $\exp(-j2\pi Tf)$. For practical reasons we may typically restrict the upper limit of the summation in Eq. (3.10) to a finite value; but, in certain cases, even with an infinite number of terms, a "closed form" may exist. For example, if $x(k) = a^k$, the transform may be summed using the (infinite) geometric sum formula, giving us

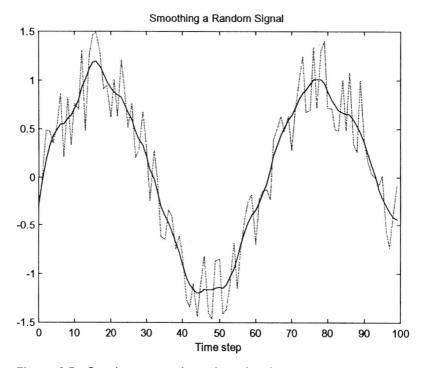

Figure 3.5 Creating a smooth random signal.

$$X(f) = \frac{1}{1 - a\exp(-j2\pi Tf)} \tag{3.11}$$

whenever $|a| < 1$. Many other closed form transforms may be found in the literature in the form of z-transforms, where $z = \exp(j2\pi Tf)$. It is clear that the Fourier representation of a DT signal is periodic in f, since it is a function of $\exp(j2\pi Tf)$. This is the dual statement to the fact that a periodic time signal has a discrete frequency (Fourier series) representation.

With $z = \exp(j2\pi Tf)$, reasonably complex signals can be generated by the simple procedure of passing a signal through a system. In Chapter 5 transfer functions are discussed in the context of an alternate means of describing a (linear) system; and, in fact, techniques are presented which addressed the conversion between state space and transfer function models. For present purposes we will illustrate that a narrowband signal can be obtained from a unit pulse using a system transformation, as presented in the previous subsection.

Example 3.4

Let us consider that a unit pulse is the input to a system with a transfer function given by:

$$H(z) = \frac{z^2}{z^2 - 1.7554z + 0.9025}$$

The output signal will be represented by $H(z)$, whose frequency plots (magnitude and phase) are shown in Fig. 3.6; these plots were calculated from $H(z)$ using $z = \exp(j2\pi Tf)$ with $T = 1$. Note that the magnitude and phase of $H(z)$ constitute the complex value at each frequency. The plots are shown as a function of "normalized" frequency, $2\pi Tf$, which has the dimensionless units of radians. The plots are symmetric about π (with the magnitude having even symmetry and the angle having odd symmetry). These plots are normally plotted for only a half-period of the frequency response

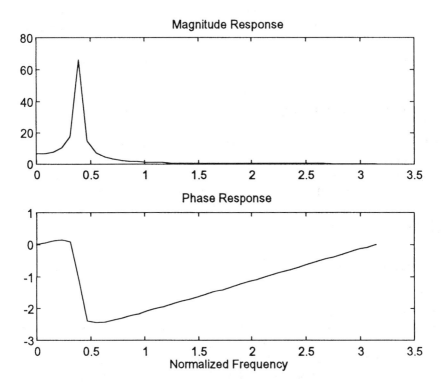

Figure 3.6 Frequency plots of the signal of Example 3.4.

function, since the only non-redundant information is for normalized frequency between 0 and π.

Computer Code
```
function [mag,ang] = dtfreq(num,den,N)
[w = 0:pi/N:2*pi; m = length(w); na = length(den); nb = length(num);
ta = na-1:-1:0; tb = nb-1:-1:0, Ta = ta(ones(1,m),:);
Tb = tb(ones(1,m),:); wa = w(ones(1,na),:); wb = w(ones(1,nb),:);
Wa = wa.*Ta'; Wb = wb.*Tb'; Ca = cos(Wa); Cb = cos(Wb);
   Sa = sin(Wa);
Sb = sin(Wb); Rb = num*Cb; Ra = den*Ca; Ib = num*Sb;
   Ia = den*Sa;
Num = Rb + i*Ib; Den = Ra + i*Ia; mag = abs(Num./Den);
ang = angle(Num./Den);
```

The frequency (z) domain provides a concise method of describing signals, and viewing the signal as a function of real frequency can be useful to designers. For example, the signal of Fig. 3.6 is clearly dominant in a "narrow band" of frequencies and is therefore called a *narrowband* signal. The next section discusses the FFT as a means of efficiently computing frequency response functions.

3.5 FAST FOURIER TRANSFORMS

The *discrete Fourier transform* (DFT) was developed as a technique for approximating Fourier transforms using digital calculations. *Fast Fourier transforms* are simply ultra-efficient algorithms for calculating the DFT. Let us first define the DFT. Assume a time sequence of N samples given by $g(kT)$ for $k = [0, N-1]$. Then, the DFT of this "time" sequence is the "frequency" sequence

$$G\left(\frac{n}{NT}\right) = \text{DFT}\{g(kT)\} = \sum_{k=0}^{N-1} g(kT) \exp\left(\frac{-j2\pi \, nk}{N}\right) \quad (3.12)$$

for $n = 0, 1, 2, \ldots, N-1$. Letting $F^{-1} = NT$, it may be noted that there are only N distinct values of $G(nF)$ and that F represents the sample spacing in frequency. The reader should keep in mind that to emphasize the discrete nature of both domains, k is the index being used in the time domain and n is the index for the frequency domain. The *inverse DFT* (denoted IDFT) is similarly defined as

$$g(kT) = \text{IDFT}\{G(nF)\} = \frac{1}{N}\sum_{n=0}^{N-1} G(nF)\exp\left(\frac{j2\pi\, nk}{N}\right) \quad (3.13)$$

for $k = 0, 1, 2, \ldots, N-1$. To simplify notation, we will define $W = \exp(-j2\pi/N)$. Thus,

$$G(nF) = \text{DFT}\{g(kT)\} = \sum_{k=0}^{N-1} g(kT)W^{nk} \quad (3.14)$$

for $n = 0, 1, 2, \ldots, N-1$. Equation (3.14) can be translated into a vector/matrix equation given by $G = Wg$, or in expanded form,

$$\begin{bmatrix} G_0 \\ G_1 \\ \vdots \\ G_{N-1} \end{bmatrix} = \begin{bmatrix} W^{0\cdot 0} & W^{0\cdot 1} & \cdots & W^{0\cdot (N-1)} \\ W^{1\cdot 0} & W^{1\cdot 1} & \cdots & W^{1\cdot (N-1)} \\ \vdots & \vdots & \cdots & \vdots \\ W^{(N-1)\cdot 0} & W^{(N-1)\cdot 1} & \cdots & W^{(N-1)^2} \end{bmatrix} \begin{bmatrix} g_0 \\ g_1 \\ \vdots \\ g_{N-1} \end{bmatrix} \quad (3.15)$$

The entries of $W = \{W^{nk}\}$ can be determined by noting that W^m is a periodic function of m; these values are the complex values found at angular intervals of $-2\pi/N$ radians around the unit circle of the complex plane.

The idea of the *fast Fourier transform* can be presented in the algebraic development of an $N = 4$ point algorithm. Let the indices k and n be represented in binary (2-bit) numbers:

$$k = \{0, 1, 2, 3\} = \{00, 01, 10, 11\} = (k_1 k_0)_2 \quad (3.16a)$$

$$n = \{0, 1, 2, 3\} = \{00, 01, 10, 11\} = (n_1 n_0)_2 \quad (3.16b)$$

Beginning with "level 0," the time samples $g_0(k) = g(k)$, each subsequent level of calculation will be made until at level $(\log_2 N)$ the final DFT signal $G(n)$ is available.

$$G(n) = \text{DFT}\{g(k) = g_0(k)\} = \sum_{k=0}^{N-1} g_0(k)W_N^{nk} \quad (3.17a)$$

Using the "positional bit notation," e.g. $(n)_{10} = (n_1 n_0)_2 = 2n_1 + n_0$,

$$G(n_1 n_0) = \sum_{k_0=0}^{1}\sum_{k_1=0}^{1} g_0(k_1 k_0) W_N^{(2n_1+n_0)(2k_1+k_0)} \quad (3.17b)$$

Since $N = 4$, we can expand the exponent into

$$(2n_1 + n_0)(2k_1 + k_0) = 4n_1 k_1 + 2n_1 k_0 + 2n_0 k_1 + n_0 k_0 \quad (3.18a)$$

$$W_4^{nk} = W_4^{4n_1 k_1} W_4^{2n_0 k_1} W_4^{(2n_1+n_0)k_0} \quad (3.18b)$$

The first factor is always equal to 1, and the "decimation" step is:

$$G(n_1 n_0) = \sum_{k_0=0}^{1} \left[\sum_{k_1=0}^{1} g_0(k_1 k_0) W_4^{2n_0 k_1} \right] W^{(2n_1+n_0)k_0} = \sum_{k_0=0}^{1} [g_1(n_0 k_0)] W^{(2n_1+n_0)k_0} \quad (3.19)$$

$$G(n_1 n_0) = g_2(n_0 n_1) = \sum_{k_0=0}^{1} g_1(n_0 k_0) W^{(2n_1+n_0)k_0} \quad (3.20)$$

The inside summation of Eq. (3.19) represents the calculation for the first level, given by the following four equations.

$$g_1(00) = g_0(00)W^0 + g_0(10)W^0, \quad g_1(01) = g_0(01)W^0 + g_0(11)W^0$$

$$g_1(10) = g_0(00)W^0 + g_0(10)W^2, \quad g_1(11) = g_0(01)W^0 + g_0(11)W^2$$

Similarly, the summation in Eq. (3.20) represents the calculation for the second level. The following first-level calculation illustrates the simplification obtained in the number of complex multiplications.

$$\begin{bmatrix} g_1(00) \\ g_1(01) \\ g_1(10) \\ g_1(11) \end{bmatrix} = \begin{bmatrix} 1 & 0 & W^0 & 0 \\ 0 & 1 & 0 & W^0 \\ 1 & 0 & W^2 & 0 \\ 0 & 1 & 0 & W^2 \end{bmatrix} \begin{bmatrix} g_0(00) \\ g_0(01) \\ g_0(10) \\ g_0(11) \end{bmatrix} \quad (3.21)$$

The reordering of the calculated values, as indicated in Eq. (3.20), is necessary to obtain the proper order of the calculated DFT values; this procedure is called "bit reversal," because reversing the binary description of the indices regains the correct order.

Having understood the basic method, consider the extension to $N = 8$:

$$G(n_2 n_1 n_0) = \sum_{k_0} \sum_{k_1} \sum_{k_2} g_0(k_2 k_1 k_0) W_8^{nk} \quad (3.22)$$

where $n = 4n_2 + 2n_1 + n_0$ and $k = 4k_2 + 2k_1 + k_0$. Proceeding as above,

$$g_1(n_0 k_1 k_0) = \sum_{k_2} g_0(k_2 k_1 k_0) W^{4n_0 k_2} \quad (3.23)$$

$$g_2(n_0 n_1 k_0) = \sum_{k_1} g_1(n_0 k_1 k_0) W^{(2n_1+n_0)2k_1} \quad (3.24)$$

$$g_3(n_0 n_1 n_2) = \sum_{k_0} g_2(n_0 n_1 n_0) W^{(4n_2+2n_1+n_0)k_0} \quad (3.25)$$

Algorithms Used in Signal Analysis

Finally, $G(n_2 n_1 n_0) = g_3(n_0 n_1 n_2)$ provides the necessary bit reversal. The complete *decimation in frequency* DFT developed above is represented as a signal flow graph in Fig. 3.7, illustrating these calculations. Note the "butterfly" patterns shown in heavy lines, and the final stage of reordering. Butterfly patterns are important in simplifying DFT computations; each one can be reduced to a single complex multiplication. It can be shown that within a butterfly pattern the complex factors are always W^p and $W^{(p+n/2)}$, and that these factors are the same except for a negative sign.

We can identify a butterfly pattern with upper-left node k and lower-left node q. In Fig. 3.7, for instance, we see nodes (0,5) between level 0 and level 1, or (1,3) between level 1 and level 2. Determining the power p of a particular pattern involves a manipulation of the bit string representing the upper node, k [1]. One butterfly calculation (from level $(l-1)$ to level l can be summarized as (the *dual* node, $q = k + N/2^l$):

$$g_l(k) = g_{l-1}(k) + g_{l-1}(q)$$
$$g_l(q) = [g_{l-1}(k) + g_{l-1}(q)]W_N^p \qquad (3.26)$$

There are many variations of FFT algorithms, but little insight would be gained from their study here. Given that efficient methods exist for calculating the DFT, we will now consider some applications.

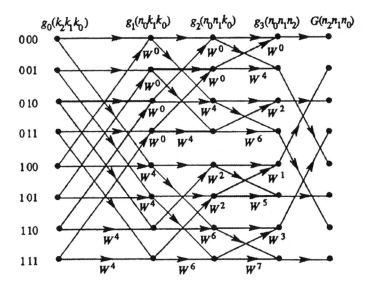

Figure 3.7 Signal-flow graph illustrating an FFT.

3.5.1 Fourier Series

If a signal is periodic and band-limited, we may use the DFT to calculate the Fourier series representation which is a sine/cosine expansion of the signal. In order to use the DFT for this function, we must be careful to provide samples of the original (time) signal that are "close enough" for a good representation. (The sampling frequency should be several times the highest frequency of the signal.) In addition, the samples must represent exactly one period of the signal. Finally, the DFT output must be scaled by the number of samples and truncated to avoid the repetition inherent in the output. The following example will illustrate this application.

Example 3.5 (Fourier Series Calculation)

This signal is comprised of three sinusoids. The fundamental frequency is 20 Hz; the other components are the third and fifth harmonics. The amplitudes are $1, -\frac{1}{3}, \frac{1}{5}$, respectively. In addition, a d.c. component of amplitude 2 is present. The sampling frequency is 800 Hz. Figures 3.8 and 3.9 illustrate the results.

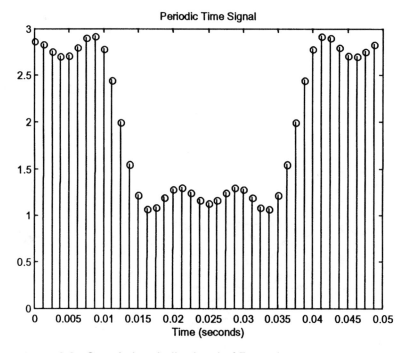

Figure 3.8 Sampled periodic signal of Example 3.5.

Algorithms Used in Signal Analysis

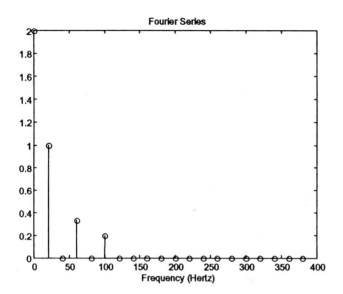

Figure 3.9 Fourier series coefficients for the signal of Fig. 3.8.

Computer Code
```
f0 = 20; F = [1*f0; 3*f0; 5*f0]; fs = 40*f0; T = 1/f0;
A = [1; -1/3; 1/5]; t = 0:1/fs:(T-1/fs); N = length(t);
[y,B] = sumsig(A,F,T,fs); y = y(1:N) + 2*ones(1:N);
Y = fft(y); Y = Y(1:N/2); F0 = 1/(N/fs); f = 0:F0:(N*F0/2)-F0;
if Y(1) ~= 0, Y(1) = Y(1)/2; end
```

Other applications of the DFT include harmonic synthesis (converting from the discrete Fourier coefficients to periodic data) and approximating (continuous) direct and inverse Fourier transforms. We will illustrate the Fourier transform approximation.

3.5.2 Fourier Transform Approximation

The direct Fourier transform is given by

$$G(f) = \int_{-\infty}^{\infty} g(t) \exp(-j2\pi ft) dt \qquad (3.27)$$

The method will be illustrated by using a simple example. We will take $g(t) = \exp(-t)$ for $t \geq 0$ and 0 for $t < 0$. For comparison with the DFT results, it is well known that the corresponding (true Fourier) transform is

$$G(f) = \frac{1}{1+j2\pi f} = \frac{1}{1+(2\pi f)^2} + j\frac{-2\pi f}{1+(2\pi f)^2} \quad (3.28)$$

The DFT will be taken on a 32 point uniform sampling of $g(t)$ with a sampling frequency of 4 Hz; however, as required by the Fourier transform, at the point of discontinuity ($t = 0$) the value is the mean of the upper and lower points of continuity, i.e. $g(0) = 0.5$. The 32 point sampling with an interval of 0.25 s fairly represents the function in that the amplitudes are down to quite small values ($\exp(-8) = 3.35 \times 10^{-4}$). The DFT output requires an additional factor of T, the sample interval when used to approximate the integral transform. As we know, for an N point DFT the useful range of output corresponds to $n \leq N/2$ with a frequency resolution given by $F = (NT)^{-1}$. Figure 3.10 shows the error between the real and imaginary parts of the DFT and that of the true (non-periodic curves) values given in Eq. (3.28). The following computer code was used to plot Fig. 3.10. Note that the code reflects the above discussion.

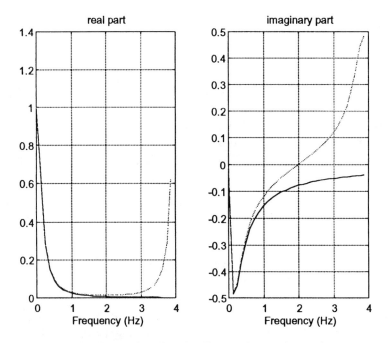

Figure 3.10 Approximating the Fourier integral transform.

Computer Code

```
N = 32; n = [0:1:N-1]; T = 0.25; t = T*n; g = exp(-t); g(1) = 0.5;
F = 1/(N*T); f = F*n; D = 1 + (2*pi*f).^2; R = ones(1,32)./D;
I = -2*pi.*f./D; G = T*fft(g); clg
subplot(121),plot(f,R,f,real(G),':')
title('real part') xlabel('Frequency (Hz)') grid
subplot(122),plot(f,I,f,imag(G),':') title('imaginary part')
xlabel ('Freqency (Hz)') grid
```

Summary

In this section the DFT (and one of its "fast" FFT algorithms) was discussed. After a brief introduction to the basic DFT, the computational details of an FFT were presented. The FFT algorithms are widely available for use in obtaining frequency representations of signals. Two important specific uses were illustrated: (1) using the DFT to calculate Fourier series information, and (2) using the DFT as an approximation to the Fourier integral transform. In the first case care must be taken to sample the time-domain signal over exactly one period. Both cases require proper scaling to perform their respective operations.

3.6 FILTERING ALGORITHMS

In this section we discuss systems which have a certain frequency selectivity. Such systems are called *filters*. They are used to provide some predictable effect on signals. For example, a *low-pass filter* can be used to smooth signals which have added random noise.

Filters are generally called low-pass, bandpass, bandstop or high-pass according to their action on an input signal; e.g. a bandpass filter will pass signals within a certain frequency band. The techniques of filter design, either analog or digital, are based on a "prototype low-pass filter"; from this design any of the other filter types listed above can be obtained by a suitable transformation. Analog filter design has been fully developed for many years. Consequently, we will take a particular CT design and convert this design to a corresponding digital filter, called digital "redesign." This process is equivalent to the approximation of an analog filter by a digital filter. The approximation is then typically verified by comparing the (real) frequency responses, i.e. $C(s) = C(j\omega)$ for the CT system and $D(z) = D(\exp(j\omega T))$ for the DT system (over some range of frequencies ω).

3.6.1 Bilinear Transformation

The fundamental relation between the Laplace domain and the DT (z) domain is that $z = \exp(sT)$. An approximation to this relation is given by

$$s = \frac{2(z-1)}{T(z+1)} \tag{3.29}$$

One way to implement a conversion from an analog filter $C(s)$ to a digital filter $D(z)$ is to make the substitution indicated in Eq. (3.29). We will carry this *bilinear transformation* one step further in order to make it more practical. The basic bilinear transformation maps the imaginary axis of the s-plane into the unit circle of the z-plane, but in doing so it causes a warping of the frequencies.

Recall that both analog and digital filters have "continuous" frequency responses. As a consequence, the degree of similarity of a digital filter to an analog filter is usually measured by a direct comparison of frequency plots. Let us consider this warping effect more closely by looking at the real frequency mapping. From Eq. (3.29)

$$j\omega_c = \frac{2(\exp(j\omega T - 1))}{T(\exp(j\omega T + 1))} \tag{3.30}$$

where ω_c is the true frequency and ω (without a subscript) is the achieved value as mapped by the digital filter. We can modify Eq. (3.30) as follows:

$$j\omega_c = \frac{2\exp(j\omega T/2)\{\exp(j\omega T/2) - \exp(-j\omega T/2)\}}{T\exp(j\omega T/2)\{\exp(j\omega T/2) + \exp(-j\omega T/2)\}} \tag{3.31}$$

Thus, we obtain the desired relationship between the normalized CT frequency and the normalized DT frequency:

$$\omega_c T = 2\tan\left(\frac{\omega T}{2}\right) \tag{3.32}$$

This result tells us that a true frequency ω_c will be warped into a frequency $\omega < \omega_c$ as a natural effect of the bilinear mapping. This warping effect is displayed graphically in Fig. 3.11. Note the severity of the distortion as the curve approaches the asymptote line at $\omega T = \pi$. Figure 3.11 is an excellent indicator of how fast the sampling should be done so that the frequencies of interest will lie in the nearly linear part of the curve.

To counteract the effect of this warping, it is possible to anticipate this effect on certain critical frequencies and "pre-warp" them before substituting the bilinear transformation in Eq. (3.29). With the correct *prewarping* the DT filter frequencies will be the same as those of the desired CT filter. The following steps outline the procedure:

Algorithms Used in Signal Analysis

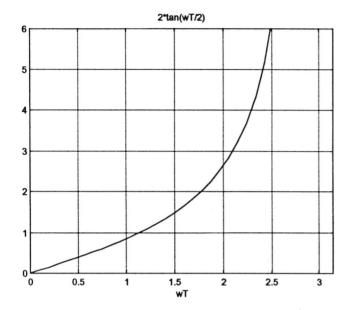

Figure 3.11 Frequency warping.

1. Prewarp critical frequencies of the CT system $C(s)$ according to Eq. (3.32). This will result in higher initial frequencies which the subsequent design will warp back to the desired values.
2. Design a CT system (filter) to meet the specifications at the *prewarped* frequencies.
3. Transform $C(s)$ to the designed DT filter $D(z)$ using Eq. (3.29).

A simple example will help us to understand digital redesign using the bilinear transformation with prewarping.

Example 3.6 (Digital Redesign)

Using a sampling frequency of 10 Hz ($T = 0.1$ s), we wish to perform a "digital redesign" of the CT filter

$$C(s) = \frac{\omega_0}{s + \omega_0} = \frac{1}{1 + (s/\omega_0)} \tag{3.33}$$

where $\omega_0 = 16$ rad/s. In the real frequency domain ($s = j\omega$)

$$C(j\omega) = \frac{1}{1 + (j\omega/16)} \Rightarrow \frac{1}{1 + (j\omega/20.6)} = \frac{20.6}{j\omega + 20.6} \tag{3.34}$$

The last step represents the prewarping of $\omega_0 = 16$ to according to Fig. 3.11 (with $T = 0.1$). The final step is to substitute the bilinear expression of Eq. (3.29) for $s = j\omega$. The result is the DT filter

$$D(z) = \frac{0.5074(z+1)}{z + 0.0148} \tag{3.35}$$

Within the accuracy of the calculations, it is easily checked that the two filters match at the prewarped frequency, i.e.

$$|C(j16)| = |D(\exp(j1.6))| \tag{3.36}$$

In Example 3.6 the sampling frequency is barely sufficient. Ideally, one would want a faster sampling frequency (smaller T) so that there is less difference between the original and warped frequencies.

3.6.2 Finite Impulse Response (FIR) Filter Design

Another useful and general method is to generate the (finite) pulse response sequence of an approximating filter. The desired frequency response is assumed to be periodic since it is represented by a DT frequency response. The Fourier coefficients are then obtained for this periodic function and properly truncated to obtain the required pulse response. An FIR filter may be specified by its finite (im)pulse response of N samples. The transfer function is given by

$$H(z) = \sum_{n=0}^{N-1} h(n) z^{-n} \tag{3.37}$$

where $h(n)$ is the unit pulse response sequence of length N. Given an ideal (desired) DT filter function $H_d(z)$, the real frequency response is periodic with period T^{-1} Hz. The Fourier series representation for this periodic frequency function is

$$H_d(f) = \sum_{n=-\infty}^{\infty} h_d(n) \exp(-2\pi n T f) \tag{3.38}$$

where

$$h_d(n) = T \int_0^{1/T} H_d(f) \exp(j2\pi n T f) df$$

The DT signal $h_d(n)$ is the unit pulse response of the desired filter. Since this response is likely to be infinite and non-causal, i.e. non-zero for n

negative, we will consider a method for approximation. Assuming that only $(2m + 1)$ terms are taken, the approximating filter is

$$H(z) = \sum_{n=-m}^{m} h_c(n)z^{-n} \qquad (3.39)$$

For realizability (causality) we must consider the first element to correspond to $n = 0$. Therefore, with this modification we have

$$H(z) = \sum_{n=0}^{2m} h_d(n - m)z^{-n} \qquad (3.40)$$

A brief aside will be taken to discuss the options for data truncation.

Window Functions
The simplest window function is the rectangular function, $w(t) = \text{rect}(t/2m)$. Windowing a (long) data set $x(t)$ is equivalent to multiplying $x(t)$ by a window function $w(t)$ to obtain $y(t) = w(t) x(t)$. Multiplying by $w(t)$ is the same as convolving in frequency by $W(f)$; and, since a minimum amount of distortion is desirable, $W(f)$ should be as close to an impulse as possible. For this reason data window functions are compared in two ways: (1) narrowness of the main frequency lobe, and (2) the level of the highest side lobe. The first comparison relates to frequency resolution and the second to undesirable rippling (the Gibbs phenomenon). For example, a Hamming window function has a main lobe width of 1.46 times that of a rectangular window, but has side lobes which are smaller by 30 dB! We can now summarize the FIR filter design steps.

Design Steps
From a selected "ideal" (periodic) filter response $H_d(z)$:

1. Calculate the "Fourier coefficients" $h_d(n)$.
2. Apply a window function (symmetrically about the origin).
3. Shift the resulting sequence for realizability.
4. Test the FIR design (by comparing the filter's frequency response to that of $H_d(z)$).

The Fourier coefficients in Step 1 refer to the double-sided "exponential" coefficients of Eq. (3.38). The DFT (FFT) can be used effectively to calculate these values. We conclude this section with an example to approximate the ideal low-pass filter characteristic shown in Fig. 3.12.

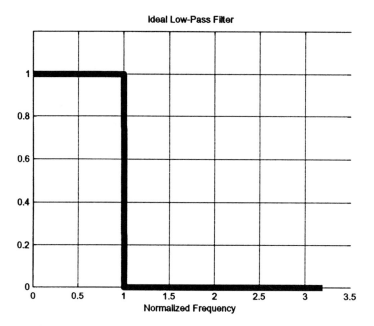

Figure 3.12 An ideal low-pass DT filter.

Example 3.7 (FIR Filter Design)

In this case the ideal coefficients can be calculated analytically using Eq. (3.38). They are

$$h_d(n) = \frac{\sin(n)}{n\pi}$$
$$= \{\cdots, -0.060, 0.015, 0.145, 0.268, 0.318, 0.268, 0.145, 0.015, -0.060, \cdots\}$$

which exhibit even symmetry about the origin. We will compare two designs, one with a rectangular window (which uses ±10 coefficients listed above, and another which uses the same coefficients, but with a Hamming window, i.e.

$$w(n) = 0.54 + 0.46 \cos\left(\frac{n\pi}{N+1}\right), |n| \leq N$$

Figure 3.13 illustrates both frequency responses versus normalized radian frequency, i.e. π is one-half of the period of the frequency response. Note that using the raw data values (rectangular truncation) creates a filter

Algorithms Used in Signal Analysis

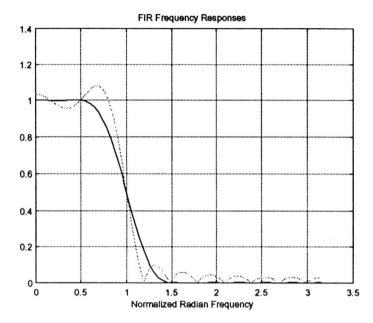

Figure 3.13 Low-pass filter approximations.

that has a sharper transition between passband and stopband, but at the expense of more ripple in the response. This is a direct effect of the large sidelobes of the frequency response of the rectangular window. For most applications the filter response of the (Hamming) windowed data is preferable.

Computer Code

```
r = 10; n = -r:r; n(r + 1) = eps; m = ceil(length(n)/2);
x = sin(n)./n; x = x/pi; w = .54 + .46*cos(n*pi/m);
h = x.*w; den = [1 zeros(1,2*r)]; N = 200; om = 0:pi/N:pi;
[mh,ah] = dtfreq(h,den,N); [mx,ax] = dtfreq(x,den,N);
figure(1); plot(om,mx,'g:',om,mh,'r-');
title('FIR Frequency Responses');
xlabel('Normalized Radian Frequency'); grid
```

Summary

In this section two methods of designing digital filters were presented: one, which is a "redesign" of a given analog filter, is useful because of existing analog filter designs available in the literature; and, another, an FIR filter design, which is attractive in that the design can use the FFT. An additional comment about the second design method is that the phase of the filter is linear, i.e. there is no "phase distortion," which can be a problem with some applications.

3.7 EIGENSYSTEMS AND JORDAN FORM DECOMPOSITION

A standard tool used in array processing is the *similarity transformation*. For a square matrix, A, the matrix B is said to be similar to A if there is a nonsingular *transformation matrix* P such that

$$B = P^{-1}AP \tag{3.41}$$

Perhaps the most useful transformation is one for which the resulting matrix B is diagonal, or block diagonal, i.e. a *Jordan form*.

The eigenvalues of a (square) matrix A are those "characteristic" numbers d that satisfy that for some non-zero vector x

$$Ax = dx \tag{3.42}$$

Assuming that A is an $n \times n$ array, let us write Eq. (3.42) as

$$(A - dI)x = 0 \tag{3.43}$$

It is clear that the $\det(A - DI)$ must be zero, for otherwise, the only solution would be $x = 0$. The determinant of $(A - DI)$ is, then, an nth-order polynomial in d. The n roots of this polynomial are the *eigenvalues* of A. The vector(s) x_i that satisfy Eq. (3.43) for a given eigenvalue d_i are *eigenvectors* corresponding to that eigenvalue. Together, the set of eigenvalues and corresponding eigenvectors comprise the *eigensystem* for the matrix A.

Jordan Form

In the next discussion a specific transformation matrix will be determined to obtain the Jordan form. The key characteristic of the Jordan form state model is that the transformed state coefficient matrix $P^{-1}AP$ is in the Jordan matrix form. The Jordan matrix, J, is a block diagonal matrix, each block of which is associated with a single eigenvalue of A. Specifically, J is of the form

$$J = \begin{bmatrix} J_1 & 0 & \cdots & 0 \\ 0 & J_2 & \cdots & 0 \\ 0 & 0 & \ddots & 0 \\ 0 & 0 & \cdots & J_v \end{bmatrix} \tag{3.44}$$

Algorithms Used in Signal Analysis

where the individual J-blocks have the following structure:

$$J_k = \begin{bmatrix} \lambda & 1 & \cdots & 0 \\ 0 & \lambda & \ddots & 0 \\ 0 & 0 & \ddots & 1 \\ 0 & 0 & \cdots & \lambda \end{bmatrix} \qquad (3.45)$$

i.e. a diagonal of the same (repeated) eigenvalue and a super-diagonal of ones. The transformation P in Eq. (3.41) that creates this form is called the *modal matrix*, since the resulting system model is thereby decomposed into its *modes*. The special case where A has no repeated eigenvalues results in a Jordan form with each J-block being a scalar (eigenvalue).

3.7.1 Algorithm

It is assumed that the square matrix A and its eigenvalues are given and that some criterion has been employed to decide which of these eigenvalues (which may all be at least slightly different) should be grouped as multiple values. Following this process, we may assume a list of distinct eigenvalues, E, each of which may have a certain, but unknown, multiplicity.

Ordinary eigenvectors of an eigenvalue e must lie in the null space of $Q = (A - eI)$; and, since all columns of the transformation matrix P must be linearly independent, we can only use a number of vectors equal to the dimension of this null space. The remaining vectors for this eigenvalue are called generalized eigenvectors. A *generalized eigenvector of order k* may be defined as a vector in the null space of $Q^k = (A - eI)^k$, but in the range space of $Q^{k-1} = (A - eI)^{k-1}$. Thus, an ordinary eigenvector is a "generalized eigenvector of order 1." A generalized eigenvector generates a *chain* of eigenvectors. Given an eigenvector of order k, p_k, a chain of k vectors is obtained as follows:

$$p_{k-i} = Q^i p_k \quad \text{for } i = [1, k-1] \qquad (3.46)$$

Every chain ends with an ordinary eigenvector. Thus, the number of J-blocks equals the nullity of Q.

The following is a description of the algorithm:

1. Determine the range space R and null space N of the matrix Q.
2. The nullity, $v = \dim(N)$, is the number of chains to be found.
3. Beginning with $k = 1$, calculate Q^k and determine those vectors which are in R but not in N_k, the null space of Q^k. If there are q non-zero vectors, say the columns of M, each one is used to generate a chain as in Eq. (3.46). If $q = 0$, increment k and repeat.

4. Each time there are non-zero columns in Step 3, new chains must be kept linearly independent. This is accomplished by concatenating R with M each time through. The process terminates when the accumulation of the number of vectors (columns of the M matrices) equals v.
5. The previous steps are carried out for each (distinct) eigenvalue of A.

The complete details of this algorithm as well as the necessary utility algorithms involved are contained in [4].

Example 3.8 (Jordan Form)

The following system has eigenvalues at 2 and 3 with multiplicities 2 and 1, respectively. By implementing the above algorithm, the transformation (modal) matrix P and the Jordan form state model are presented. Only the dynamic state equations are shown (no output equations) to save space. (the reader is referred to Chapter 5).

Original model:

$$A = \begin{bmatrix} 2.6000 & -1.1528 & 0.5944 \\ -0.4000 & 2.1389 & -0.3222 \\ 0.6000 & 1.6806 & 2.2611 \end{bmatrix} \quad B = \begin{bmatrix} -0.2056 \\ -0.1222 \\ 0.4611 \end{bmatrix}$$

Transformation matrix:

$$P = \begin{bmatrix} 1.2357 & 0.0231 & 0.9168 \\ -0.1901 & -0.9907 & -0.3806 \\ -1.6159 & 0.1343 & -0.1211 \end{bmatrix}$$

Jordan form model:

$$A = \begin{bmatrix} 2 & 1 & 0 \\ 0 & 2 & 0 \\ 0 & 0 & 3 \end{bmatrix} \quad B = \begin{bmatrix} -0.2877 \\ 0.1169 \\ 0.1606 \end{bmatrix}$$

Summary

The algorithm presented in this section generates the Jordan form state model by simultaneously computing the eigenstructure and associated eigenvector-generalized eigenvector chains, starting with the shortest chains and increasing the length until the invariant subspaces are exhausted. This method avoids difficulties in generalization found in other methods [5].

3.8 SINGULAR VALUE DECOMPOSITION

Singular value decomposition (SVD) is a general and very useful method of decomposing a rectangular matrix into a special form that can be used for solving a linear system of equations and much more. SVD can be thought of as a generalization of the eigensystem calculation for matrices that are not necessarily square.

We will assume that A is a real matrix of dimensions $m \times n$ where $m \geq n$. The objective is to represent A as

$$A = UWV^T \qquad (3.47)$$

where W is an $m \times n$ diagonal matrix with rank r, and U and V are $m \times m$ and $n \times n$ matrices, respectively. Noting that AA^T and A^TA are positive semi-definite (symmetric) $m \times m$ and $n \times n$ matrices, respectively,

$$AA^T u_i = \sigma_i^2 u_i \text{ and } A^T A v_j = \lambda_j^2 v_j \qquad (3.48)$$

for $i = [1, m]$ and $j = [1, n]$. The use of "squares" of the singular values is justified since both matrix products are positive semi-definite. This development is based on the attractive properties of symmetric matrices; namely, that they always possess a full set of eigenvectors which are, or can be made to be, mutually orthogonal. In addition, it can be shown that the non-zero eigenvalues of AB and BA are equal, so that we can use the symbol σ^2 for both equations of (3.48). Since the vectors $z_i = A^T u_i$ can be shown to be eigenvectors of $A^T A$, it follows that

$$A z_i = \sigma_i^2 u_i \qquad (3.49)$$

follows from Eq. (3.48). Since the norm of $z_i = \sigma_i$ it is also seen that $z_i = \sigma_i v_i$. Dividing Eq. (3.49) by σ_i and concatenating all the resulting equations,

$$A \begin{bmatrix} v_1 v_2 & \cdots & v_n \end{bmatrix} = \begin{bmatrix} u_1 & u_2 & \cdots & u_m \end{bmatrix} \begin{bmatrix} \sigma_1 & 0 & \cdots & 0 \\ 0 & \sigma_2 & \cdots & 0 \\ \vdots & \vdots & \ddots & \vdots \\ 0 & 0 & \cdots & \sigma_n \\ 0 & 0 & \cdots & 0 \\ 0 & 0 & \cdots & 0 \\ \vdots & \vdots & \ddots & \vdots \\ 0 & 0 & \cdots & 0 \end{bmatrix} \qquad (3.50)$$

Finally, by post-multiplying by V^T, we obtain the desired form of Eq. (3.47). The last columns of U and the corresponding zeros of W can be deleted, in which case the dimensions of U, W, and V^T become $m \times n$,

$n \times n$, and $n \times n$, respectively. If A is less than full rank, the final rows of V^T and the corresponding zero columns of W can be omitted, leaving the dimensions of U, W, and V^T as $m \times r$, $r \times r$, and $r \times n$, respectively.

Example 3.9 (Singular Value Decomposition—Least-Squared Error Solution)

Consider the linear vector/matrix equation, $Ax = b$, with A and b given:

$$\begin{bmatrix} 1 & 1 \\ -1 & 1 \\ 0 & 1 \end{bmatrix} \begin{bmatrix} x_1 \\ x_2 \end{bmatrix} = \begin{bmatrix} 2 \\ -2 \\ 3 \end{bmatrix} \qquad (3.51)$$

The SVD of the matrix A is given by $A = USV^T$:

$$U = \begin{bmatrix} 0.5774 & 0.7071 & -0.4082 \\ 0.5774 & -0.7071 & -0.4082 \\ 0.5774 & 0 & 0.8165 \end{bmatrix}, S = \begin{bmatrix} 1.7321 & 0 \\ 0 & 1.4142 \\ 0 & 0 \end{bmatrix}, V = \begin{bmatrix} 0 & 1 \\ 1 & 0 \end{bmatrix}$$

Defining $U^T b = w$ and $V^T x = v$ establishes the new equation to be solved as $Sv = w$. In this instance the equations prove to be inconsistent. Even so, a *least-squared error* solution can be found by neglecting the last element of w. Solving yields $v = [1 \ 2]^T$; so that our solution is $x = Vv = [2 \ 1]^T$. The code that follows shows the SVD algorithm for the case where the number of columns of A is greater than or equal to the number of rows of A. If this is not the case, the decomposition can be performed on A^T.

Computer Code

```
function [U,S,V] = svd(A)
[m,n] = size(A); p = max(m,n); if n = = p, aat = A*A';
[u,d] = eig(aat); s = real(diag(d)); [y,i] = sort(s);
j = i(max(i):-1:min(i)); ss = s(j); I = find(ss); r = max(I);
sig = sqrt(ss(1:r)); SS = diag(sig); UU = real(u(:,j));
U = UU(:,1:r); V = A'*U*inv(SS); S = diag([sig' zeros(1,(n-r))]);
S = real(S(1:m,:)); ata = A'*A; [E,D] = eig(ata); sss = diag(D);
[yy,ii] = sort(sss); jj = ii(max(ii):-1:min(ii)); E = E(:,jj);
V = real([V E(:,(r+1):n)]); if r~=m,
U = real([U UU(:,(r+1):m)]); end
```

Summary

The singular value decomposition (SVD) of a matrix was discussed. This algorithm is numerically robust and has many diverse applications, particularly related to approximation. An example of solving an inconsistent set of equations was presented.

3.9 SPECIAL RANDOM NUMBER SEQUENCES

In Sect. 3.2 we saw some examples of signals. Random signals were presented as having some degree of variation which cannot be predicted exactly. So-called *white noise* sequences are completely uncorrelated which means that knowledge of the sequence value at one time provides absolutely no indication of what the next value will be. This is in contrast to signals which are *time correlated*. We will discuss how time correlation can be incorporated into a random signal, as well as the generation of special random numbers [6].

Recall that a system can "transform" one signal (the input) into another signal (the output). We must first consider the "averaging" operator, E, that ideally provides the (*ensemble*) *average* of a random signal. Given a random signal $x(k)$, averages provide a means of describing the signal, for example, the *mean value* of $x(k)$, denoted

$$m_x = E[x(k)] \tag{3.52}$$

Similarly, the *autocorrelation* function of $x(k)$ and the *cross-correlation* between $x(k)$ and $y(k)$, respectively, can be written as (the complex conjugate operation denoted by the asterisk can be disregarded for real-valued signals):

$$R_{xx}(\alpha) = E[x^*(t)x(t+\alpha)] \quad \text{and} \quad R_{xy}(\alpha) = E[x^*(t)y(t+\alpha)] \tag{3.53}$$

Here, we have assumed that the random signals are scalar-valued, but the principles carry over to vector-valued signals. The parameter α in both cases represents a time difference, which gives these averages the ability to "measure" the similarity of one signal (at time t) to another (or the same signal) at a time differing by α seconds. At one extreme, *white noise* is a type of signal which has an autocorrelation function that is identically zero for non-zero α. We can think of a *deterministic* signal as being the other extreme in that knowing the functional relation $f(t)$, the value of the signal at one time is completely related to the value at any other time. It is often important to be able to model a signal that is time-correlated in a specific way. We will discuss signals with a *first-order* correlation.

Time-Correlated Signals

Although the subject of random signals goes far beyond this brief introduction, a simple development can easily provide the reader with a basic understanding of the subject. For present purposes let us investigate a DT scalar random signal that is "processed" through a first-order linear system.

$$x(t+1) = a\,x(t) + w(t) \tag{3.54}$$

where $w(t)$ is a white noise sequence described by

$$E[w(t)] = 0 \quad \text{and} \quad R_{ww}(\tau) = b^2 \delta(\tau) \tag{3.55}$$

Thus, the discrete-time sequence $w(t)$ has a zero average and a variance of b^2. (The distribution of w values is not specified; they could be from a uniform, normal, or other distribution.) We can simulate $w(t)$ by calling (independent) random numbers in sequence and multiplying by $b > 0$ (to achieve the correct variance). In turn, $x(t)$ is constructed from $w(t)$ using Eq. (3.54) for some specific values of a and b, starting with, say, $x(0)$ initialized randomly; this will be used later.

Taking the expectation (average) of Eq. (3.54), term by term, we see that

$$E[x(t+1)] = a\,E[x(t)] \tag{3.56}$$

so that $E[x(t)] = 0$ for all t, since we are assuming that $E[x(0)] = 0$. To continue the analysis, the variance V can be calculated from

$$V = R_{xx}(0) = E[x^2(t+1)] = E[a^2 x^2(t) + w^2(t) + 2ax(t)w(t)] \tag{3.57}$$

Taking the average term by term

$$V = a^2 V + b \quad \text{or} \quad V = \frac{b^2}{1-a^2} \tag{3.58}$$

When $x(0)$ is initialized with this variance, the random sequence will be in a "stochastic steady state." We can now calculate the autocorrelation function of $x(t)$. From Eq. (3.54) the general expression for $x(t+k)$ can be derived as

$$x(t+k) = a^k x(t) + \sum_{n=0}^{k-1} a^{k-n-1} w(t+n) \tag{3.59}$$

The autocorrelation funcion can then be determined to satisfy

$$R_{xx}(k) = E[x(t)x(t+k)] = a^k R_{xx}(0) \tag{3.60}$$

Since $R_{xx}(0)$ is known from Eq. (3.58),

$$R_{xx}(k) = a^k V = \frac{a^k b^2}{1-a^2} \tag{3.61}$$

Algorithms Used in Signal Analysis

Assuming that a has a magnitude less than one, the correlation of $x(t)$ with itself is a decreasing exponential function of the time lag k. By using a higher-order difference equation to generate $x(t)$ from the independent samples of $w(t)$, the degree of correlation can be increased. In the following discussion a technique for "creating" the proper distribution for random numbers is considered.

Generating Special Distributions of Random Numbers

It is occasionally desirable to be able to modify the pseudo-random numbers available from a utility program to form a more complex distribution. Specifically, we will assume that random numbers uniformly distributed between 0 and 1 are available. Such a distribution is said to have a constant *density*, $f(x)$, between 0 and 1, zero elsewhere, i.e.

$$f(x) = \begin{cases} 1, & \text{for } 0 < x < 1 \\ 0, & \text{elsewhere} \end{cases} \tag{3.62}$$

For any density function we define the *cumulative distribution function* (CDF),

$$F(x) = \int_{-\infty}^{x} f(u) du \tag{3.63}$$

so that the probability of getting a value of x between a and b is $F(b) - F(a)$.

Suppose that we want to generate random numbers from a distribution that has a CDF given by the function $G(x)$, with associated density function $g(x)$, which is the derivative of $G(x)$. The result is that from uniform [0, 1] numbers, y, we apply the inverse of G, i.e.

$$x = G^{-1}(y) \tag{3.64}$$

This result is easily shown in reverse as follows. Given that $y = G(x)$, and letting $F(y)$ be the CDF of y,

$$\begin{aligned} F(y_0) &= \Pr\{y \leq y_0\} = \Pr\{G(x) \leq y_0\} \\ &= \Pr\{x \leq G^{-1}(y_0)\} = G(G^{-1}(y_0)) = y_0 \end{aligned} \tag{3.65}$$

This string of equalities begins with the definition of the CDF of y, the defined relation, leading to the definition of the CDF of x, and the simple result. Since a CDF is always monotone increasing from 0 to 1, the result of Eq. (3.65) indicates that the CDF of y is linear for y between 0 and 1, whence the derivative $g(y)$ is constant. Turning this around, if y is uniformly distributed, then x from Eq. (3.64) is distributed according to $G(x)$, and equivalently has the density $g(x)$, as desired.

Example 3.10 (Generating Exponentially Distributed Random Numbers)

From uniform [0,1] random numbers we will generate numbers that have the density function $g(x) = \exp(-x)$ for non-negative values of x. (Density functions must integrate to unity.) The desired CDF is therefore $G(x) = [1 - \exp(-x)]$ for x non-negative. Using Eq. (3.64), the numbers y are called from the utility program and are subsequently transformed into the x values. Since $y = [1 - \exp(-x)]$, then $x = \ln[(1-y)^{-1}]$. Figure 3.14 illustrates the histogram of the generated numbers.

Computer Code
```
N=5000; bars=30; y=rand(1,N); x=-log(1-y);
[ny,y]=hist(y,bars); [yy,nny]=bar(y,ny); scaley=N/bars;
[nx,x]=hist(x,bars); [xx,nnx]=bar(x,nx); scalex=N/4;
subplot(2,1,1),plot(yy,nny/scaley); title('Uniform Dist.')
subplot(2,1,2),plot(xx,nnx/scalex); title('Exp. Dist.')
```

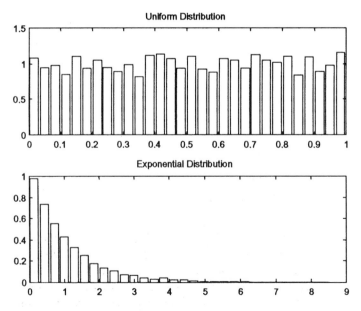

Figure 3.14 Histogram of generated random numbers.

Summary

In this section a brief discussion was made of incorporating first-order time correlation into a random sequence. The method involved the use of a linear system model to transform a white (uncorrelated) input sequence. A second topic presented was a general technique for generating non-standard distributions of random numbers. This procedure is simple and effective, but does require analytic knowledge of the desired cumulative distribution function.

REFERENCES

1. E. O. Brigham (1998), *The Fast Fourier Transform and Its Application*, Prentice-Hall, Inc., Englewood Cliffs, NJ.
2. C. S. Burrus *et al.* (1994), *Computer-Based Exercises for Signal Processing Using Matlab*, Prentice-Hall, Inc., Englewood Cliffs, NJ.
3. A. V. Oppenheim and R. W. Schafer (1989), *Discrete-Time Signal Processing*, Prentice-Hall, Inc., Englewood Cliffs, NJ.
4. S. Bingulac and H. F. VanLandingham (1993), *Algorithms for Computer-Aided Design of Multivariable Control Systems*, Marcel Dekker, Inc., NY.
5. W. L. Brogan (1991), *Modern Control Theory*, 3rd edn, Prentice-Hall, Inc., Englewood Cliffs, NJ.
6. H. Stark and J. W. Woods (1994), *Probability, Random Processes, and Estimation Theory for Engineers*, 2nd edn, Prentice-Hall, Inc., Englewood Cliffs, NJ.

4
Communication Systems

Timothy Pratt
Virginia Polytechnic Institute and State University, Blacksburg, Virginia

4.1 ANALOG MODULATION

Analog modulation techniques are widely used in broadcasting and for point-to-point radio links. AM broadcast stations use amplitude modulation (AM), television broadcasting uses vestigial sideband modulation (VSB), stereo radio broadcasts employ double sideband suppressed carrier AM (DSB-SC) of a sub-carrier to carry stereo information, and HF radio links use single sideband AM. Frequency modulation (FM) is widely used for high-quality sound broadcasting, using large deviation ratios and wide bandwidth signals. Narrow bandwidth small deviation FM is used in cellular telephones which employ analog modulation (AMPS), and also for handheld radios used for voice communications. Wideband FM (WBFM) is also used for distribution of television signals via geostationary satellites.

Analog modulation is generally simpler to implement than digital modulation, especially where the signal to be transmitted is analog, such as voice or television signals. AM requires less bandwidth than FM, but requires more transmitter power for a given signal-to-noise ratio (S/N) at the receiver. FM transmissions using wideband techniques always provide better S/N at the receiver output than AM, but always require more RF bandwidth. FM is less susceptible to interference from adjacent radio stations or noise spikes than AM.

4.1.1 Amplitude Modulation Waveforms and Power

Broadcast AM signals have a large carrier component which serves only to operate the envelope detector in the radio receiver. The general equation of an AM wave is:

$$V_{AM}(t) = [A + f(t)] \cos \bar{\omega}_m t \quad (4.1)$$

where A is the amplitude of the carrier, $f(t) = \cos \bar{\omega}_c t$, with $\bar{\omega}_c$ is the angular frequency of the carrier and $f(t)$ is the modulation signal. For simplicity, $f(t)$ is often modeled as a sine wave of the form $V_m \cos \bar{\omega}_m t$. The AM signal can then be written as

$$V_{AM}(t) = A \cos \bar{\omega}_c t + V_m \cos \bar{\omega}_m t \cos \bar{\omega}_c t \quad (4.2)$$

The modulation index of the AM wave is defined as

$$m = V_m/A \quad (4.3)$$

with the restriction than $V_m < A$ and therefore $m < 1$. The AM wave can then be rewritten as

$$V_{AM}(t) = A[\cos \bar{\omega}_c t + m \cos \bar{\omega}_{mt} \cos \bar{\omega}_c t] \quad (4.4)$$

or

$$V_{AM}(t) = A[\cos \bar{\omega}_c t + \frac{1}{2} m \cos(\bar{\omega}_c + \bar{\omega}_m)t + \frac{1}{2} m \cos(\bar{\omega}_c - \bar{\omega}_m)t] \quad (4.5)$$

The magnitude of the RF signal will vary between a maximum of $(A + m)$ and a minimum of $(A - m)$. The spectrum consists of a carrier at angular frequency $\bar{\omega}_c$ and two sidebands spaced evenly about the carrier by a frequency separation $\bar{\omega}_m$. This is the most basic form of amplitude modulation, known as double sideband large-carrier AM or DSB-LC.

In broadcast AM, the average value of the modulation index, m, is typically about 0.2. This leads to little power being transmitted in the sidebands, where the information is located, and most of the power residing in the carrier, which contains no information. More efficient forms of AM remove some or all of the power in the carrier. DSB-SC (suppressed carrier) and SSB (single sideband) AM both remove all the carrier power, leaving only one sideband in SSB and both sidebands in DSB. Some form of carrier must be regenerated at the receiver in order to demodulate DSB and SSB signals.

Analog television signals are transmitted using vestigial sideband (VSB) in which the high-frequency part of the lower sideband of a DSB-LC signal is removed as a bandwidth reduction measure. Since most of the energy in the baseband TV signal spectrum is in the lower frequencies, total power in

Communication Systems

the VSB signal differs little from the full AM signal, but a smaller RF bandwidth is required and the receiver can employ an envelope detector.

4.1.2 AM Transmitted Power Example

A radio transmitter has a peak envelope power (PEP) rating of 10 kW. What is the maximum power radiated in the sidebands when the modulation index is 1 and 0.2 when large-carrier, DSB-LC, and SSB AM are employed?

For large-carrier AM, the r.m.s. power in the carrier is $\frac{1}{2}A^2$, and the power in the sidebands is $2 \times \frac{1}{2}(\frac{1}{2}Am)^2 = 0.25\ A^2m^2$, normalized to an impedance of one ohm. Peak envelope power describes the maximum r.m.s. power that the transmitter can transmit, which will occur when $m = 1$. For any modulation index, power in the carrier is $\frac{1}{2}A^2$. For $m = 1$, power in the sidebands is $0.25A^2$. Thus the transmitter carrier power must be set to 6.667 kW, so that when $m = 1$ the sideband power is 3.333 kW, and the total power transmitted in the RF signals is 10 kW. When $m = 0.2$, power in the carrier remains at 6.667 kW, and power in the sidebands is $\frac{1}{2} \times m^2 \times 6.667$ kW $= 0.133$ kW. This represents only 1.33% of the total transmitted power. Thus, most of the time, an AM transmitter operates inefficiently sending little of its total output in the information content part of the RF signal.

If the transmitter used SSB instead of full AM, no carrier is transmitted and the RF waveform with sine wave modulation would be

$$V_{\text{SSB}}(t) = Am \cos \bar{\omega}_m t \cos \bar{\omega}_c t$$

For a maximum allowable value of $m = 1$, the transmitted power is $\frac{1}{2}A^2 = 10$ kW. With an average modulation index $m = 0.2$, the transmitted power is $0.5\ A^2 m^2 = 0.4$ kW. All of the transmitted power is sent as an information carrying signal, and peak power is radiated only when the modulating signal reaches its maximum permitted value and $m = 1$.

DSB-SC AM is used in FM sound broadcasting to carry stereo information on a sub-carrier, and in television broadcasting to carry the chrominance (color) part of the TV signal, again on a sub-carrier. In both cases, DSB was chosen to minimize the power required in the sub-carrier to convey the additional information, since the subcarrier is added to the audio or TV baseband signal. DSB-SC AM is used rather than single sideband AM because it is easier to demodulate and less prone to distort the signal.

4.1.3 Signal-to-Noise Ratio for Amplitude Modulation

The S/N ratio for the recovered baseband signal at the output of a demodulator depends on several factors: the carrier-to-noise ratio (C/N) of the received RF signal, the form of modulation employed, the demodulation technique, and the modulation index.

Full AM signals with both sidebands present are invariably demodulated with an envelope detector, followed by a low-pass filter which removes the carrier frequency components to leave just the baseband signal. The resulting S/N ratio at the output of the low pass filter is given by

$$S/N_{DSB-LC} = C/N \times [2m^2/(2+m^2)] \qquad (4.6)$$

When $m = 1$, S/N = 2/3 × C/N, and with $m = 0.2$, S/N = 0.0392 C/N.

Thus the best achievable S/N (when $m = 1$) is 1.8 dB below the RF signal C/N, and the typical S/N for an average modulation depth of 0.2 is 14.1 dB below the RF signal C/N. The poor S/N ratio of the demodulated baseband signal when the modulation index is small reflects the fact that most of the transmitted power in the RF signal is in the carrier component and very little in the information carrying sidebands.

As an example, suppose we want an S/N ratio of 50 dB in a baseband audio signal transmitted using full AM. The C/N at the input to the demodulator, with maximum depth of modulation $m = 1$, must be 51.8 dB. When the modulation depth is $m = 0.2$, the S/N will be 50 − 14.1 = 35.9 dB.

DSB-SC (suppressed carrier) AM signals do not have a carrier. The S/N at the output of the demodulator, which must multiply the DSB-SC signal by a regenerated carrier, is given by

$$S/N_{DSB-SC} = 2 \times C/N \qquad (4.7)$$

The factor of two is present because there are two sidebands, each of which carries a replica of the baseband signal. The multiplicative demodulation process superimposes the energy from each sideband, doubling the output S/N relative to the input C/N.

A single sideband AM signal has only one sideband, and the demodulator simply translates the RF signal to baseband. Thus for SSB

$$S/N_{SSB} = C/N \qquad (4.8)$$

In both DSB-SC and SSB forms of AM, signal-to-noise ratios are dependent on the modulation index because all of the transmitted energy in the RF signal lies in the sidebands, and the sideband energy is dependent on the depth of modulation.

As an example, consider the case of an SSB signal in which a modulation index of $m = 1$ causes the transmitter to broadcast an RF power of 10 kW. At a particular location, the RF C/N = 50 dB at the input to the demodulator of the receiver. The demodulated S/N = 50 dB also. If the average modulation index of the baseband signal is 0.2, the average

RF power from the transmitter is 400 W, and the receiver C/N is 50 − 14 = 36 dB. The S/N at the demodulator output is also 36 dB.

Signal-to-noise ratio is usually specified for the maximum amplitude of a modulating signal, so in the SSB example above, the S/N for that receiver would be quoted as 50 dB, despite the fact that the average S/N for the listener is only 36 dB.

4.1.4 Frequency Modulation

Frequency modulation is widely used for high-quality audio broadcasting, for analog cellular telephony, and for analog satellite television transmission. FM offers better interference and impulsive noise rejection than AM, and also the possibility that the S/N ratio at the output of a demodulator can be substantially higher than the C/N at the demodulator input. This latter characteristic of FM is used in satellite transmission of TV signals, where the C/N at the demodulator may be as low as 10 dB and the weighted S/N at the demodulator output can be 45 dB. The improvement in S/N ratio is achieved at the expense of a much larger RF bandwidth than required for equivalent AM transmission of the same baseband signal. Improvement of S/N relative to C/N always requires an increase in RF bandwidth – bandwidth is traded for S/N improvement in a wideband FM (WBFM) system. In narrowband FM (NBFM) systems, the S/N at the demodulator output may be lower than the C/N at the demodulator input; however, FM is preferred over AM for its superior noise and interference rejection properties.

The bandwidth occupied by an FM signal is defined by Carson's rule

$$B = 2(\Delta f_{pk} + f_m) \tag{4.9}$$

where Δf_{pk} is the peak frequency deviation of the FM signal and f_m is the maximum modulating frequency. The bandwidth defined by Carson's rule encompasses most of the energy in the FM signal spectrum, leading to minimal distortion of the baseband signal at the demodulator output.

The peak frequency deviation of an FM signal is given by

$$\Delta f_{pk} = kV_{max} \tag{4.10}$$

with maximum permissible value for the modulating signal at V_{max}. The ratio of peak frequency deviation to maximum modulating frequency is called the "deviation ratio," D, and is given by

$$D = \Delta f_{pk}/f_{max} \tag{4.11}$$

A deviation ratio of one or less defines the signal as NBFM, and a deviation ratio of two or more defines wideband FM.

The spectral content of an FM signal is sometimes analyzed with a baseband sinusoidal test signal at a frequency f_t. The result is an infinite set of sidebands offset from the carrier frequency by all possible integer multiples of the test signal frequency, $\pm f_t, \pm 2f_t \cdots$. The magnitudes of the sidebands are determined by Bessel function coefficients dependent on the deviation ratio D. In practice, the RF bandwidth required to transmit an FM signal is nearly always determined from Carson's rule.

4.1.5 FM Bandwidth Examples

A single audio frequency baseband signal is transmitted using FM. The FM signal is defined by a peak frequency deviation of $\Delta f_{pk} = 5$ kHz and a maximum modulating frequency $f_{max} = 5$ kHz. The deviation ratio is

$$D = 5/5 = 1$$

which defines a narrowband FM signal. The bandwidth occupied by this signal is given by Carson's rule as

$$B = 2(5+5) = 20 \text{ kHz}$$

This is somewhat larger than the bandwidth occupied by an equivalent DSB-AM signal. The bandwidth defined by Carson's rule encompasses most of the energy in the FM signal spectrum, leading to minimal distortion of the baseband signal at the demodulator output.

The same audio signal could be transmitted over a link using wideband FM. If the peak frequency deviation is increased to $\Delta f_{pk} = 25$ kHz with $f_{max} = 5$ kHz, then

$$D = 25/5 = 5$$

and

$$B = 2(25+5) = 60 \text{ kHz}$$

The 60 kHz RF bandwidth will not be completely filled all the time, since the width of the spectrum depends on the frequency content of the signal. If most of the baseband energy lies at the lower frequencies, the FM spectrum tends to have most of its energy concentrated around the carrier frequency.

Television signals intended for cable TV distribution are usually sent out via geostationary satellite to many cable television companies. In the United States, there are over 10,000 such companies receiving these signals, and several million households also receive these signals using 8 ft or 10 ft dish antennas. The video portion of a standard television signal in the United States, known as NTSC, has a baseband bandwidth of 4.2 MHz.

Communication Systems

The TV signal is modulated onto a 6 GHz or 11 GHz carrier using WBFM and a frequency deviation of $\Delta f_{pk} = 10$ MHz. The resulting RF bandwidth is

$$B = 2(10 + 4.2) = 28.4 \text{ MHz}$$

This signal is transmitted through a single satellite transponder with bandwidth 36 MHz.

4.1.6 S/N for FM Transmissions

The baseband S/N ratio for frequency modulated signals depends on the bandwidth of the receiver, B_{IF}, the maximum baseband frequency, f_{max}, and the peak frequency deviation, Δf_{pk}, of the FM signal. In wideband FM, the baseband S/N can be made much larger than the C/N at the demodulator input by making the deviation ratio much larger than one. The baseband S/N ratio is further improved by several weighting factors, which will be discussed shortly.

The baseband S/N of an unweighted FM signal is given by

$$S/N = C/N \times [3/2 \times B_{IF}/f_{max} \times (\Delta f_{pk}/f_{max})^2] \tag{4.12}$$

The S/N expression is often given in dB form as

$$S/N = C/N + [\log_{10}(B_{IF}/f_{max}) + 20 \log_{10}(\Delta f_{pk}/f_{max}) + 1.8] \text{ dB} \tag{4.13}$$

where C/N is in dB. The term in the square brackets is known as the "FM improvement" in WBFM systems, since it increases the baseband S/N above the C/N. If the deviation ratio is less than one, the FM improvement factor may be less than unity, or negative in dB, so that S/N < C/N. In systems where C/N is very high, such as line-of-sight microwave links, the deviation ratio may be made small to conserve RF bandwidth, and S/N < C/N.

The S/N can be improved by using de-emphasis in the receiver, which reduces the total noise power in the baseband at the demodulator output. Baseband noise power increases as the square of the baseband frequency, but can be reduced by employing a simple RC low-pass filter. A single RC low-pass section has an output voltage which falls linearly with frequency above its cutoff frequency, defined by $f_c = 1/(2\pi RC)$ or the corresponding time constant $t_c = RC$. The noise power at the output of the single RC section falls in proportion to the square of the frequency, above the cutoff frequency f_c, so this simple low-pass filter can be used to flatten the rising baseband noise power characteristic of the FM demodulator. A complementary RC section must be added at the transmitter to provide pre-emphasis of the baseband signal, with the same time constant used in the receiver.

The exact S/N improvement which can be achieved with pre-emphasis and de-emphasis in an FM link depends on the signal being transmitted. The improvement is subjective, since low-frequency noise in the baseband is not changed but high-frequency noise is reduced. In audio systems, high-frequency noise is more annoying than low-frequency noise, so de-emphasis makes FM audio sound better. In television signals, the snow on a TV screen created by analog noise is less noticeable when de-emphasis is used. Consequently, standard factors for de-emphasis improvement are used, denoted by the addition of a factor P dB in the S/N equation.

Thus for a FM audio signal, the baseband S/N is

$$\text{S/N} = \text{C/N} + [\log_{10}(B_{\text{IF}}/f_{\text{max}}) + 20\log_{10}(\Delta f_{\text{pk}}/f_{\text{max}}) + P + 1.8]\,\text{dB} \quad (4.14)$$

Sound broadcast using FM employs a standard pre-emphasis and de-emphasis time constant of 75 μs, equivalent to an RC section cutoff frequency of 2.1 kHz, which gives $P = 13$ dB. This value of P assumes that the addition of pre-emphasis to the baseband signal at the transmitter does not increase its total power significantly, which is the case most of the time.

FM television has a second subjective improvement factor, denoted by Q dB, which accounts for the sensitivity of the human eye to noise in a video signal, and some other factors related to the definition of S/N for a video signal. Thus for an FMTV signal, the baseband S/N after the FM demodulator is given by

$$\text{S/N} = \text{C/N} + [\log_{10}(B_{\text{IF}}/f_{\text{max}}) + 20\log_{10}(\Delta f_{\text{pk}}/f_{\text{max}}) + P + Q + 1.8]\,\text{dB}$$
$$(4.15)$$

The value of Q for a standard NTSC signal is typically 9 dB, and depends on the exact shape of the filters used in the video sections of the transmitter and receiver. Pre-emphasis and de-emphasis are used to obtain a further subjective improvement of about 8 dB, so for NTSC television signals transmitted by FM, $P + Q \approx 17$ dB.

4.1.7 FM S/N Examples

FM broadcasting in the United States employs standard frequency deviation and de-emphasis parameters. The peak deviation is $\Delta f_{\text{pk}} = 75$ kHz, the maximum baseband audio frequency is $f_{\text{max}} = 15$ kHz, giving a deviation ratio $D = 5$. A 75 μs pre/de-emphasis time constant is used, giving a subjective improvement factor $P = 13$ dB. From Carson's rule, the bandwidth required to transmit the FM signal is

$$B = 2(\Delta f_{\text{pk}} + f_{\text{max}}) = 2 \times 90 \text{ kHz} = 180 \text{ kHz}$$

Broadcast FM signals in the United States are spaced 200 kHz apart in the VHF band, and receivers have IF filters with $B_{IF} = 180$ kHz. The S/N at the receiver for a broadcast FM audio signal is

S/N = C/N + [FM improvement factor] dB

The FM improvement factor is

[FM improvement factor] = $10 \log_{10}(B_{if}/f_{max}) + 20 \log_{10}(\Delta f_{pk}/f_{max})$
$+ P + 1.8$ dB
$= 10.8 + 14.0 + 13.0 + 1.8 = 39.6$ dB

Thus with a C/N = 20 dB, the corresponding baseband S/N = 59.6 dB. An audio signal with S/N of 60 dB is generally regarded as high quality, so good quality audio can be obtained with an FM broadcast provided the C/N is above 20 dB. By comparison, a DSB-LC AM signal with C/N = 20 dB would give S/N < 20 dB, which is very poor audio quality. The FM improvement will be maintained for all C/N ratios down to the FM threshold. Below threshold, typically at a level of C/N = 9 to 13 dB, noise spikes from the demodulator rapidly reduce the average S/N in the baseband.

The S/N calculated from the improvement factor above represents the maximum achievable value. To obtain the full 39.6 dB improvement, the audio signal must have maximum amplitude to generate peak frequency deviation, and also be at $f_{max} = 15$ kHz. This is an unlikely occurrence, since loud signals are usually at the lower frequencies. The calculated S/N is therefore most useful for comparison purposes, and does not represent the actual S/N achieved in practice, especially when FM is used for telephony. When the amplitude of the signal is small, and the frequency deviation is also small, the improvement factor is much reduced.

For example, a "quiet talker" using a telephone may produce an audio signal 30 dB below the maximum permitted value, giving a peak frequency deviation 15 dB below the maximum. (Frequency deviation is proportional to signal voltage, which is the square root of signal power, so in decibels the reduction in Δf_{pk} is one half of the reduction in signal power.) The FM improvement is then reduced by 30 dB, making the S/N much lower than the ideal case. Telephone systems are designed to achieve S/N = 50 dB for the maximum voice signal level, so the "quiet talker" is heard by the listener at a 20 dB signal-to-noise ratio if the link uses FM. Companding is often used to overcome this problem, both in telephony and broadcasting. Companding compresses the dynamic range of the audio signal before transmission, so that the signals applied to the FM modulator have a small range of magnitudes. At the receiver, the inverse process restores the correct dynamic range. Companding is the analog equivalent of non-linear A/D encoding used in digital speech links.

The distribution of NTSC television signals for cable companies in the United States is almost entirely via satellite. When analog transmission through a single transponder is employed, the modulation is FM with a peak deviation around 10 MHz and an RF bandwidth of 28 MHz. Dish antennas 8 to 10 ft in diameter can be used to receive the FMTV signals, and will produce a C/N of 10 to 12 dB in clear air from a satellite transponder with an EIRP of 26 dB. Most FM satellite television receivers use a bandwidth less than 28 MHz to increase the C/N further above threshold, at the expense of a slight increase in distortion in the picture.

The baseband video S/N of the received TV signal can be calculated for a standard NTSC video signal with $f_{max} = 4.2$ MHz and $\Delta f_{pk} = 10$ kHz, and a receiver IF bandwidth of 25 MHz, as follows.

The FM improvement using standard de-emphasis and subjective improvement factors is given by

$$S/N = C/N + [\log_{10}(B_{IF}/f_{max}) + 20\log_{10}(\Delta f_{pk}/f_{max}) + P + Q + 1.8]\,\text{dB}$$

where $P + Q = 17$ dB. Hence

$$S/N = C/N + [7.8 + 7.5 + 17 + 1.8]\,\text{dB} = C/N + 34.1\,\text{dB}$$

If the C/N = 12.0 dB, the video S/N = 48.1 dB. This is regarded as an acceptable video S/N in cable TV systems. (S/N = 45 dB is the lowest value considered acceptable, and 55 dB is rated as "studio quality" video.) The FM threshold for a satellite TV receiver is typically at 9.0 dB, using a threshold extension demodulator. At this C/N, the FM improvement is reduced by 1 dB – the definition of where threshold occurs. Thus in a 3 dB rain fade on the satellite link, the C/N will fall to 9.0 dB, and the video S/N will fall by 4 dB to 44.1 dB. As the carrier falls below threshold, noise spikes appear at the demodulator output which cause white flecks ("sparklies") on the TV screen. The video signal still produces a good quality picture, but further reduction in C/N ratio swamps the picture with impulsive noise.

4.2 DIGITAL TRANSMISSION

Communication systems are increasing moving to digital transmission in place of analog. The conversion is almost complete for the terrestrial telephone network in the United States, where transmission of voice and data signals is entirely by digital techniques, mainly using optical fibers. Only for the "last mile" between the local telephone exchange and the subscriber's house or office is the signal in analog form. Terrestrial broadcasting remains the last domain of analog modulation in radio transmission. Direct broad-

Communication Systems

cast satellite television is digital, direct broadcast satellite audio is digital, and high-definition television is also digital. Almost all audio reproduction is now from compact discs, which store information in optical digital form.

There are two steps in the transmission of digital signals. If the baseband signal is in analog form, it must first be converted to a digital bit stream. The signal must then be encoded and transmitted directly over an optical fiber, or modulated onto a carrier for RF transmission. The encoding process may include the addition of redundant bits to the data stream for error detection and/or forward error correction (FEC).

4.2.1 Conversion of Analog to Digital Signals

Analog signals are converted to digital form by sampling and digitizing the analog signal waveform. Sampling must be done at a frequency above the Nyquist rate, defined as twice the highest frequency present in the analog signal. The samples are then converted to digital words with an analog to digital converter (ADC). The conversion process results in quantization noise in the recovered baseband signal, with a level dependent on the number of bits in the digital words. Sampling an analog signal at the Nyquist rate allows the signal to be converted back to its analog form using only a low-pass filter, but Nyquist rate sampling requires perfect low-pass filters with infinitely fast roll-off in their stopbands.

Digital signals can be multiplexed onto a single transmission line by sending the digital words sequentially. One widely used system, known as T1 or DS1, sends signals from 24 analog channels as a continuous 1.544 Mbps bit stream. Multiple T1 bit streams can be combined into higher rate T2, T3, etc., bit streams for trunk transmission. For transmission on a radio link, the digital signals must modulate a carrier, giving a "shift keying" modulation. Possible modulations are amplitude (ASK), frequency (FSK), or phase (PSK). PSK is the most widely used digital modulation. It gives the best bit error rate performance for a given carrier power, but at the expense of considerable complexity in the radio receiver.

4.2.2 Baseband Digital Transmission

Consider as an example the process of digitizing an analog telephone channel. The frequency range of a telephone signal in the United States is specified to extend from 300 Hz to 3400 Hz. Thus the highest frequency present, f_{max}, is 3.4 kHz, and the Nyquist rate for this signal, f_N, is

$$f_N = 2 \times f_{max} = 6.8 \text{ kHz} \tag{4.16}$$

In practice, analog telephone channels are always sampled at 8 kHz to allow for the use of real filters. A NTSC television video signal extends from close to 0 Hz to 4.2 MHz. This signal must be sampled at a rate higher than 8.4 MHz. Typically, 10 MHz sampling of video signals is used. The process of creating a digital signal from an analog signal is sometimes called PCM, pulse code modulation, a misnomer since it is neither a coding nor a modulation process.

The sampled analog signal is converted to a bit stream by an ADC. Each sample of the signal creates one digital word of N bits. In telephony, the standard value for N is 8 bits, so a digital telephone channel carries a 64 kbps bit stream (8 bits per word × 8 kHz samples). It is standard practice in telephone systems to use non-linear A/D conversion, to improve the signal to noise ratio for the "quiet talker". An NTSC video signal sampled at 10 MHz with an 8 bit ADC gives a bit stream at 80 Mbps. Because of the wide bandwidth of this digital signal, compression techniques such as MPEG-2 are widely used to reduce the bit rate of the transmitted bit stream. MPEG-2 will reduce an 80 Mbps digital video signal to an average rate around 6 Mbps. With some additional processing, movies can be digitized and stored for replay at 1.5 Mbps.

4.2.3 S/N of Baseband Digital Signals

Recovery of sampled signals using a low-pass filter leaves some quantization noise present in the baseband analog waveform. The S/N radio depends on the number of bits, N, used for transmission of each sample. If the signal is assumed to have a uniform amplitude probability distribution, so that any signal level is as likely to occur as any other, the quantization signal to noise ratio for the maximum magnitude signal is

$$(S/N)_Q = 2^{2N} = 6N \text{ dB} \tag{4.17}$$

Thus an 8 bit uniformly distributed and uniformly quantized digital signal will have a quantization S/N of 48 dB.

If a sine wave test signal is used, the amplitude distribution is not uniform and the quantization S/N is lower

$$(S/N)_{Q \text{ sine wave}} = 6N - 7.3 \text{ dB} \tag{4.18}$$

Speech signals do not have uniform amplitude distributions, nor are they sine waves. The "quiet talker" using a telephone may have a speech power level 30 dB below the maximum permitted level in the speech channel, resulting in a $(S/N)_Q$ ratio of 18 dB rather than 48 dB. Effectively, the quiet talker is using only 3 bits of the ADC, resulting in much more quantization noise, relative to the small signal, than will occur with a louder talker. Quantization noise is then very noticeable to the listener. Non-linear encoding, known as non-uniform quantization, is used to overcome this problem. The speech signal is processed through an amplifier with a logarithmic gain curve and then applied to a conventional (linear) 8 bit ADC. Small signals are amplified more than large signals, compressing the dynamic range of the incoming signals. The signal from the quiet talker then uses more bits in the ADC, and the quantization signal to noise ratio improves. At the receiving end of the link, an amplifier with an inverse log (antilog) gain characteristic compensates for the non-linear amplifier at the transmit end. When a loud talker generates a maximum amplitude speech signal, the quantization S/N is actually worse in the non-linear system that in a linear system. However, a feature of human hearing is that we can listen to only one loud sound at a time, so the increased quantization noise acompanying a loud speech signal is not noticeable.

The non-linear logarithmic characteristic used in telephony systems complies with one of two published curves, either "A law" or "μ Law." The μ Law characteristic is used mainly in the United States, and A law is used everywhere else. The difference in performance for the two systems is small, provided they are not cross-connected. The improvement in $(S/N)_Q$ obtained from non-uniform quantization is assumed as 17 dB for μ Law used in the United States. Hence, the resulting S/N is

$$(S/N)_{Q \text{ non-uniform quantization}} = 6N + 17 \text{ dB} = 48 + 17 = 63 \text{ dB}$$

assuming a uniform amplitude distribution of the signal and maximum signal amplitude.

4.2.4 CD Audio Reproduction

Compact discs record digital information by changing the reflectivity of a spiral track just below the transparent surface of the disk. Each bit is only a micron long, and the track is a micron wide, allowing a standard 5 inch disk to store over 10 Gbits of data. Audio signals are stored as a stereo pair using 44 kHz sampling. Each sample is converted to a 16 bit word using linear A/D conversion. The resulting bit stream has a rate

$$R_b = 2 \times 44 \times 16 \text{ kHz} = 1.408 \text{Mbps}$$

A double layer of Reed Solomon forward error correction (FEC) code is applied to the data stream and additional information is added giving a stored bit rate on the CD of over 3 Mbps.

4.2.5 Bit Error Rate in Digital Data Links

All binary data transmission systems must recover the transmitted bits at a receiver which samples the received waveform and decides whether a 1 or a 0 was transmitted. If the decision is incorrect, a bit error occurs. The probability of a bit error, P_b, known as the bit error rate (BER), determines the quality of the transmission system. Generally, speech transmission requires a BER better than 10^{-6}, and video transmission requires BER $< 10^{-7}$ to provide acceptable signals. When lower BER is required, error control techniques may be used.

Bit errors are caused by noise, either thermal or impulsive, by external interference, or by intersymbol interference (ISI). Impulsive noise which exceeds the magnitude of a bit at the receiver has a 50% chance of causing a bit error, since the noise may increase or decrease the magnitude of the sampled signal. Thermal noise has a Gaussian probability distribution and is specified by its r.m.s. value. The probability of a bit error due to a given thermal noise level can be calculated using statistical methods. ISI occurs when the waveform in the receiver allows signal from one bit to be present when the next bit is sampled, which will happen if the sampling clock suffers timing jitter. In combination with thermal noise, ISI increases the BER.

When analog signals are converted from digital binary form (PCM) to a baseband analog waveform, the S/N ratio for Gaussian distributed errors caused by thermal noise is given by

$$(S/N)_{th} = \frac{1}{4} P_b \qquad (4.19)$$

When combined with quantization noise for an N bit word, the resulting S/N for a uniform probability analog signal is

$$S/N = 2^{2N}/(1 + 4P_b \times 2^{2N}) \qquad (4.20)$$

Example

A digital speech link uses 8 bit words and 8 bit sampling to generate a 64 kbps digital speech signal. At the receiver, the BER under normal operating conditions is 10^{-7}. Find the S/N ratio in the recovered baseband audio signal.

The actual transmission rate of 64 kbps is irrelevant – bit error rate is not a rate, but an error probability which does not depend on the number of bits transmitted each second. For $P_b = 10^{-7}$ and $N = 8$, the S/N ratio is

Communication Systems

$$S/N = 2^{2N}/(1 + 4P_b \times 2^{2N}) = 65{,}536/(1 + 4 \times 10^{-7} \times 65{,}536)$$
$$= 65{,}536/1.0262 = 63{,}862 \text{ or } 48.0 \text{ dB}$$

The result does not differ from S/N = 48 dB for quantization noise alone. The quantization noise is dominant when the BER < 10^{-6}.

If thermal noise increases relative to the signal, the BER will fall. Suppose BER = 10^{-4}. Then the S/N will be

$$S/N = 65{,}536/(1 + 4 \times 10^{-4} \times 65{,}536) = 65{,}536/27.21 = 2408 \text{ or } 33.8 \text{ dB}$$

The thermal noise errors now dominate the quantization noise.

4.2.6 BER in Optical Fiber Links

Optical fibers transmit digital signals as pulses of light, usually generated with a laser which is modulated by a binary bit stream. A photodiode at the receiving end of the link converts the pulses of light back to a voltage waveform which is sampled by the bit clock.

Thermal noise in the first stage of the receiver is the primary cause of bit errors in optical fiber links when the optical signal is small. For the case of a pn photodiode driving a low-noise amplifier with a noise figure F, the S/N at the receiver output is approximately.

$$S/N = 0.5 \text{ m } R_0^2 P_r^2/(4kT_0 B(F-1)/R) \qquad (4.21)$$

where m is the modulation index of the laser (ratio of ON state output to OFF state output); R_0 is the photodiode conversion efficiency in A/W; P_r is the received optical power at the photodiode; k is Boltzmann's constant (1.38×10^{-23} J/K); T_0 is the standard noise temperature of 290 K; B is the receive amplifier bandwidth in hertz; F is the noise figure of the receiver (not in dB); and R is the resistive load on the photodiode.

The noise figure is usually expressed in decibels, and the product $T_0(F-1)$ is the noise temperature of the receiver in kelvins. For many systems $m \approx 1$. The bandwidth of the receiver must match the bit rate of the transmission system. For ideal raised cosine filters, the bandwidth should be one half the bit rate R_b, so $B = 0.5 R_b$.

The bit error rate for a baseband binary signal which has $V_t = 0$ when a binary 0 is transmitted and $V_t = A$ when a binary 1 is transmitted is given by

$$P_b = \text{erfc}[(0.5 \times S/N)^{1/2}] \qquad (4.22)$$

where erfc is the complimentary error function. Table 4.1 shows values for erfc(x) for $0 < x < 6$.

Table 4.1 Function erfc(x)

x	erfc(x)	x	erfc(x)	x	erfc(x)	x	erfc(x)
0.0	1.00000	1.00	0.15730	2.0	5.167E-3	4.0	1.587E-8
0.05	0.94363	1.05	0.13776	2.1	3.267E-3	4.1	1.880E-9
0.10	0.88754	1.10	0.11979	2.2	2.029E-3	4.2	2.932E-9
0.15	0.83200	1.15	0.10388	2.3	1.237E-3	4.3	1.224E-9
0.20	0.77730	1.20	0.08969	2.4	7.408E-4	4.4	5.012E-10
0.25	0.72367	1.25	0.07710	2.5	4.357E-4	4.5	2.013E-10
0.30	0.67137	1.30	0.06599	2.6	2.515E-4	4.6	7.925E-11
0.35	0.62062	1.35	0.05624	2.7	1.426E-4	4.7	3.060E-11
0.40	0.57161	1.40	0.04771	2.8	7.932E-5	4.8	1.159E-11
0.45	0.52452	1.45	0.04030	2.9	4.331E-5	4.9	4.303E-12
0.50	0.47950	1.50	0.03389	3.0	2.321E-5	5.0	1.567E-12
0.55	0.43668	1.55	0.02838	3.1	1.220E-5	5.1	5.596E-13
0.60	0.39614	1.60	0.02363	3.2	6.297E-6	5.2	1.959E-13
0.65	0.35797	1.65	0.01692	3.3	3.187E-6	5.3	6.727E-14
0.70	0.32220	1.70	0.01621	3.4	1.583E-7	5.4	2.265E-14
0.75	0.28884	1.75	0.01333	3.5	7.713E-7	5.5	7.476E-15
0.80	0.25790	1.80	0.01091	3.6	3.687E-7	5.6	2.420E-15
0.85	0.22933	1.85	0.00889	3.7	1.729E-7	5.7	7.680E-16
0.90	0.20309	1.90	0.00721	3.8	7.951E-8	5.8	2.390E-16
0.95	0.17911	1.95	0.00582	3.9	3.587E-7	5.9	7.291E-17

4.2.7 Optical Link Example

A fiber optic link transmits binary data at 20 Mbps. The receiver uses a pn diode with a 10 kΩ load and an amplifier with a noise figure of 6 dB. The receiver bandwidth is 10 MHz. Find the BER at the receiver output for a received power of −40 dBm (10^{-7} W).

The first step in calculating BER is to find the receiver S/N. The receiver noise temperature is

$$T = T_0 \times (F - 1) - 290 \times (4 - 1) = 870 \text{ K} \qquad (4.23)$$

If we assume that $m = 1$ (transmitting laser diode is off when the input is a binary 0) and that the pn diode has a conversion efficiency of 0.5 A/W:

$$\begin{aligned} S/N &= 0.5 m R_0^2 P_r^2 / (4 k T_0 B (F-1)/R) \\ &= 0.5 \times 0.25 \times 10^{-14} / (4 \times 1.38 \times 10^{-23} \times 10^7 \times 870/10^4) \\ &= 0.125 \times 10^{-14} / 4.80 \times 10^{-17} \\ &= 26.04 \text{ or } 14.2 \text{ dB} \end{aligned}$$

Communication Systems

The BER is given by

$$P_b = \text{erfc}[(0.5 \times S/N)^{1/2}]$$
$$= \text{erfc}[5.10] = 5.6 \times 10^{-13}$$

In practice, the error rate would be slightly higher than 5×10^{-13} because no account has been taken of ISI caused by sample clock jitter and non-ideal filters in the receiver. Note that even if the bit error rate were 10^{-10}, with a bit rate of 20 Mbps a bit error occurs once every nine minutes. Such transmission is said to be "essentially error free".

4.2.8 Digitial Radio Links

Digital radio links require a digitally modulated carrier. Options are amplitude shift keying (ASK), normally used with envelope detection, frequency shift keying (FSK), also using envelope detection, or phase shift keying (PSK). PSK requires coherent detection, either by regenerating a local carrier in the receiver, or by using differential detection in which the previous bit is used as the phase reference for the current bit. Although coherent detection could be used with ASK and FSK, the complexity required in the receiver for carrier regeneration justifies the better BER performance of PSK. ASK is not often used on its own, but is often used in conjunction with four-phase PSK (QPSK) to increase the number of bits per transmitted symbol. Such systems are known as QAM (quadrature amplitude modulation) digital links.

FSK with envelope detection is used in low data rate links such as low-speed modems for telephone lines, because of the ease with which an FSK signal can be generated and recovered.

The BER formulas for ASK, FSK, and PSK depend on the type of demodulation employed at the receiver (coherent or non-coherent). All digital radio links require a receiver noise bandwidth equal to the symbol rate, so $B_{RF} = R_s$. The RF bandwidth occupied by the radio signal, B_{occ}, is given by

$$B_{occ} = R_s(1 + \alpha) \tag{4.24}$$

where α is the roll-off factor of the filters used in the transmitter and receiver and $0 < \alpha < 1$.

For envelope detection of ASK and FSK, the BER with zero ISI and envelope detection at the receiver is

$$P_{b-ASK} = 0.5 \exp[-(R_b/2B) \times (E_b/N_0)] \tag{4.25}$$

$$P_{b-FSK} = 0.5 \exp[-(R_b/2B) \times (E_b/N_0)] \tag{4.26}$$

where R_b is the transmitted bit rate, B is the equivalent noise bandwidth of the filter before the envelope detector, and E_b/N_0 is the ratio of the average energy per bit to the single-sided noise power spectral density.

If there are ideal Nyquist (square root raised cosine) filters in both the transmitter and receiver, the predetection bandwidth B is equal to the bit rate R_b, so BER becomes

$$P_{b-ASK} = P_{b-FSK} = 0.5 \exp[-0.5(E_b/N_0)] \tag{4.27}$$

FSK transmits energy for both binary 1 and binary 0 bits, so the average transmitted power (normalized to a circuit resistance of 1 ohm) is $0.5A^2$, where A is the amplitude of the sine wave. With ideal Nyquist filters, the receiver bandwidth $B = R_b = 1/T$. Hence for FSK

$$[E_b/N_0]_{FSK} = ST/(N/B) = S/N \tag{4.28}$$

The corresponding BER for FSK is

$$P_{b-FSK} = 0.5 \exp[-0.5(S/N)] \tag{4.29}$$

In ASK, energy is transmitted only for a binary 1, so the transmitter is on only half the time. Compared with FSK, ASK requires twice as much transmitted power. With ideal Nyquist filters, the receiver predetection bandwidth B is equal to $R_b = 1/T$. Hence for ASK

$$[E_b/N_0]_{ASK} = 0.5ST/(N/B) = 0.5 \, S/N \tag{4.30}$$

and then the BER for ASK is

$$P_{b-ASK} = 0.5 \exp[-0.25(S/N)] \tag{4.31}$$

Because it takes more power at the transmitter to achieve a given BER with ASK than with FSK, binary ASK is rarely used in radio links.

Coherent detection of binary PSK achieves a lower BER than envelope detection of ASK or FSK. The BER with binary PSK (BPSK) and with four-phase PSK (QPSK) is approximately

$$P_{b-BPSK} = 0.5 \, \text{erfc}[(E_b/N_0)^{1/2}] \tag{4.32}$$

As with FSK, the transmitter is on all the time with BPSK, so with ideal square root raised cosine filters in the transmitter and receiver

$$[E_b/N_0]_{BPSK} = ST/(N/B) = S/N \tag{4.33}$$

Hence the BER for BPSK is

$$P_{b-BPSK} = 0.5 \, \text{erfc}[(S/N)^{1/2}] \tag{4.34}$$

Communication Systems

QPSK sends two bits with each transmitted symbol and effectively has two BPSK transmitters operating in parallel sending orthogonal BPSK waveforms in a common channel with bandwidth B Hz, where $B = 0.5R_b$. As a result, the transmitter power required to send a QPSK signal is twice the power required to send a BPSK signal in the same bandwidth, which requires twice the signal-to-noise ratio for QPSK to obtain the same BER as for the equivalent BPSK signal sent in bandwidth B Hz. Hence for QPSK, with ideal Nyquist filters,

$$P_{b-QPSK} = 0.5 \, \text{erfc}[(0.5 \, S/N)^{1/2}] \tag{4.35}$$

Ideal Nyquist filters do not exist, so in a PSK radio link with real filters, the theoretical BER is never achieved. ISI caused by the real filters and sample clock timing jitter increases the bit error rate. This is accounted for in the BER calculations by the addition of an "implementation margin" to the theoretical S/N required for a given BER. The implementation margin is typically 0.5 dB for a carefully designed low bit rate system ($R_b < 2$ Mbps), and up to 2 dB in high bit rate systems.

4.2.9 Digital Radio BER Examples

Example 1

A digital radio link transmits binary data at a rate 10 Mbps. The receiver has a 3 dB bandwidth of 10 MHz. At the input to the demodulator in the receiver, the S/N is 14 dB. Find the BER for non-coherent detection of FSK and for coherent detection of BPSK, with a receiver implementation margin of 1.0 dB. What is the bit rate and BER if the same link is used to send QPSK?

For non-coherent FSK, the BER for a system with ideal filters is

$$P_{b-FSK} = 0.5 \exp[-0.5(S/N)]$$

The actual S/N is 14 dB with a 1 dB implementation margin, so for the calculation of BER we need to use S/N = 13 dB = 20. Then for FSK with envelope detection

$$P_{b-FSK} = 0.5 \exp[-0.5(20)] = 2.27 \times 10^{-5}$$

For BPSK with coherent detection, the BER is obtained from the erfc tables with S/N = 20:

$$P_{b-BPSK} = 0.5 \, \text{erfc}[(S/N)^{1/2}] = 0.5 \, \text{erfc}[4.47] = 1.5 \times 10^{-10}$$

It would require approximately 3 dB more transmitter power using non-coherent FSK to achieve the same BER as BPSK with coherent detection.

If we use QPSK in the radio link, keeping the 10 MHz bandwidth, we would send data at a rate of 20 Mbps. The BER with a receiver S/N = 14 dB and 1 dB implementation margin would be

$$P_{b-QPSK} = 0.5 \, \text{erfc}[(0.5 \, S/N)^{1/2}] = 0.5 \, \text{erfc}[3.16] = 5 \times 10^{-6}$$

The higher transmission rate results in a higher bit error rate. Again, we would need 3 dB more transmitter power to restore the 10^{-10} BER obtained with BPSK.

Example 2

A radio link is needed to send binary data at 30 Mbps with a BER at the receiver of 1×10^{-8}. Find the S/N required at the receive end of the link allowing a receiver implementation margin of 2 dB, the receiver 3 dB bandwidth required, and the bandwidth occupied by the RF signal for filters with $\alpha = 0.5$, for the following cases:

1. FSK with non-coherent detection.
2. BPSK with coherent detection.
3. QPSK with coherent detection.

In each case, the filter 3 dB bandwidth will be equated to the noise bandwidth required for the given bit rate.

FSK

The filter bandwidth required is equal to the bit rate of 30 Mbps, so $B_{3\,dB} = 30$ MHz. The bandwidth occupied by the RF signal is $R_b(1 + \alpha) = 30 \times 1.5$ MHz $= 45$ MHz.

To achieve a BER of 10^{-8} we require that

$$10^{-8} = 0.5 \exp[-0.5(S/N)]$$

By trial and error we find S/N = 35.4 or 15.5 dB. The actual receiver S/N required is 17.5 dB to account for the implementation margin of 2 dB.

BPSK

The filter bandwidth required is equal to the bit rate of 30 Mbps, so $B_{3\,dB} = 30$ MHz. The bandwidth occupied by the RF signal is $R_b(1 + \alpha) = 30 \times 1.5$ MHz $= 45$ MHz, as with FSK. To achieve a BER of 10^{-8} we require that

$$P_{b-QPSK} = 10^{-8} = 0.5 \, \text{erfc}[(S/N)^{1/2}]$$

This requires erfc$(x) = 2 \times 10^{-8}$, and from the erfc tables, $x = 3.98$. Hence

$S/N = x^2 = 15.8 = 12.0$ dB (no implementation margin)

With a 2 dB implementation margin, the receiver S/N required for BPSK transmission of the 30 Mpbs data stream is 14.0 dB.

QPSK

QPSK symbols each carry two bits, so the symbol rate on a QPSK link with a bit rate of 30 Mbps is $R_s = 15$ Msps. The 3 dB bandwidth of the transmitter and receiver filters is equal to R_s, so $B_{3\,dB} = 15$ MHz. The RF bandwidth occupied by the radio signal using filters with roll-off factor $\alpha = 0.5$ is 15×1.5 MHz $= 22.5$ MHz.

The S/N required to achieve a BER of 10^{-8} with QPSK is given by

$$P_{b-QPSK} = 0.5 \operatorname{erfc}[(0.5\, S/N)^{1/2}]$$

This requires $\operatorname{erfc}(x) = 3.98$, where $x^2 = 0.5(S/N)$. Hence for QPSK we need $S/N = 2 \times 3.98^2 = 2 \times 15.8$ or 15 dB without implementation margin. Thus QPSK requires $S/N = 17.0$ dB with a 2 dB receiver implementation margin.

This example illustrates the trade-offs available in a digital radio link. FSK is the easiest modulation to implement, but requires the highest S/N at the receiver. BPSK requires the same bandwidth as FSK, but 3.5 dB lower S/N at the receiver for the same BER. QPSK needs 3 dB more S/N than BPSK, 0.5 dB less S/N than FSK, but transmits the same volume of data as the other two links in the half the RF bandwidth. QPSK is the most complex system to implement.

4.3 RADIO LINKS

Radio links require a transmitter, a receiver, and a path between the two. At higher frequencies, above about 500 MHz, a clear path (called *line-of-sight* or *LOS*) is needed if severe attenuation of the signals is to be avoided. Calculations of link performance are usually made first with a line-of-sight path to establish received signal levels under the best conditions, and then attenuation factors are added for obstructions in the path, absorption or scattering by the atmosphere, or other attenuating effects.

A radio link operates by modulating the baseband signal, which contains the information to be conveyed by the link, onto an RF carrier which can be radiated by an antenna. The basic measure of performance in a radio link is carrier to noise ratio, C/N, at the receiver. The C/N ratio must exceed a minimum value in the receiver IF bandwidth for successful recovery of the baseband signal. The required minimum level depends both on the baseband signal being sent over the link and the form of modulation used. The latter topic is discussed in Sects. 4.1 and 4.2. This section is concerned with the calculation of the received carrier power in a radio frequency link; the techniques can be applied to all forms of radio link from terrestrial broadcasting through satellite communication and radar.

4.3.1 Basic Equations

We will start by assuming that the transmitter of a radio link transmits a constant power level, P_t watts, into an antenna with a gain G_t. The product of transmit power and transmit antenna gain is called *Effective Isotropic Radiated Power* or *EIRP*. At the receiving end of the link, the radio signal is received by an antenna with a gain G_r. The power at the antenna output is given by

$$P_r = P_t G_t G_r / L_p L_m \text{ watts} \tag{4.36}$$

where L_p is the free space path loss and L_m is the sum of all other RF losses in the link.

Free space path loss is calculated as

$$L_p = [4\pi R/\lambda]^2 \text{ or } 20\log_{10}(4\pi R/\lambda) \tag{4.37}$$

Usually, these calculations are made using decibel values:

$$P_r = P_t + G_t + G_r - L_p - L_m \text{ dBW} \tag{4.38}$$

4.3.2 Example 4.1 (Radio Link)

A 6 GHz microwave line of sight link has a transmitter output power of 5 W, transmit and receive antennas with 40 dB gain, and a single hop of length 50 km. The antennas are mounted at the top of towers and connect to the transmitter and the receiver through 50 meters of waveguide. Loss in this waveguide at 6 GHz is 0.05 dB/meter. The system noise temperature of the receivers used in the LOS link is 1000 K including the effect of the waveguides. The atmosphere introduces a loss of 0.1 dB per kilometer along the LOS path in clear air conditions. Calculate the power received in clear air conditions, and also when rain causes an additional 15 dB attenuation on the path between the transmitter and the receiver. Find the receiver C/N ratio when a signal with an RF bandwidth of 40 MHz is transmitted.

The calculation is usually performed in a tabular fashion known as a link power budget.

Communication Systems

First calculate the path loss and the sum of all the other losses. At 6 GHz the wavelength is 0.05 m, so path loss for a 50 km = 5×10^4 m path is

$$L_p = 20 \log_{10}(4\pi R/\lambda) \text{ dB} = 20 \log_{10}(4\pi \times 5 \times 10^4/(0.05)^2) \text{ dB}$$
$$= 20 \log_{10}(12.57 \times 10^6) = 142.0 \text{ dB}$$

The waveguide loss at each end of the link is 0.05 dB/m × 50 m = 2.5 dB. In clear air the path loss is 0.1 dB/km, so total atmospheric loss over a 50 km path is 0.1 db/km × 50 km = 5 dB.

Hence the total of "other" losses in clear air conditions is

$$L_m = 2.5 + 2.5 + 5.0 = 10.0 \text{ dB}$$

The transmission power of 5 watts must be converted to dB watts (dBW):

$$P_{t \text{ in dB}} = 10 \log_{10}(P_t \text{ in watts}) = 10 \log_{10}(5) = 7 \text{ dBW}$$

The received power can now be calculated as P_r, where

$$P_r = P_t + G_t + G_r - L_p - L_m \text{ dBW}$$
$$= 7 + 40 + 40 - 142 - 10 = -65 \text{ dBW}$$

The noise power referred to the input of the receiver is given by $N = kT_sB$ watts. For a system noise temperature $T_s = 1000$ K = 30 dBK, and a bandwidth of 40 MHz = 76 dBHz,

$$N = -228.6 + 30 + 76 = -122.6 \text{ dBW}$$

Hence the C/N ratio for a single hop is

$$C/N = P_r - N = -65 + 122.6 = 57.6 \text{ dB}$$

When there is an additional 15 dB of rain attenuation on the LOS path, the received power, P_r, will fall by 15 dB to $-65 - 15 = -80$ dBW. The C/N falls to 42.6 dB. This received power level and C/N can easily support a TV channel or hundreds of voice and data channels using either analog or digital modulation.

4.3.3 Multiple Hop LOS and Satellite Links

A satellite link always has two LOS paths – one up to the satellite (the *uplink*) and one down from the satellite (the *downlink*). At the satellite a receiver and transmitter are connected in series to form a transponder. The transponder receives the uplink signal, amplifies it, changes its frequency, and then retransmits the signal back to Earth on the downlink. The power received from the downlink can be calculated if the gain of the transponder is known; or if the satellite transponder is to be operated at a specified output power, the required gain in the transponder can be calculated. A similar approach can be used for multiple hop LOS links. Note that the gain calculation ignores the change in frequency. Transponder gain is simply the ratio of power at the transponder output to power at the transponder input.

4.3.4 Example 4.2 (Link Power Budget)

The radio link in Sect. 4.3.2 was for a single hop. If the link is extended another 50 km using identical equipment, a total distance of 100 km can be tranversed. A transponder will be required to join the two LOS links together. The gain of the transponder, G_{tr}, is calculated for clear air conditions from the received signal power and the required transmit power

$$G_{tr} = P_t - P_r \text{ dB} = +7 \text{ dBW} - (-65 \text{ dBW}) = 72 \text{ dB}$$

It is assumed here that the transponder is a linear amplifier with a frequency shift, and does not demodulate and remodulate the signal. The latter design is called a regenerative transponder.

The link power budget can now be developed for the two hop LOS link. Suffices 1 and 2 refer to the first and second hops of the link. The noise power at the input to the satellite receiving system is

$$N = kT_s B_n \text{ watts or } N = k + T_s + B_n \text{ dBW} \tag{4.39}$$

where k is Boltzmann's constant $= 1.38 \times 10^{-23}$ J/K, T_s is the system noise temperature in Kelvins, and B_n is the noise bandwidth of the receiver at the station receiving the signal.

Path loss is calculated from the formula in Eq. (4.37):

$$L_p = [4\pi R/\lambda]^2 \text{ or } 20\log_{10}(4\pi R/\lambda) \text{ dB}$$

where R is the distance in meters between the transmitting and receiving antennas in the link and λ is the wavelength in meters. Path loss accounts for the reduction in power received as the signal spreads out in free space.

The C/N ratio at each receiver is calculated from

$$C/N = P_r - Nd \tag{4.40}$$

where P_r and N are in dB units (dBW). The overall C/N ratio $(C/N)_o$ at the end of a multi-hop link must be found by combining the individual receiver C/N values using

$$1/(C/N)_o = 1/(C/N)_1 + 1/(C/N)_2 + \ldots \tag{4.41}$$

If the gain of the transponder is set correctly, and each relay station has an identical transmitter and receiver, the second link has the same parameters as the first, and $(C/N)_1 = (C/N)_2 = 57.6$ dB or a power ratio of 575,430. The overall C/N is then

$$1/(C/N)_o = 1/575{,}430 + 1/575{,}430 = 1/287{,}772$$

Hence $(C/N)_o = 287{,}772$ or 54.6 dB.

If the first hop in the link is affected by rain, the C/N in the first relay falls, and the output power also falls. For example, a 15 dB rain fade on the first hop makes $(C/N)_1 = 42.6$ dB and P_t from the transmitter of the first relay station becomes $+7 - 15 = -8$ dBW. The second hop is assumed to operate in clear air, because heavy rain is unlikely to occur on both hops at the same time. The reduction in transmit power from the first relay station means that the received power at the second station is lower by 15 dB, at -80 dBW. This gives $(C/N)_2$ at the second station of $57.6 - 15 = 42.6$ dB. The resulting overall C/N is 39.6 dB.

4.3.5 Regenerative Transponder

A regenerative transponder can be used to advantage in a digital LOS link. The received signal is demodulated and the baseband data stream recovered. The data stream is then used to modulate a new RF carrier. The advantage of this technique is that bit errors occur with equal probability on any hop in the link, so the BER after N hops is simply $N \times$ BER for one hop. With linear transponders, the noise from the first hop is amplified by every successive hop, so noise builds up along the link and C/N steadily falls. This increases the BER at the end of the link much faster than when regenerative repeaters are used.

4.4 GEOSTATIONARY SATELLITE COMMUNICATIONS

4.4.1 Introduction

Geostationary satellites orbit the Earth over the equator at an altitude above the Earth's surface of 35,872 km. The angular rotation rate of the satellite exactly matches that of the Earth so the satellite is always above a particular point on the equator. The location of the satellite is defined by the longitude of the Earth directly below the satellite – the sub-satellite point. Earth stations can communicate through geostationary satellites with fixed antennas, since the satellite appears stationary in the sky. The position of the satellite in the sky for a given Earth station is defined by its look angles. The calculation of look angles for a geostationary satellite is discussed in Sect. 4.6.

Almost all geostationary satellites carry wideband transponders which receive signals from Earth stations, amplify the signals and shift their frequencies to a different band, and then retransmit the signals back to Earth. These are known as "bent pipe" satellites, since the received signal is effectively bent back towards Earth by the satellite. Most transponders have bandwidths of 36, 52, or 72 MHz and can be used for single or multiple transmissions of voice, data, or television signals. A typical large geostationary satellite might carry 24 transponders, each with a bandwidth of 36 MHz.

Geostationary satellites operate in C-band, Ku-band, and Ka-band. C-band satellites generally receive signals from Earth in the 5.925–6.425 GHz frequency range, and transmit them back to Earth in the band 3.7–4.2 GHz. Ku-band satellites receive in the 14.0–14.8 GHz band and retransmit at 11.7–12.5 GHz. Ka-band satellites are just beginning to be used commercially at the time of writing (1998), and receive signals from Earth in the 28–31 GHz band and retransmit in the 18–21 GHz band. C-band has proved the most popular frequency range for geostationary satellites because it suffers the lowest attenuation through rain. C-band satellites are in geostationary orbit every two or three degrees around the equator, providing television distribution and long-distance voice and data links throughout the world. The geostationary orbit is rapidly filling up with Ku-band satellites, especially where it serves heavily populated regions of the Earth.

The critical parameter in any radio link is the C/N ratio at the receiving station. For a geostationary satellite link, the RF power at the receiving Earth station is always low because of the long path lengths from Earth to the satellite. High-gain dish antennas and low-noise receiving systems are nearly always required for geostationary satellite links. In high-capacity systems, the Earth station antennas may be as large as 30 m in diameter to achieve the very high gains that are required.

There are always two radio paths in a geostationary satellite communications system: the uplink from Earth to the satellite and the downlink from the satellite to the Earth. The C/N ratio at the receiving Earth station must exceed a minimum value for communications to be successfully established. This minimum value ranges between 6 dB and 25 dB depending on the modulation and error control coding used in the link. The S/N or BER for the received baseband signal can be calculated using the methods described in Sects. 4.1 and 4.2

4.4.2 Geostationary Satellite Link Calculations

The uplink and downlink C/N ratios for a geo-satellite link are calculated separately, and then combined to give the overall C/N on the link. The

Communication Systems

procedure used is described in Sect. 4.3, with the equivalent of a two-hop link. The calculation steps will be reviewed again here.

For either link, the power received is given by Eq. (4.36):

$$P_r = P_t G_t G_r / L_p L_m \text{ watts} \tag{4.42}$$

which is usually expressed in decibels as

$$P_r = P_t G_t + G_r - L_p - L_m \text{ dBW} \tag{4.43}$$

where P_r is the power received at the output port of the antenna, $P_t G_t$ is the transmitting Earth station (or satellite) EIRP in dBW, G_r is the gain of the receiving antenna in dB, L_p is the path loss in dB, and L_m accounts for any other losses in the link.

The noise power referred to the receiving antenna port is given by Eq. (4.39):

$$N = kT_s B \text{ dBW} \tag{4.44}$$

where k is Boltzmann's constant, 1.38×10^{-23} watts/kelvin, or -228.6 dBW/K/Hz, T_s is the receiving system noise temperature in kelvins, and B is the receiver bandwidth in hertz.

The C/N ratio at the receiving end of the radio link is simply C/N = P_r/N, usually quoted in decibels as

$$C/N = P_r - N \text{ dB} \tag{4.45}$$

A bent-pipe geostationary satellite link consists of two parts, the uplink and the downlink, each with a different C/N. The overall C/N at the receiving Earth station is denoted $(C/N)_0$, given by

$$1/(C/N)_0 = 1/(C/N)_{up} + 1/(C/N)_{dn} \tag{4.46}$$

where the C/N values are ratios, not in decibels. The suffices up and dn correspond to the C/N ratios for the up- and downlinks. The overall C/N combines noise transmitted by the satellite with noise at the receiving Earth station.

The uplink and downlink C/N ratios are usually calculated using link budgets. The link budgets are tabulated forms of Eqs. (4.43) and (4.44), which give the link C/N using Eq. (4.45). Setting out the equations in tabular form allows a link designer to make trade-offs between parameters in the link very easily, since the decibel values of parameters can be added and subtracted. The following example for a C-band geostationary satellite carrying a single FM-TV signal illustrates the process.

4.4.3 Geostationary Satellite Link Example

This example is for a C-band geostationary satellite with a standard 36 MHz bandwidth transponder. The uplink is at a carrier frequency of 6.00 GHz. All C-band satellites use a 2225 MHz downshift in frequency between uplink and downlink, so the satellite retransmits this signal on the downlink at 3.775 GHz. One FM-TV signal is transmitted through the transponder and received by an Earth station with a bandwidth of 27 MHz.

The parameters of the satellite and the uplink and downlink Earth stations are given in Tables 4.2, 4.3, and 4.4.

Table 4.2 Geostationary Satellite Parameters

Saturated transponder output power	20 W
Uplink frequency	6000 MHz
Downlink frequency	3775 MHz
RF signal bandwidth	33.9 MHz
Receive antenna gain (uplink)	26 dB
Transmit antenna gain (downlink)	26 dB
Satellite receiver system noise temperature	500 K
Maximum range to edge of coverage zone	39,000 km

Table 4.3 Uplink Earth Station Parameters

Transmitter output power (maximum available)	300 W
Antenna gain (transmit)	56 dB
Transmitter RF bandwidth	30 MHz
Clear air atmospheric loss (uplink)	0.5 dB
Miscellaneous losses (antenna mispointing, etc.)	0.5 dB

Table 4.4 Downlink Earth Station Parameters

Antenna gain (receive)	45 dB
Receiving system noise temperature	75 K
Receiver IF bandwidth	33.9 MHz
Required overall C/N in clear air	17 dB
Clear air atmospheric loss (downlink)	0.3 dB
Miscellaneous losses (antenna etc.) mispointing	0.5 dB

4.4.4 Preliminary Calculations

The path losses for the uplink and downlink need to be calculated from Eq. (4.37):

$$L_p = (4\pi R/\lambda)^2 = 20\log_{10}(4\pi R/\lambda) \text{ dB}$$

The calculation is usually made for the longest possible path length, to ensure that the worst case of a station located at the edge of the coverage zone is considered. If the uplink station is at a known location the specific range from that station to the satellite can be used. In this example, the maximum range to the satellite of 39,000 km will be used for both uplink and downlink.

For the uplink at 6.000 GHz, the wavelength λ is 0.05 m and the path loss is 199.8 dB. For the downlink at 3.775 GHz, the wavelength is 0.0795 m and the path loss is 195.8 dB. Note that the two path losses differ by $20\log_{10}(6.000/3.775)$, the decibel ratio for the square of uplink to downlink frequencies.

The transmitter powers, the Earth station receiver bandwidth, and the satellite and Earth station noise temperatures need to be converted to dBW units, rounded to the nearest 0.1 dB. For the uplink, the Earth station maximum transmit power is 300 W = 24.8 dBW. For the downlink, the satellite maximum transmit power is 20 W = 13 dBW. The satellite receiving system noise temperature is 500 K = 27 dBK, and the receiving Earth station noise temperature is 75 K = 18.8 dBK. The Earth station receiver IF bandwidth is 33.9 MHz = 75.3 dBHz. We can now tabulate the link budgets.

The transmitter at the Earth station is rarely operated at its maximum power level. The exact setting depends on the gain of the satellite antennas and transponder. Let's suppose that in this example the Earth station transmit power is 100 W = 20 dBW. Combined with the transmit Earth station antenna gain of 56 dB, the EIRP is $20 + 56 = 76$ dBW.

The uplink achieves a high $(C/N)_{up}$ ratio in the transponder because the Earth station has a large, high-gain antenna and transmits 100 W. Note that the calculation of $(C/N)_{up}$ is made in the receiving Earth station's IF bandwidth, not the transponder bandwidth. This is because the noise radiated by the satellite which affects the overall C/N of this link is restricted to the receiving Earth station's IF bandwidth regardless of the actual transponder bandwidth.

The satellite transponder is rarely operated at its saturated output power level because of non-linearity problems. Output back-off is used to reduce intermodulation products caused by the non-linearity. For a single access to the transponder, as in this case, a typical transponder output back-off value would be 1 dB, so the transponder transmits at $P_t = 13 - 1 = 12$ dBW.

With multiple accesses to the transponder in FDMA, back-off values up to 3 dB may be needed.

Combined with the satellite transmit antenna gain of 26 dB, the satellite EIRP into the downlink is $12 + 26 = 38$ dBW. The downlink power budget is given in Table 4.6.

The overall C/N at the receiving Earth station is calculated by combining the uplink C/N and downlink C/N, since both the transponder and the Earth station receiver add noise to the signal. Equation (4.46) is used to calculate overall carrier to noise ratio $(C/N)_o$ at the receiving Earth station, using straight C/N ratios, not in dB:

$$1/(C/N)_o = 1/(C/N)_{up} + 1/(C/N)_{dn}$$

In this example, taking the uplink and downlink C/N values from Tables 4.5 and 4.6:

$(C/N)_{up} = 28.0$ dB $= 631.0$ as a ratio. $(C/N)_{dn} = 21.4$ dB $= 138.0$

Hence the overall C/N is given by

$$1/(C/N)_o = 1/631 + 1/138 = 0.00883$$

and thus

$$(C/N)_o = 113.2 = 20.5 \text{ dB in clear air}$$

The overall C/N ratio at the receiving station exceeds the minimum required value of 17 dB by 3.5 dB. This is a system margin, which can be used to provide a guarantee that the minimum required value of overall C/N of

Table 4.5 Uplink Power and Noise Budget

Parameter	Symbol	Value	Units
EIRP of transmit station	$P_t G_t$	76.0	dBW
Gain of receiving antenna	G_r	26.0	dB
Path loss at 6.000 GHz	L_p	−199.8	dB
Miscellaneous losses	L_m	−0.5	dB
Received power at satellite	P_r	−98.3	dBW
Boltzmann's constant	k	−228.6	dBW/K/Hz
Satellite noise temperature	T_s	27.0	dBK
Receiver bandwidth	B	75.3	dBHz
Noise power at satellite	N	−126.3	dBW
C/N in satellite transponder	$(C/N)_{up}$	28.0	dB

Communication Systems

Table 4.6 Downlink Power and Noise Budget

Parameter	Symbol	Value	Units
EIRP of satellite	$P_t G_t$	38.0	dBW
Gain of receiving antenna	G_r	45.0	dB
Path loss at 3.775 GHz	L_p	−195.8	dB
Miscellaneous losses	L_m	−0.3	dB
Received power	P_r	−113.1	dBW
Boltzmann's constant	k	−228.6	dBW/K/Hz
Earth station noise temperature	T_s	18.8	dBK
Receiver bandwidth	B	75.3	dBHz
Noise power	N	−134.5	dBW
C/N in Earth station receiver	$(C/N)_{down}$	21.4	dB

17 dB will always be maintained. Note that the gain of the transponder, between receive and transmit antenna ports, must be set to $12 + 98.3 = 110.3$ dB to achieve the correct transponder back-off setting. This is because the transponder receives a signal at −98.3 dBW and retransmits it at +12 dBW.

4.4.5 Rain Attenuation

Under conditions when rain affects the links, the overall C/N can be lowered by rain causing attenuation on the uplink, or by rain causing attenuation on the downlink and also an increase in system noise temperature. The latter effect is caused by increased radiation of noise by the atmosphere. Typically, under clear sky conditions the atmosphere absorbs only a small amount of the RF signals at C-band frequencies. In the example above, the 4 GHz receive band absorption was given as 0.3 dB. When heavy rain is present on the receive path, the attenuation in the atmosphere could increase by 1 dB to 1.3 dB total. This will cause the sky temperature to increase, which in turn causes the receive antenna noise temperature and the system noise temperature to increase.

The sky noise temperature T_{sky} is given by the equation

$$T_{sky} = T_m(1 - L) \tag{4.47}$$

where T_m is the medium temperature of the rain, usually assumed to be 270 K, and L is the loss in the atmosphere as a ratio. The value of L is obtained from the decibel value of attenuation in the atmosphere, A dB, as

$$L = 1/10^{A/10} \tag{4.48}$$

In the example given here, the loss in the atmosphere increases from 0.3 dB in clear air to 1.3 dB when heavy rain is present in the downlink. The clear air sky temperature is 18 K. With 1 dB of rain attenuation, the sky temperature becomes 70 K. The increase in sky temperature of 52 K does not translate directly to an increase in system noise temperature of 52 K. Only about 90% of the sky noise translates to antenna noise temperature, so we should expect the system noise temperature to increase by about 47 K to 122 K. This corresponds to a receiving system noise temperature increase of $10 \log_{10} (122/75) = 2.1$ dB, which will cause a similar increase in system noise power and reduction in downlink C/N. Thus the 1 dB of downlink rain attenuation in this example causes a 3.1 dB reduction in $(C/N)_{dn}$ from 21.4 dB to 18.3 dB.

The overall C/N falls from its clear air value of 20.5 dB, giving a 1 dB downlink rain fade value

$$1/(C/N)_o = 1/631 + 1/67.6 = 0.0164$$

and thus

$$(C/N)_o = 61.1 = 17.9 \text{ dB}$$

Rain affects Ku-band links much more severely than C-band links. Attenuation in excess of 10 dB can occur in thunderstorms, causing C/N to fall below its threshold value and resulting in a link outage. Rain attenuation at C-band is sufficiently small that outages caused by rain are unlikely if the system has a 3 dB margin in overall C/N. Ka-band links suffer frequent attention in excess of 10 dB in rain, and are therefore less reliable than C-band and Ku-band links. Services using Ka-band satellites must either be designed to accept outages in rain, or an alternative Earth station must be available in an area not affected by heavy rain. Heavy rain always occurs in thunderstorms, which rarely exceed 10 km in the region where rain is heaviest. Two stations separated by 10 km, with switching of the signal between the two whenever one is affected by a rain outage, can significantly improve the reliability of Ka-band links. This is known as station diversity. It is expensive to implement because two Earth stations are needed, with a full bandwidth link between the two, and therefore has not been widely employed.

4.4.6 Receiving FM-TV Signals With a Small Earth Station

In the example above, the receiving Earth station had a gain of 45 dB at a frequency of 3.775 GHz. This corresponds to an antenna diameter of about 5.7 m or 20 feet. Antennas of this diameter are used at cable TV head-ends to obtain a high C/N and guarantee good reception of the signal during rain fades. The video signals intended for distribution by cable TV companies are all sent out by geostationary satellite and can be received with smaller dishes by private individuals, with adequate C/N ratio to permit recovery of the video. For example, an 8 ft diameter dish will have a gain 8 dB below the gain of the 20 ft dish, or 37.0 dB at 3.775 GHz. With this value for G_r, the downlink C/N in the example becomes 13.4 dB in clear air, ad the overall $(C/N)_o$ is 13.2 dB. In a rain fade which reduces $(C/N)_{dn}$ by 3.1 dB, the $(C/N)_o$ is 10.2 dB, still above the threshold of the FM video demodulator, typically at 9.5 dB.

4.4.7 Video S/N for the Satellite Signal

The S/N ratio for the recovered video is calculated using the equations in Section 1 of this chapter. Assuming a US standard NTSC signal is transmitted, the video S/N is given by Eq. (4.15), repeated here as Eq. (4.49).

$$S/N = C/N + 10 \log_{10}(B_{IF}/f_{max}) + 20 \log_{10}(\Delta f_{pk}/f_{max}) + P + Q + 1.8 \text{ dB} \tag{4.49}$$

In clear air conditions with the cable TV company's 20 ft receiving antenna, C/N = 20.5 dB.

We must first find Δf_{pk} and f_{max} for this signal. The value of f_{max} for an NTSC video signal is 4.2 MHz, and Δf_{pk} is found from Carson's rule (Eq. 4.9)

$$B = (2\Delta f_{pk} \text{ and } f_{max}) \tag{4.50}$$

The video transmission uses an occupied RF bandwidth of 30 MHz, so substituting in Eq. (4.49) gives $\Delta f_{pk} = 10.8$ MHz. However, at the receiver, the IF bandwidth is restricted to 27 MHz, which increases the $(C/N)_{dn}$ at the expense of some distortion of the video signal, since the receiver bandwidth is less than the bandwidth occupied by the FM signal.

Using the preceding values in Eq. (4.49), and the standard value $P + Q = 17$ dB, gives

$$S/N = C/N + 8.1 + 8.2 + 18.8 = 35.1 \text{ dB}$$

With the 20 ft dish, the video S/N ratio is 20.5 + 35.1 = 55.3 dB in clear air. This provides a high-quality video signal, which falls by only 3 dB in heavy rain at the receiving Earth station. The 8 ft dish gives a lower overall C/N of 13.2 dB, producing a video S/N = 48.3 dB in clear air and S/N = 45.2 dB downlink if affected by heavy rain. A video S/N ratio of 55 dB is rated as "studio quality" with noise barely perceptible in the picture. At S/N = 45 dB, the video signal is rated as "cable TV quality", with some noise visible in the picture. There is a degradation of S/N and picture quality in the coaxial cable distribution networks of cable TV systems. As a result, a signal which is received at the head end with video S/N of 55 dB may be degraded to S/N of 45 dB by the time it reaches some distant customers.

4.4.8 Multiple Access Techniques

The example above showed how a transponder on a geostationary satellite can be used to relay a single video signal using frequency modulation. A transponder can be shared by several signals so that more than one Earth station can use the transponder for different signals. Some method of sharing the transponder between different communication links is required – a technique known as multiple access. There are three primary methods of multiple access. Frequency division multiple access (FDMA), time division multiple access (TDMA), and code division multiple access (CDMA). All three techniques are in use in different classes of satellite system. An example of FDMA follows here, and TDMA is considered in Sect. 4.6.

FDMA divides up the frequency band of the transponder in much the same way that broadcast radio stations share the VHF spectrum. However, the signals must also share the available power of the transponder, and more back-off of the transponder output power is required as the number of accesses increases. It is also important that each signal should maintain approximately the same power spectral density in the transponder bandwidth to maintain adequate suppression of third-order intermodulation products caused by non-linearity of the transponder characteristic.

In the example that follows, five transmitting stations with RF signal bandwidths of 10, 5, and 2.5 MHz share the transponder described in the previous example. There are two stations with transmitted RF bandwidths of 10 MHz, two with bandwidth of 5 MHz, and one at 2.5 MHz. Back-off of the transponder is set to 3 dB, so that its output power is 10 dBW, or 10 W.

The 10 watts of transponder power must be shared in proportion to the bandwidth occupied by each signal.

$$P_t = P_{total} \times B_{signal}/B_{total} \tag{4.51}$$

Thus the 10 MHz signals each have 3.08 watts, the 5 MHz signals have 1.54 watts each, and the 2.5 MHz signal has 0.77 watts of transponder power. In decibel units these values are 4.9 dBW, 1.9 dBW, and -1.1 dBW. We can now set up new downlink budgets to find the overall $(C/N)_o$ for each of the signals. Taking a 10 MHz signal as an example, the transponder EIRP is $4.9 + 26 = 30.9$ dBW. The receiver IF bandwidth will be 10 MHZ $=$ 70 dBHz. All other parameters in the downlink budget remain the same.

Communication Systems

The downlink C/N is approximately 2 dB lower than for the single access case. This is because the transponder output power was backed off by 2 dB. Thus using FDMA with proper power sharing does not lead to a reduction in overall C/N relative to the single access case except for the extra back-off needed in the transponder. Similar analysis at the transmitting Earth station shows that the transmitter power required for each signal is a total 100 watts shared out according to the RF bandwidths of the signal.

The RF bandwidth of the multiple access FDMA signals can be used to carry compressed video, data, or voice signals, employing analog (FM) or digital (PSK) modulations. Once the overall $(C/N)_o$ ratio has been calculated, the baseband S/N for analog signals or BER for digital signals can be found using the procedures described in Sects. 4.1 and 4.2.

4.5 PERSONAL COMMUNICATION SYSTEM USING LOW EARTH ORBIT SATELLITES

4.5.1 Description of Satellite Communication System

This satellite communication system consists of 24 low Earth orbit (LEO) satellites in 1000 km orbits, a large hub station, and many handheld transceivers. The handheld transceivers are similar to cellular telephone handsets, and provide two-way voice communications through a central hub station. They operate in L-band (the 1500 and 1600 MHz bands), which is allocated for mobile satellite communications. The hub station connects to the public switched telephone networks (PSTN), so a handset user can dial a regular telephone number from anywhere in the satellite system's coverage zone and connect to that number through whichever LEO satellite happens to be in view at that time. Some LEO systems use inter-satellite links so a user can connect to any point in the world through a series of satellite links.

Low Earth orbit satellites have orbital periods around 100 minutes, so they move across the sky in 5 to 15 minutes. As one satellite disappears over the horizon, another satellite comes into view over another part of the horizon, permitting continuous connection between the user and the hub station. A hand-off process is required similar to that used in cellular telephone networks. The hub station usually contains all the computers and software that figure out how to keep the user connected to the hub as the satellites pass overhead and disappear over the horizon.

This problem is concerned with the link between the user and the hub station. The RF signal received at the handheld unit is always weak, typically less than one picowatt (10^{-12} W). At the receiver, the signal must be greater than the noise generated in the receiver by about 10 dB for reliable communications to be possible. If forward error correction coding is applied to a digital bit stream, carrier-to-noise ratios (C/N) down to 5 dB can be used. Thus calculation of the carrier-to-noise ratio is a first step in the analysis of such a system. A second problem is to share the available bandwidth of the satellite transponder among a number of users. The choice of a multiple access technique and the number of simultaneous users sharing one transponder are important parameters in the system design.

In this example, the handheld units transmit to a transponder using frequency division multiple access (FDMA). In FDMA, each transmitter is allocated its own frequency, just like radio stations in the AM and FM bands. The following subsections establish the parameters of the satellite and the handset.

Satellite Parameters

Saturated output power	10 W
Uplink frequency for handheld transceiver	1650 MHz
Downlink frequency for handheld transceiver	1550 MHz
Transponder bandwidth	1 MHz
Antenna gain (one beam) uplink	20 dB
Antenna gain (one beam) downlink	20 dB
Satellite receiver system noise temperature	500 K
Maximum range to edge of coverage zone	2000 km

Note that the satellite has multiple beams serving different parts of its coverage zone. A single beam from a LEO satellite serving all of the coverage zone would have very low gain. A useful approximate relationship between gain and beamwidth for aperture antennas is:

$G = 30{,}000/(3 \text{ db beamwidth})^2$, G not in dB.

For an antenna with a gain of 20 dB, $G = 100$ and the beamwidth is $(30{,}000/100)^{1/2} = 17°$. A single beam covering all the visible Earth from a satellite at 1000 km altitude would have a beamwidth of 120°, and a gain of only 3 dB. It would be very difficult to design a link from a handset with a low-gain antenna to a satellite 2000 km away which also had a low-gain antenna. By using a multiple beam antenna on the satellite, the antenna gain per beam can be increased and this increases the C/N ratio.

Handheld Transceiver Parameters

Transmitter output power	0.5 W
Antenna gain (transmit and receive)	0 dB

Communication Systems

Receiver system noise temperature	200 K
Transmit bit rate	5000 bps
Receive bit rate	100 kbps
Required maximum bit error rate	10^{-5}

Note that antenna gain at the handset is low, with a minimum value of 0 dB, because the antenna coverage must be very broad. If the handset antenna gain were increased, its beam would narrow and the user would have to point the handset at the satellite. By using an omni-directional antenna (with inherent low gain) the user is free to move around without having to point the antenna in a specific direction. The voice signal from the handheld transmitter has been heavily compressed to a 5000 bps digital bit stream using vocoder and linear predictive encoding techniques.

The 100 kbps signal received from the hub is a TDM sequence of 20 digital voice channels each at 5000 bps. Since there is only one hub station and many handheld units, the hub can transmit a wideband signal using TDM.

The handheld transceiver sends and receives binary phase shift keyed (BPSK) modulation. We will assume that ideal Nyquist (square root raised cosine) filters with $\alpha = 0.5$ are used throughout the system. For any RF digital link with ideal (Nyquist) filters, the noise bandwidth in the receiver, B_n, will be $B_n = R_s$, where R_s is the symbol rate for the digital signal. Note that the narrowband filter which limits the noise to B_n is located in the intermediate frequency (IF) section of the receiver – either in the hub for transmissions from the handheld unit or in the handheld unit for transmissions from the hub. The transponder in the satellite typically has a bandwidth much wider than B_n in the receiver, but this is ignored since only transponder noise within the bandwidth B_n reaches the receiver's demodulator.

Hub Station Parameters

The hub station–satellite link uses Ku-band and a large antenna at the hub.

Transmitter output power (maximum)	100 W
Antenna gain (transmit 14.0 GHz)	57 dB
Antenna gain (receive 11.6 GHz)	56 dB
Receive system noise temperature	150 K
Transmit bit rate	500 kbps
Receive bit rate	5000 bps
Required maximum bit error rate	10^{-5}

4.5.2 Calculation of Parameters for Handheld Unit to Hub Station Link (Inbound)

The handheld units transmit BPSK signals at 5000 bps. Each handheld unit is allocated a different frequency, so signals in the transponder are separated by their frequencies. This is Frequency Division Multiple Access (FDMA). At the hub station, the antenna feeds many identical IF receivers tuned to the individual frequencies of the handheld transmitters. Each IF receiver has a noise bandwidth of 5000 Hz, set by a square root raised cosine filter with $\alpha = 0.5$. The bandwidth occupied by each BPSK signal is given by:

$$B_{\text{occ}} = R_b(1+\alpha) = 5000(1+0.5) = 7500 \text{ Hz} \tag{4.52}$$

If the transponder is bandwidth limited, the number of handheld transceivers, N, that can share one transponder is given by:

$$N = B_{\text{tr}}/(B_{\text{occ}} + B_g) \tag{4.53}$$

where B_{tr} is the bandwidth of the transponder and B_g is a guard band that separates the RF signal in the transponder so that the individual channel filters in the receiver can separate the signals without crosstalk. Typically, B_g is 10 to 20% of B_{occ}.

For a transponder with bandwidth $B_{\text{tr}} = 1$ MHz $= 1000$ kHz, RF channels with $B_{\text{occ}} = 7.5$ kHz, and a guard band of 1.5 kHz between channels:

$$N = 1000/(7.5 + 1.5) = 111$$

(Numbers of channels must be integers, so any decimal remainder in N is discarded.) The value of $N = 111$ represents the maximum number of 5000 bps BPSK channels that can be sent simultaneously through the 1 MHz filter under the stated conditions. The actual number of channels that can be sent may be lower if the transponder is power limited.

At the receiving end of the link, the C/N at the input to the BPSK demodulator must be high enough to provide an acceptable bit error rate. Thus the starting point for C/N calculation is the BER specification. BER is found from tables or graphs of BER versus E_b/N_0, where E_b is the energy per bit and N_0 is the noise power spectral density. In a link with ideal Nyquist filters, $E_b/N_0 = $ C/N for BPSK, and $E_b/N_0 = \frac{1}{2}$ C/N for QPSK. Ideal Nyquist filters do not exist, and Butterworth or Chebyshev or similar practical bandpass filters must be used in real links. As a result, a performance degradation occurs, and to achieve a particular bit error rate always requires a higher C/N ratio than the corresponding E_b/N_0 ratio would indicate. To account for this difference, a factor called *implementation margin* is introduced. The C/N ratio required to achieve a specified BER is then

Communication Systems

higher than the theoretical E_b/N_0 ratio by the implementation margin. For low bit rate BPSK links, an implementation margin of 0.5 dB is typical. At Mbps bit rates, implementation margin may be as much as 2 dB.

In the LEO satellite communication system example, we require a maximum BER of 10^{-5}. A BER of 10^{-5} with BPSK requires $E_b/N_0 = 9.6$ dB. So under ideal conditions (no implementation margin and ideal Nyquist filters) C/N = 9.6 dB. Allowing an implementation margin of 0.6 dB, we need a minimum C/N = 10.2 dB to mee the BER specification. Now we can design the satellite link to achieve the minimum C/N.

4.5.3 Earth–Satellite Link Design

Link design is done using power and noise *link budgets*. A link budget is simply a table of factors that can be summed to give C for the power budget of N for the noise budget. We first analyze the uplink from the handheld unit to the satellite. What we need to find is the carrier to noise power ratio in the satellite transponder, $(C/N)_{tr}$. The calculation of C/N is made at the output of the receiving antenna, where $C = P_r$, the power received by the antenna, and N is the noise power of a noiseless receiver with system noise temperature T_s.

The received power at the output of the uplink antenna on the satellite is P_r where:

$$P_r = P_t G_t G_r / L_p L_m \text{ watts} \tag{4.54}$$

Usually, these calculations are made using decibel values:

$$P_r = P_t + G_t + G_r - L_p - L_m \text{ dBW} \tag{4.55}$$

where P_t is the Earth station transmit power in dBW, G_t is the Earth station (transmit) antenna gain, G_r is the satellite receive antenna gain, L_p is the path loss of the link, and L_m accounts for any other losses. The product $P_t G_t$, called the effective isotropically radiated power, or EIRP, is often used to characterize transmitters since P_t and G_t always appear together in a link budget.

The noise power at the input to the satellite receiving system is:

$$N = kT_s B_n \text{ watts or } k + T_s + B_n \text{ dBW} \tag{4.56}$$

where k is Boltzmann's constant = 1.38×10^{-23} J/K, T_s is the system noise temperature in Kelvins, and B_n is the noise bandwidth of the receiver at the Earth station receiving the signal.

Path loss is calculated from the formula

$$L_p = [4\pi R/\lambda]^2 \text{ or } 20\log_{10}(4\pi R/\lambda) \text{ dB} \tag{4.57}$$

where R is the distance in meters between the transmitting and receiving antennas in the link and λ is the wavelength in meters. Path loss accounts for the reduction in power received as the transmitted RF signal spreads out as it propagates away from the transmitting antenna.

The Earth station and satellite specifications provide the parameters for the link budget.

In the LEO link example, the handheld unit has $P_t = 0.5$ W $= -3$ dBW and $G_t = 0$ dB, so EIRP $= -3$ dBW. At the satellite, $G_r = 20$ dB, and the maximum range for the link is 2000 km. The uplink frequency is 1650 MHz, giving $\lambda = 0.1818$ m. Then path loss is:

$$L_p = 20 \log_{10}(4\pi \times 2 \times 10^6 / 0.1818) = 162.8 \text{ dB}$$

We will assume that there are miscellaneous losses in the link of 0.5 dB, caused by polarization misalignments, antenna mispointing, gaseous absorption in the atmosphere, etc. The calculation of C/N is usually made for an Earth station located on the -3 dB contour of the satellite antenna beam, so a 3 dB reduction in satellite antenna gain is applied. This reduction is included in the L_m loss value. We can now set out the link budget:

Parameter	Symbol	Value	Units
EIRP of handheld unit	$P_t G_t$	−3	dBW
Gain of receiving antenna	G_r	20	dB
Path loss at 1650 MHz	L_p	−162.8	dB
Miscellaneous losses	L_m	−3.5	dB
Received power at satellite	P_r	−149.3	dBW

The noise power budget for the satellite transponder is calculated in a similar way. The transponder noise temperature is 500 K, and the calculation is made in the hub receiver noise bandwidth of 5000 Hz:

Parameter	Symbol	Value	Units
Boltzmann's constant	k	−228.6	dBW/K/Hz
System noise temperature	T_s	27.0	dBK
Noise bandwidth	B_n	37.0	dBHz
Noise power	N	−164.6	dBW

The C/N ratio in the transponder can now be calculated directly from the power and noise budgets:

$$(C/N)_{tr} = P_r/N = -149.3 \text{ dBW} - (-164.6 \text{ dBW}) = 15.3 \text{ dB}$$

Communication Systems 197

Note that this is the lowest C/N that should occur in the transponder in clear conditions, since the calculation was made for an Earth station at the longest range from the satellite and at the edge of a satellite antenna beam. The handset antenna gain has also been set to its minimum value of 0 dB. If the satellite were directly overhead the range would be 1000 km instead of 2000 km, and the miscellaneous losses would be 3 dB lower at the center of the satellite antenna beam. The power received at the transponder would then be 9 dB greater, and $(C/N)_{tr} = 24.3$ dB.

4.5.4 Link from Satellite to Hub Station

The next step in calculating the C/N for the inbound link is to calculate the C/N in the hub receiver. We are operating the transponder in FDMA, so the individual handset signals must share the output power of the transponder. We had calculated that a maximum of 111 handsets signals could share the 1 MHz transponder bandwidth. Let's assume that a maximum of 100 handsets access the transponder simultaneously. Then the output power of the transponder must be shared equally between the 100 signals. We cannot use the saturated output power of the transponder when operating in FDMA. The non-linearity of the input–output characteristics of the transponder will cause intermodulation between signals. We must reduce the output power until the characteristics are sufficiently linear to produce acceptable intermodulation levels. This is known as *quasi-linear* operation of the transponder. The ratio of saturated output power to operating output power is called *back-off*.

Transponder operation in FDMA with a large number of SCPC signals requires relatively large output back-off. Let's set a figure of 3 dB, so that the output power from the transponder is 5 W or +7 dBW. Each 5 kbps BPSK signal at the transponder output is allocated 1/100 of this power, or 0.05 W = −13 dBW. We can now establish a link budget for the satellite–hub downlink. We will use the same worst case conditions as for the uplink – maximum path length and minimum satellite antenna gain. Since the satellite to hub link operates in Ku-band at 11.8 GHz, path loss is substantially higher than at L-band. The satellite antenna gain is low, +3dB, because a single broad beam is used for communication with the hub station. The hub station uses a large, high-gain antenna to ensure high C/N on the downlink. Because the gain of the antenna is high, 56 dB, the beamwidth is narrow, about 0.3 degrees, and the hub station must track the satellite as it crosses the sky.

The downlink power budget is:

Parameter	Symbol	Value	Units
EIRP per channel	$P_t G_t$	−10	dBW
Gain of receiving antenna	G_r	56	dB
Path loss at 11.6 GHz	L_p	−179.1	dB
Miscellaneous losses	L_m	−3.5	dB
Received power at satellite	P_r	−136.6	dBW

The hub receiver noise power budget is calculated in the same way as the transponder's:

Parameter	Symbol	Value	Units
Boltzmann's constant	k	−228.6	dBW/K/Hz
System noise temperature	T_s	21.8	dBK
Noise bandwidth	B_n	37.0	dBHz
Noise power	N	−169.8	dBW

The C/N in the 5 kHz noise bandwidth of a hub receiver is given by:

$$(C/N)_{dn} = P_r/N = -136.6 - (-169.8) = 33.2 \text{ dB}$$

Note that the C/N for the downlink is higher than for the uplink because of the high gain of the hub station antenna.

The C/N at the hub is calculated by combining the uplink C/N and downlink C/N, since both the transponder and the hub station receiver add noise to the signal: the result is called the *overall C/N* for the link. The values used in the formula are ratios, i.e. C/N is not in dB:

$$1/(C/N)_o = 1/(C/N)_{tr} + 1/(C/N)_{dn} \qquad (4.58)$$

Hence for the inbound link with $(C/N)_{tr} = 15.3$ dB $= 33.9$ and $(C/N)_{dn} = 33.2$ dB $= 2089.3$

$$(C/N)_o = 1/(1/33.9 + 1/2089) = 33.4 \text{ or } 15.2 \text{ dB}$$

The C/N of 15.2 dB at the hub station receiver guarantees that under the stated conditions the BER for the received signal will be much lower than 10^{-5}, since that BER requires only C/N = 10.2 dB, allowing a 0.6 dB implementation margin. The 5.0 dB difference between the expected overall C/N value of 15.2 dB and the minimum allowed value of 10.2 dB is called the *system margin*. The margin is needed to allow for additional losses in the links which have not been taken into account in the initial calculations. On the L-band uplink, *vegetative shadowing* by trees is the main cause of additional loss of power at the satellite. On the Ku-band downlink, heavy rain can cause significant attenuation.

We can calcualte the uplink and downlink margins from the overall C/N formula, by allowing the uplink C/N and the downlink C/N to fall to a value which reduces the overall C/N to its minimum allowable value of 10.2 dB. For example, for the L-band uplink, the minimum $(C/N)_{up,\ min}$ value is given by:

$$(C/N)_{up,min} = 1/[1/(C/N)_{o,min} - 1/(C/N)_{dn}]$$
$$= 1/(1/10.47 - 1/2089) = 10.22 \text{ dB}$$

Thus the uplink C/N can fall to 10.2 dB before the BER at the hub receiver increases above 10^{-5}.

This provides a margin on the uplink of $15.0 - 10.2 = 4.8$ dB for vegetative shadowing. Note that this is under worst case uplink conditions, when the satellite is at maximum range and the user is at the edge of a satellite antenna beam.

Repeating the analysis for the downlink, where $(C/N)_{dn} = 33.2$ dB under clear air conditions, we find that the downlink minimum C/N is:

$$(C/N)_{dn,min} = 1/[1/(C/N)_{o,min} - 1/(C/N)_{up}]$$
$$= 1/(1/10.47 - 1/32.4) = 11.9 \text{ dB}$$

The downlink fade margin is then $33.2 - 11.9 = 21.3$ dB. This is large enough to ensure that the downlink will continue to operate in rain for 99.99% of the year. Typically, rain attenuation levels on satellite–earth paths in the United States exceed 10 dB for less than 0.01% of a year – about one hour per year. These attenuation levels occur only in the heaviest of thunderstorms, for a few minutes at a time. However, since all the signals from every handset are lost when the downlink fails, a large link margin is provided for the downlink to ensure that this happens infrequently. When a single uplink fails because of blockage by trees or buildings, only one user is affected. When the downlink fails, every user is disconnected.

The calculation for allowable downlink attenuation was made for clear conditions on the uplink and rain on the downlink. If tree shadowing occurs on the uplink at the same time as heavy rain is attenuating the downlink, the BER will increase above the 10^{-5} design objective. A final check can be made using 10 dB of downlink attenuation in heavy rain to determine the available uplink fade margin. With 10 dB downlink rain attenuation, $(C/N)_{dn} = 23.2$ dB. Hence:

$$(C/N)_{up,min} = 1/[1/(C/N)_{0,min} - 1/(C/N)_{dn}]$$
$$= 1/(1/10.47 - 1/208.9) = 10.4 \text{ dB}$$

This allows an uplink margin of $15.1 - 10.4 = 4.7$ dB for vegetative shadowing when the downlink is affected by 10 dB rain attenuation.

4.5.5 Hub Station to Handheld Unit Link (Outbound)

The outbound link from the hub station to the handheld transceiver uses a bandwidth of 1 MHz to send a continuous 500 kbps TDM bit stream using BPSK modulation. The bit stream is a series of packets addressed to active handheld units. At an average data rate of 5000 bps to each handheld transceiver, 100 transceivers can be served. This matches the 100 inbound channels which share a separate 1 MHz wide transponder using FDMA. If ideal Nyquist filters with $\alpha = 0.5$ are used in the outbound link, the RF bandwidth occupied by the TDMA signal will be 750 kHz. The noise bandwidth of the handheld receiver must be 500 kHz.

The uplink and downlink C/N values are calculated in exactly the same way as for the inbound link. In this example, the power and noise budgets are combined to give C/N directly from a single calculation. The hub station transmitter output power is set at 20 W = 13 dBW.

At the uplink frequency of 14 GHz, clear air atmospheric attenuation of 1.0 dB is included in the miscellaneous losses. The satellite antenna edge of beam loss of 3 dB is also included there.

The uplink C/N budget is:

Parameter	Symbol	Value	Units
EIRP	$P_t G_t$	70	dBW
Gain of receiving antenna	G_r	3	dB
Path loss at 14.0 GHz	L_p	−180.7	dB
Miscellaneous losses	L_m	−4.5	dB
Received power at satellite	P_r	−112.2	dBW
Boltzmann's constant	k	−228.6	dBW/K/Hz
System noise temperature	T_s	27.0	dBK
Noise bandwidth	B_n	57.0	dBHz
Noise power	N	−144.6	dBW
Uplink C/N	$(C/N)_{tr}$	32.4	dB

Communication Systems

The satellite transponder carrying the 500 kbps outbound signal can be operated close to saturation. We will allow 0.8 dB back-off to avoid saturating the transponder, so $P_t = 9.2$ dBW. The downlink C/N budget is:

Parameter	Symbol	Value	Units
EIRP of satellite	$P_t G_t$	29.2	dBW
Gain of receiving antenna	G_r	0	dB
Path loss at 1550 MHz	L_p	−162.3	dB
Miscellaneous losses	L_m	−3.5	dB
Received power at satellite	P_r	−136.6	dBW
Boltzmann's constant	k	−228.6	dBW/K/Hz
System noise temperature	T_s	23.0	dBK
Noise bandwidth	B_n	57.0	dBHz
Noise power	N	−148.6	dBW
Downlink C/N	$(C/N)_{dn}$	12.0	dB

C/N at the handheld receiver must be calculated by combining the uplink C/N and downlink C/N, to give the overall C/N for the link, $(C/N)_o$. Converting the C/N values from dB,

$$(C/N)_{tr} = 32.4 \text{ dB} = 1737.8, \quad (C/N)_{dn} = 12.0 \text{ dB} = 15.8$$

Hence, the overall C/N for the outbound link is:

$$(C/N)_o = 1/[1/(C/N)_{tr} + 1/(C/N)_{dn}]$$
$$= 1/[0.000575 + 0.0757] = 13.11 \text{ or } 12.0 \text{ dB}$$

Note that the downlink C/N is so much lower than the uplink C/N that the overall C/N is equal (within the 0.1 dB precision used for C/N calculations) to the downlink C/N.

The clear air C/N is 1.8 dB above the minimum allowed for BER = 10^{-5} on the outbound link, leaving a 1.8 dB margin for vegetative shadowing on the downlink, or for uplink rain attenuation which lowers the power output of the satellite. The latter problem can be partially overcome by the use of uplink power control. When attenuation is detected on the uplink (by observation of the signal received from the satellite at the hub station), the transmitter power is increased to bring the received signal level back up to the clear air value. To overcome a maximum uplink attenuation of 10 dB using uplink power control (UPC), the transmit power must be increased by 10 dB, from the normal output power of 100 W in clear air to 1000 W in rain. This means that a 1 kW transmitter must be installed at the hub station, but the maximum output power will be needed only one or two hours per year.

4.5.6 Conclusion

The personal communication system in this example uses a network of low Earth orbit satellites to link a user anywhere in the system's coverage zone to a hub station, and then to the public switched telephone network. The user's handset is similar to a cellular telephone, with a low gain, omnidirectional antenna, and operates in L-band. The transmissions are digital, using speech compression to achieve a bit rate of 5 kbps per speech channel, and BPSK modulation. The inbound link from the user to the hub station achieves a minimum bit error rate of 10^{-5} with a margin of 4.8 dB for tree shadowing of the uplink to the satellite, and 10 dB rain attenuation on the downlink to the hub. Up to 100 users can simultaneously share a 1 MHz wide transponder on the satellite using FDMA.

The outbound link from the hub to the satellite uses a continuous TDM bit stream at 500 kbps, and has a 1.8 dB tree shadowing margin on the downlink from the satellite. Uplink power control is used to overcome rain attenuation on the Ka-band link between the hub and satellite.

4.6 PROBLEM ON LOOK ANGLES FOR GEOSTATIONARY SATELLITES

4.6.1 Introduction: The Geostationary Earth Orbit

A geostationary satellite is in a circular orbit in the Earth's equatorial plane. The radius of the orbit, measured from the center of the Earth, is 42,164.2 km. The Earth's mean radius at the equator is 6378.16 km, so the mean altitude of a geostationary satellite above the equator is 35,786 km. For this particular altitude, the angular velocity of the satellite is identical to the angular velocity of the Earth, with a period of one sidereal day. A sidereal day is 23 hours, 56 minutes, 4.09 seconds. It differs from the clock day of 24 hours because the Earth orbits the Sun every 365 days, adding a small time increment (3 minutes 55.91 seconds) to the average time between sunrises through the year.

A satellite is truly geostationary only when the above conditions are met exactly. Then the satellite will appear to be stationary above a point on the equator, defined by a longitude on the Earth's surface. A line from the satellite to the center of the Earth intersects the Earth's surface at the sub-satellite point. A sub-satellite point is defined by its latitude and longitude. Geostationary satellites have latitude 0 degrees when in the Earth's

Communication Systems

equatorial plane. However, gravitational effects from the Moon and the Sun tend to tilt the satellite's orbit out of the equatorial plane, giving it an inclination. The satellite will then have a sub-satellite point that changes sinusoidally in latitude through each sidereal day, with a peak value equal to the inclination of the orbital plane to the equatorial plane.

All geostationary satellites share the same orbit, with different longitudes. Currently (1998) there are several hundred geostationary satellites in orbit around the Earth's equator, providing communication links, weather observations, and broadcast transmissions for millions of people. Satellites are allocated to orbital "slots" under international agreements, with longitude spacing of two or three degrees. Several satellites can share one orbital slot if they operate at different frequencies.

Geostationary satellites do not remain stationary over one point on the Earth's surface for long. A number of small forces tend to accelerate the satellite away from its desired position, so periodically the orbit must be reset to maintain a nearly geostationary orbit and prevent the satellite moving away from its slot in the geostationary orbit. The orbit is changed by firing small rocket motors on the satellite to increase or decrease its velocity along each of its three axes.

Communication with geostationary satellites requires high gain antennas with narrow beams at the user's Earth station because of the large transmission distances. The advantage of a geostationary satellite, compared with a satellite in any other orbit, is that Earth station antennas can be pointed in a fixed direction instead of having to track the satellite across the sky.

This problem illustrates the calculation of the look angles for Earth station antennas that transmit signals to and receive signals from geostationary satellites.

4.6.2 Calculating Look Angles

Look angles define where to point an Earth station antenna beam in order to communicate with a geostationary satellite. The angles are defined as azimuth (Az) and elevation (El). An azimuth angle is measured in the horizontal plane at the Earth station location starting at true North and measuring in degrees through east. Thus east is 90°, south is 180° and west is 270°. Elevation angle is measured up from the horizontal plane to the satellite. Thus an elevation angle of 0° points the antenna horizontally and an elevation angle of 90° points the antenna at the local zenith. Look angles are given as the angle pair (Az, El).

Calculation of look angles requires three formulas and starts with specification of the geostationary satellite's sub-satellite point and the Earth station's latitude and longitude. The satellite location is L_s, l_s, where L_s is the sub-satellite latitude and l_s is the sub-satellite longitude, in degrees. The Earth station location is specified as L_e, l_e, where L_e is the Earth station latitude and l_e is the Earth station longitude, in degrees. A longitude of 0° corresponds to the Greenwich Meridian, which is a line on the Earth's surface from the North Pole to the South Pole through Greenwich, UK. Longitudes are usually given as east or west, ranging from 0° to 180°. Lines of latitude are circles round the Earth's suface orthogonal to lines of longitude. A latitude of 0° corresponds to the Earth's equator, and a latitude of 90° is at a pole.

A truly geostationary satellite will have a sub-satellite latitude of 0°. However, because of the continual slow changes in orbit, the satellite may have a varying latitude caused by inclination of the orbital plane to the Earth's equator. A general solution for the look angles requires inclusion of sub-satellite latitude. If the orbital inclination becomes large compared with the Earth station antenna beamwidth, a tracking antenna may be required.

The first step in calculating look angles is to calculate the central angle, γ. The central angle is the angle at the center of the Earth between lines which go from the Earth's center to the satellite and to the Earth station. The central angle is calculated as:

$$\cos(\gamma) = \cos(L_e)\cos(L_s)\cos(l_s - l_e) + \sin(L_e)\sin(L_s) \quad (4.59)$$

If the satellite longitude and the Earth station longitude are not both specified as east or west, they are on opposite sides of the 0° or 180° lines of longitude. The sign of the $\cos(l_s - l_e)$ term must be changed to $+$ under these conditions. If the satellite is in the Earth's equatorial plane, its latitude $L_s = 0°$ and the equation for the central angle, γ, simplifies to:

$$\cos(\gamma) = \cos(L_e)\cos(l_s - l_e) \quad (4.60)$$

The elevation angle of the satellite can be found from the equation:

$$\text{El} = \tan^{-1}[(r - R_e \cos(\gamma))/R_e \sin(\gamma)] - \gamma$$

where r is the radius of the geostationary orbit and R_e is the radius of the Earth. Substituting the mean values for r and R_e gives:

$$\text{El} = \tan^{-1}[(6.610715 - \cos(\gamma))/\sin(\gamma)] - \gamma \quad (4.61)$$

The azimuth angle is found from an intermediate angle A′:

$$A' = \tan^{-1}(\tan|l_s - l_e|/\sin(L_e)) \quad (4.62)$$

Communication Systems

The azimuth look angle for the satellite is found by considering the relative positions of the satellite and the Earth station. For Earth stations in the northern hemisphere, Az is given by:

$$\text{Az} = 180° - A' \text{ if the satellite is south-east of the Earth station}$$
$$\text{Az} = 180° + A' \text{ if the satellite is south-west of the Earth station}$$
(4.63)

For earth stations in the southern hemisphere, Az is given by:

$$\text{Az} = A' \text{ if the satellite is north-east of the Earth station}$$
$$\text{Az} = 360° - A' \text{ if the satellite is north-west of the Earth station}$$
(4.64)

4.6.3 Look Angle Calculation Example

A geostationary satellite is located at longitude 25° west. Communication is established through the satellite between Earth stations near Washington DC in the United States and Cape Town, South Africa. The locations of the Earth stations are:

Washington DC, USA, latitude 39.00° N, longitude 77.20° W

Cape Town, South Africa, latitude 34.00° S, longitude 18.80° E

Find the look angles for the two Earth stations.

Washington DC Station

We begin by calculating the central angle, γ, and checking that the satellite is visible.

If the angle γ exceeds 81.3°, the satellite is below the horizon and not visible from the Earth station. For the Washington DC Earth station, $L_e = 39.00°$ N and $l_e = 77.20°$ W. The satellite is geostationary with a sub-satellite point at $l_s = 25°$ W. The central angle is given by Eq. (4.60):

$$\cos(\gamma) = \cos(L_e)\cos(l_s - l_e)$$

Evaluating the angle, $\gamma = 61.555°$. Since this is less than 81.3°, the satellite is visible.

We can now find the elevation and azimuth angles from Eqs. (4.61) and (4.62). For the elevation angle

$$\text{El} = \tan^{-1}[(6.610715 - \cos(\gamma))/\sin(\gamma)] - \gamma$$

This gives the elevation angle as El = 20.288°. The azimuth angle is found from:

$$A' = \tan^{-1}(\tan|l_s - l_e|/\sin(L_e))$$

The satellite and the Earth station both have westerly longitudes, so $|l_s - l_e| = 52.20°$.

Solving for the angle A' gives $A' = 64.981°$. The satellite is south-east of the Earth station, so from Eq. (4.63) the azimuth angle is Az $= 180° - A' = 116.019°$. The look angles show the satellite to appear at an elevation angle of 20.288° and an azimuth of 116.019°, which is 26.162° south of east, as seen from Washington DC. This is the correct portion of the sky for a geostationary satellite well to the east of a northern hemisphere Earth station.

Cape Town, South Africa Station

The Earth station near Cape Town, South Africa, has an easterly longitude of 18.8°E. The Earth station longitude must be treated as negative when evaluating the angle $(l_s - l_e)$. Thus $(l_s - l_e) = 43.80°$. The central angle for the Cape Town Earth station is found from Eq. (4.60)

$$\cos(\gamma) = \cos(L_e)\cos(l_s - l_e)$$

The central angle is 53.247°. The elevation angle is calculated in the same way as for the Washington DC station, giving El $= 20.162°$. We find the azimuth angle for the satellite from the angle A' using Eq. (4.62), remembering that l_e and l_s must have opposite signs. Then $A' = 59.753°$. The azimuth angle is found from A' using Eq. (4.64):

$$Az = 360° - A'$$

because the satellite is north-west of the Earth station. This gives the azimuth angle as Az $= 300.247°$. Thus the satellite appears at an elevation angle of 29.162°, at an azimuth angle of 300.247°, when seen from the Earth station near Cape Town, South Africa. The satellite is in the north-western sky, 30° north of west, the correct position for a geostationary satellite to the west of an Earth station in the southern hemisphere.

4.7 RADAR

4.7.1 Problem on Radar Design

A long-range surveillance radar is required for detection of aircraft. Such radars are used in air defense systems and for air traffic control. Long-range radars use either L-band or S-band frequencies because rain attenuation and scatter are tolerable at these frequencies. At higher frequencies, reflections from rain obscure targets, and attenuation in rain makes long-range opera-

Communication Systems

tion unreliable. L-band radars operate between 1 and 2 GHz. S-band radars operate between 2 and 4 GHz. The specification of the radar is as follows:

Frequency	2800 MHz
Maximum range for 2 m^2 RCS target	200 km
Range resolution	$\frac{1}{8}$ mile
Scan rate	6 r.p.m.
Antenna beamwidth	
Azimuth plane	2 degrees
Elevation plane	3 degrees
Receiver noise figure	5 dB
Total system losses	11.4 dB
Probability of detection	90%
Time between false alarms	10 h

Determine a suitable pulse repetition frequency (prf), pulse width, and receiver IF bandwidth for this radar. Find the antenna gain and dimensions, and the transmitter power to meet the target detection requirements. Find the range at which this radar can detect a target of 0.1 m^2 RCS with the same probability of detection.

4.7.2 PRF and Pulse Width

The pulse repetition frequency of a radar is set by the maximum range at which targets can be detected. The radar measures the range to a target by measuring the time it takes a pulse of RF energy to travel to the target and back to the radar. Once all the echoes have arrived from distant targets, another pulse can be transmitted. Any signals reflected from a very distant target which are received after a second pulse is transmitted will appear to be small targets close to the radar. This is called a second time around echo, at an ambiguous range. To avoid the problem of range ambiguity, the prf is set sufficiently low that all expected signals have been received before another pulse is transmitted.

The radar is required to detect a 2 m^2 RCS target at a range of 200 km. The received signal power from a target falls as range to the fourth power (R^4). The radar will be designed to guarantee detection of the 2 m^2 target at 200 km range, but a target at a longer range might also be detected. Consequently, the radar must have a prf which makes the ambiguous range for a large target much greater than the maximum range of 200 km. Let's assume that an unambiguous range of 400 km will be satisfactory in this application. EM waves travel at a velocity $c = 3 \times 10^8$ m/s, so the time, T, to go to a target at 400 km and back is:

$$T = 2R/c$$
$$= 2 \times 400 \times 10^3 \text{ m}/3 \times 10^8 \text{ m/s} = 2.66 \text{ milliseconds} \qquad (4.65)$$

The PRF is the reciprocal of T, the pulse repetition interval (pri). Hence the prf is:

prf $= 1/2.66$ ms $= 375$ Hz

The range resolution of a radar is set by the pulse width. For a resolution of d meters, the pulse width is:

$$\tau = 2 \times d/c \text{ secs} \qquad (4.66)$$

Hence for a resolution of $\frac{1}{8}$ mile $= 200$ m, we need a pulse length of 1.33 μs.

The pulse width of 1.33 μs sets the IF bandwidth of the radar receiver. For a rectangular RF pulse of duration τ seconds and a rectangular bandpass filter of bandwidth B Hz, the bandwidth which maximizes the S/N of the detected pulse is given by:

$$B = 1.4/\tau \text{ Hz} \qquad (4.67)$$

Thus for a pulse of duration 1.33 μs, the optimum bandwidth is 1.06 MHz. There is a 0.85 dB mismatch loss associated with this combination of pulse and bandpass filter shapes.

The next step in the calculation is to determine the number of "hits" on a target from the scanning antenna with the specified prf of 375 Hz. The radar antenna must rotate in the azimuth plane to provide surveillance of the surrounding airspace. The vertical beam is usually a fan beam, with a shape that corresponds approximately to cosec^2(EL), where EL is the elevation angle above the horizon. The cosecant squared elevation pattern provides a constant signal from an aircraft flying at a constant altitude as it approaches the radar. The 3° vertical plane beamwidth of the antenna is used to provide an estimate of the antenna dimensions and gain.

A scan rate of 6 r.p.m. is equivalent to 36 degrees per second, and the antenna has a 2 degree azimuth beamwidth. Thus a target lies between the 3 dB points of the antenna pattern for 2/36 seconds = 55.6 milliseconds. The prf of the radar is 375 Hz, so the target is hit by approximately 20 transmitted pulses as the antenna beam sweeps past it. This number is important because most radars use integration of the received signal to improve the probability of detecting weak targets; the more pulses that are received from a target the better the chance of it being detected.

The apparent improvement in S/N ratio after integration depends on the detection process used in the radar. Detection can be either coherent or non-coherent. A non-coherent detector is typically an envelope detector which determines the magnitude of the pulse (in its simplest form just a rectifier diode and low-pass filter). A coherent detector multiplies the received signal by a replica of the transmitted signal, usually at an intermediate frequency. For the coherent detector, the improvement in signal-to-noise ratio, called "integration gain," is equal to the number of hits on the target. For the non-coherent detector, the integration gain is equal to the number of hits per target less an integration loss. Integration loss is usually read from a graph because its calculation is difficult and depends in part on the probability of detection and the probability of false alarm. Table 4.7 shows some typical values of integration loss. If we assume non-coherent integration, integration loss is about 1.4 dB and integration gain is:

$$G_i = 10\log_{10}(20) - 2.0 \text{ dB} = 11.0 \text{ dB}$$

The next step in the analysis of this radar is to determine the single pulse S/N required to detect a specified target. The 2 m^2 radar cross-section (RCS) target is to be detected at a maximum range of 200 km. Radar cross-section is a concept used to describe any radar target in terms which allow the reflected signal to be calculated easily. For RCS values above 1 m^2, the reflected signal is equal to that received from a sphere of the same cross-sectional area.

Aircraft are complex targets at microwave frequencies because their dimensions are always a large number of wavelengths. Strong echoes can be obtained when part of the aircraft acts as a corner reflector (two surfaces at right angles). These reflections may correspond to an RCS of 10,000 m^2. An RCS of 2m^2 represents the minimum expected reflecting area for a large aircraft and the typical RCS of a single-engine general aviation aircraft. In the military environment, stealth technology is aimed specifically to reduce the RCS of an aircraft to make it more difficult for the radar to detect.

The radar receiver must have a threshold decision circuit to prevent noise spikes being misinterpreted as targets. When a noise spike crosses the threshold and is assumed to be a target, a "false alarm" is generated. The specification of this radar calls for false alarms to occur no more often

Table 4.7 Integration Loss with Non-Coherent Detection

Number of pulses	2	5	10	15	20	30	50	100	200
Integration loss (dB)	0.2	0.7	1.2	1.8	2.0	2.8	3.6	4.5	5.7

than every 10 hours. Noise in a radar receiver is assumed to be of thermal origin and therefore to have Gaussian probability statistics. After detection by an evelope detector, the noise voltage has Ricean statistics and the probability that the noise voltage V_n has a magnitude exceeding a threshold level V_T is given by:

$$P(V_n > V_T) = \exp(-V_T^2/2V_0^2) = P_{fa} \qquad (4.68)$$

where V_0 is the r.m.s. noise voltage at the input to the envelope detector and P_{fa} is the probability of a false alarm. The probability of a false alarm is simply the inverse of the product of the receiver bandwidth in hertz and the required false alarm time T_{fa} in seconds:

$$P_{fa} = 1/(BT_{fa})$$
$$P_{fa} = 1/(2.12 \times 10^6 \times 3.6 \times 10^4) = 1.3 \times 10^{-11} \qquad (4.69)$$

In most radars, false alarm probabilities are in the range 10^{-10} to 10^{-14}.

For the target to be detected, we require that the signal (after detection by the envelope detector) exceed the threshold level V_T. This means that the peak value, A volts, of the sine wave in the receiver, at the input to the envelope detector, must exceed the threshold level V_T.

The signal-to-noise (power) ratio in the receiver is given by:

$$S/N = (A^2/2)/V_0^2 = A^2/2V_0^2 \qquad (4.70)$$

Hence the minimum signal-to-noise ratio in the receiver which will result in detection of a target is $(S/N)_{min}$, where:

$$(S/N)_{min} = V_T^2/2V_0^2 \qquad (4.71)$$

and the probability of a false alarm will be:

$$P_{fa} = \exp[(-S/N)_{min}] \qquad (4.72)$$

where S/N is a power ratio (not in decibels).

The probability that a given target echo is detected depends on the threshold setting and the signal-to-noise ratio. Thus requiring a long false alarm time requires a high threshold and a higher S/N to detect the target with a given probability. The mathematics for combining probability of false alarm and probability of detection with signal-to-noise ratio is complex, so a table or graph is normally used to determine the threshold setting which provides the required probability of detection P_d and probability of false alarm P_{fa}. Table 4.8 shows some S/N_{min} values for a range of P_d and P_{fa} values.

The radar specification required a false alarm time of 10 hours with a probability of target detection of 0.9. This gave a probability of false alarm

Communication Systems

Table 4.8 S/N_{min} Values for Combinations of P_d and P_{fa}

P_{fa}	P_d	S/N_{min} (dB)
10^{-10}	0.1	12.6
10^{-10}	0.5	14.2
10^{-10}	0.9	15.6
10^{-10}	0.99	16.6
10^{-12}	0.1	13.5
10^{-12}	0.5	15.0
10^{-12}	0.9	16.4
10^{-12}	0.99	17.2
10^{-14}	0.1	14.2
10^{-14}	0.5	15.6
10^{-14}	0.9	16.8
10^{-14}	0.99	17.8

of 1.3×10^{-11}. Interpolating from Table 4.8, the minimum signal-to-noise ratio for target detection is 16.0 dB. This is the value of the S/N after integration. With a non-coherent integration gain, G_i, of 11.0 dB from 20 hits on the target, the single pulse S/N is $16.0 - 11.0 = 5.0$ dB. (Caution: if the single hit S/N is less than 3 dB, the radar may not achieve the expected integration gain with a non-coherent detector due to noise capture effects.)

Now that we know the required S/N for detection of a target with the required probabilities, we can use the radar equation to find the transmit power needed to give the this S/N ratio with a 2 m² target at a range of 200 km. The radar equation gives the power at the input to the radar receiver, P_r, from a single transmitted pulse:

$$P_r = \frac{P_t G^2 \lambda^2 \sigma}{(4\pi)^3 R^4 L} \qquad (4.73)$$

where P_t is the transmitter pulse power in watts, G is the antenna gain (as a ratio), λ is the radar wavelength in meters, σ is the target RCS in square meters, R is the range to the target in meters, and L is the total of all losses, as a ratio. Because radars existed long before calculators and computers, the radar equation is usually evaluated by converting all the terms to dB units to make the calculations easier. The decibel equivalents of the radar parameters are calculated by:

dB value $= 10 \log_{10}$ (parameter)

The dB units for quantities must be referenced to a unit of the quantity, since decibels by definition are dimensionless units of a ratio of two powers. Thus transmitted pulse power is in dBW, wavelength is expressed in terms of dBm, target RCS in dBm², R is in dBm⁴, and antenna gain and losses are in dB. The constant $(4\pi)^3$ is 33 dB. Antenna gain, G, can be calculated from the approximate relationship:

$$G = 30{,}000/(\theta_v \times \theta_h) \text{ (not in dB)} \tag{4.74}$$

where θ_v and θ_h are the 3 dB beamwidths of the antenna in the vertical and horizontal planes, in degrees. For $\theta_v = 3°$ and $\theta_h = 2°$, $G = 30{,}000/6 = 5{,}000$ or 37 dB.

The dimensions of the radar antenna can be estimated from the relationship:

$$3 \text{ db beamwidth} = 75\lambda/D \tag{4.75}$$

For the horizontal plane $\theta_h = 2°$ and $\lambda = 0.1071$ m, so $D_h = 4.02$ m, and in the vertical plane where $\theta_h = 3°$, $D_v = 2.68$ m.

Using dB units in this example gives the following result for received power, P_r:

Transmitter power	P_t	dBw
Antenna gain squared	G^2	74 dB
Wavelength squared	λ^2	−19.4 dBm²
Target RCS	σ	3 dBm²
$(4\pi)^3$		−33 dB
Range to power four	R^4	−212 dBm⁴
Losses	L	−11.4 dB
Received power	P_r	P_t − 198.8 dBW

The next step is to calculate the noise power, N, in the receiver IF bandwidth:

$$N = kT_sB \text{ watts} = -228.6 + 10\log_{10}(T_s) + 10\log_{10}(B) \text{ dBW} \tag{4.76}$$

where k is Boltzmann's constant 1.38×10^{-23} J/K = -228.6 dBW/K/Hz, T_s is the radar receiver system noise temperature in kelvins, and B is the receiver bandwidth in hertz. The relationship between system noise temperature and noise figure is:

$$T_s = 290(F-1) \tag{4.77}$$

Communication Systems

where F is the noise figure expressed as a ratio (not in dB). In this example, $F = 5$ dB. Hence:

$$T_s = 290(10^{5/10} - 1) = 627 \text{ K or } 28.9 \text{ dBK}.$$

The noise power N is given by:

Boltzmann's constant	k	−228.6	dBW/K/Hz
Receiver bandwidth	B	60.3	dBHz
Noise temperature	T_s	28.0	dBK
Noise power	N	−140.3	dBW

The signal-to-noise ratio, $S/N = P_r/N$, is required to be 5.0 dB. Thus in dB units:

$$S/N = P_r - N = P_t - 198.8 + 140.3 = 5.0 \text{ dB}$$

Solving for P_t gives:

$$P_t = 4.4 + 199.4 - 140.3 = 63.5 \text{ dBW or about 2.2 MW}$$

This is a typical transmit pulse power for a long-range surveillance radar.

The design of the radar to meet the specification is now complete. Note that the target used in this example had a constant RCS of 2 m². In practice, the echo from an aircraft or similar large target fluctuates as the target changes its orientation with respect to the radar. This is known as "target glint". The radar may be able to detect the target more easily if it glints a lot and the probability of detection is set to a low value. Alternatively, it may be more difficult to detect the target when the probability of detection is set high. Targets are classified by "Swerling case." A case 5 target has a constant RCS, as in this example. Cases 1 through 4 are fluctuating targets with different statistics.

The radar will detect a 0.1 m² target at a shorter range than the 2 m² target. The 0.1 m² target must move towards the radar until the signal reaches threshold. That requires a 13 dB (twenty-fold) increase in signal power relative to the signal at 200 km. Because power received by the radar from a target is proportional to $1/R^4$, range must decrease by 13/4 = 3.25 dB or a ratio of 2.1. Thus the maximum range at which the 0.1 m² target can be detected is:

$$R_{\text{max}} = 200/2.1 = 95 \text{ km}$$

4.8 ANTENNAS

All radio links require a transmitter, a receiver, and two antennas, one at each end of the link. Frequently, the antenna is the limiting factor in the link. This section illustrates the calculations that can be made for antennas, primarily for the gain and the beamwidth.

4.8.1 Gain and Beamwidth of Aperture Antennas

Antenna gain is a measure of an antenna's ablity to direct radiated energy in a preferred direction. The term "gain" is used to describe both the maximum gain of the antenna in its *boresight* or *axis* direction, and also the gain in any other direction.

Gain, G, is dimensionless and is defined for any antenna in its boresight direction by

$$G = 4\pi A_{\text{eff}}/\lambda^2 \tag{4.78}$$

where A_{eff} is the effective aperture of the antenna and λ is the wavelength (in meters), for an antenna operated at a frequency f Hz, where $c = f\lambda$ and c is the velocity of light in free space, in m/s. Usually c is taken to be 3×10^8 m/s. The effective aperture of an antenna is usually less than the physical aperture area, and is given by

$$A_{\text{eff}} = \eta_A A \tag{4.79}$$

where η_A is the *aperture efficiency* of the antenna and A is the physical aperture area.

The on-axis gain of the antenna is therefore

$$G = \eta_A 4\pi A/\lambda^2 \tag{4.80}$$

Gain is usually expressed in decibels, being a ratio of two powers

$$G = 10\log_{10}(\eta_A 4\pi A/\lambda^2) \tag{4.81}$$

In each case it is assumed that there is a uniform phase distribution of electric field in the antenna aperture. Non-uniform phase leads to a reduction in antenna gain and a broadening of the antenna beam.

Usually, aperture efficiency is less than unity, and the effective area of the antenna is less than the physical area. However, some low gain antennas can have effective areas considerably larger than their physical area.

Communication Systems

Reflector Antennas

A reflector antenna (often called a "dish antenna") usually consists of a parabolic reflector and a feed system. The feed and reflector create a uniform phase distribution, or plane wave, in the antenna aperture. The aperture efficiency for reflector antennas, η_A, is typically in the range 0.6 to 0.7. An antenna with a circular aperture has a gain

$$G = \eta_A 4\pi(\pi D^2/4)/\lambda^2 = \eta_A(\pi D/\lambda)^2 \tag{4.82}$$

The 3 dB beamwidth (or half power beamwidth), θ, of any antenna may be estimated from the approximate formula

$$\theta = 75\lambda/D \text{ degrees} \tag{4.83}$$

where D is the dimension of the antenna (its width) in the plane in which θ is measured. Note that a large antenna has a narrow beamwidth and a small antenna has a wide beamwidth. D and λ must have the same units.

Example 4.3

A dish antenna has a circular aperture with diameter 2.0 m. The aperture efficiency at a frequency of 6.0 GHz is 62%. Find the gain and 3 dB beamwidth at this frequency.

First calculate the wavelength for a frequency of 6.0 GHz:

$$\lambda = c/f = 3 \times 10^8/6 \times 10^9 = 0.05 \text{ m}$$

From Eq. (4.82), the antenna gain is:

$$G = \eta_A(\pi D/\lambda)^2 = 0.62 \times (\pi \times 2.0/0.05)^2 = 9790 = 39.9 \text{ dB}$$

From Eq. (4.83), the beamwidth is approximately:

$$\theta = 75\lambda/D = 75 \times 0.05/2.0 = 1.88 \text{ degrees}$$

Phased Array Antennas

A phased array antenna uses a large number of small elements, such as halfwave dipoles, open-ended waveguides, or patches, to fill an aperture. The elements are fed by power dividers and computer-controlled phase shifters, allowing the phase at each element to be adjusted. Appropriate phasing of the elements causes the beam to scan away from the array axis (defined as a line normal to the array aperture). The beam can be scanned rapidly over a volume of space in front of the array, or switched from position to position by computer command. Phased arrays have been used in radars to provide very rapid scanning and also to allow multiple targets to be tracked with a single antenna.

The gain of a phased array can be estimated from the number of elements and the element gain, provided the elements are spaced a half wavelength apart. If the element spacing exceeds one half wavelength, gain should be calculated from Eq. (4.80) using the array aperture area and aperture efficiency. For half wavelength spacing, the on-axis gain of the array is approximately

$$G = N \times G_{el} \qquad (4.84)$$

where N is the number of elements in the array and G_{el} is the gain of each element in the axis direction.

Example 4.4

A phased array has 50 rows of halfwave dipole elements, with 50 elements in each row, spaced by 0.5λ in both planes. Estimate the on-axis gain of the antenna.

The gain of a halfwave dipole is 1.64 or 2.15 dB normal to the dipole. Using Eq. (4.84), the approximate gain of the array is

$$G = 50 \times 50 \times 1.64 = 4100 = 36.1 \text{ dB}$$

The area of the aperture is $2500 \times 0.25\lambda^2 = 625\lambda^2$. If we assume an aperture efficiency of 60%, Eq. (4.80) gives the gain of the phased array antenna as

$$G = 10 \log_{10}(\eta_A 4\pi A/\lambda^2) = 0.6 \times 4\pi \times 625 = 4712 = 36.7 \text{ dB}$$

This is 0.6 dB greater than the value obtained using Eq. (4.84). However, if the elements are spaced more than a half wavelength apart, Eq. (4.84) underestimates the gain of the phased array antenna, as the following example shows.

Example 4.5

A phased array antenna consists of rows of 50 halfwave dipoles spaced 0.6λ apart, arranged as 50 rows spaced 0.7λ center to center. Find the on-axis gain and beamwidth of the array in each principal plane if the phased array has an aperture efficiency of 65%.

First calculate the width of the antenna, w, and its height, h, in wavelengths, then find the area of the aperture and hence its gain from Eq. (4.80). The antenna width is

$$w = 50 \times 0.6\lambda = 30\lambda$$

The height of the antenna is

$$h = 50 \times 0.7\lambda = 35\lambda$$

Hence the area of the antenna is $1050\lambda^2$, and the gain is

$$G = \eta_A 4\pi A/\lambda^2 = 0.65 \times 4\pi 1050\lambda^2/\lambda^2 = 8577 = 39.3 \text{ dB}$$

The principal planes of the antenna are the planes of symmetry defined by the aperture dimensions, or by polarization. In this case, the rows and columns of dipoles that make up the rectangular aperture of the phased array define the principal planes. In the width dimension of the antenna, $w = 30\lambda$, so the approximate half-power beamwidth is

$$\theta = 75\lambda/w = 75/30 = 2.50 \text{ degrees}$$

In the height dimension of the antenna, $h = 35\lambda$, so the approximate half-power beamwidth is

$$\theta = 75\lambda/h = 75/35 = 2.14 \text{ degrees}$$

If the beam is scanned away from the axis of the antenna in any plane by an angle φ, the effective aperture is reduced by a factor $\cos \varphi$. This will reduce the gain by $10 \log_{10} (\cos \varphi)$.

For example, suppose the beam is scanned off-axis by an angle of 30 degrees, the gain in that direction will be

$$G = 39.3 - 10 \log_{10}(\cos 30°) = 38.7 \text{ dB}$$

Gain, Beamwidth, and Effective Area Relationships

A useful empirical relationship between gain and beamwidth in an aperture antenna is

$$G \approx 30{,}000/(\theta_1 \times \theta_2) \tag{4.85}$$

where θ_1 and θ_2 are the 3 dB beamwidths of the antenna in its principal planes, in degrees. G is a ratio, not in dB, in Eq. (4.85). Applying this formula to the circular aperture antenna used in Example 4.3, where the beamwidth in the principal planes was 1.88 degrees

$$G \approx 30{,}000/(1.88 \times 1.88) = 8488 = 39.3 \text{ dB}$$

In Example 4.3 the gain for this antenna was found to be 39.9 dB, using the more accurate formula of Eq. (4.81) and an aperture efficiency of 62%.

The effective area of any antenna can be found from its gain, using Eq. (4.78):

$$A_{\text{eff}} = G\lambda^2/4\pi \tag{4.86}$$

Low-gain antennas, such as dipoles, slots, and some traveling wave antennas such as the Yagi and Helix antennas, can have effective areas much larger than their physical area. For example, a halfwave dipole antenna has a gain of 1.64, or 2.1 dB, broadside to the dipole. Using Eq. (4.86), the effective area of the dipole is $1.64\lambda^2/4\pi = 0.13\lambda^2$. A halfwave dipole is 0.5λ high and typically less than 0.1λ in width. Thus the physical cross-sectional area of the dipole is no more than $0.05\lambda^2$, but its effective area is $0.13\lambda^2$. Yagi antennas, consisting of an end fire array of directors which are equivalent to short-circuited halfwave dipoles, driven by a dipole or folded dipole, can have a gain of 11 dB or more, giving an effective area greater than a square wavelength, yet the cross-sectional area is no greater than that of a halfwave dipole.

Calculation of Aperture Efficiency and Antenna Patterns

The formulas for gain and beamwidth given in Sect. 4.8.1 are all approximate. They can be used when the gain or beamwidth of an antenna does not need to be known exactly. When precise calculation of the gain, beamwidth, and sidelobe pattern of an antenna is needed, diffraction theory must be used to calculate the distribution of the electric field at a large distance from the antenna. The electric field distribution is called the *antenna pattern*, and consists of the main beam and sidelobes. Antennas have identical patterns when transmitting and receiving, a principle known as reciprocity, but are usually analyzed when transmitting. As electromagnetic energy propagates away from the antenna, its distribution in space changes. This pattern is not fully developed until a distance $2D^2/\lambda$ from the antenna, where D is the largest dimension across the antenna aperture. At this distance, known as the Raleigh range, the energy is in the *far field* of the antenna and the pattern is a constant function of the angle from the antenna axis at all distances greater than the Raleigh range.

The far field pattern of an aperture antenna is calculated from the electric field distribution in the aperture. For a rectangular aperture with dimensions a in the x-direction and b in the y-direction, the far field pattern $E(\theta, \varphi)$ is given by

$$E(\theta, \varphi) = \int_x \int_y F(x, y) \exp[-jk \sin\theta(x\cos\varphi + y\sin\varphi)] \mathrm{d}x\, \mathrm{d}y \qquad (4.87)$$

where $k = 2\pi/\lambda$; θ, φ are the polar coordinates of the observation point with respect to the center of the aperture; and $F(x, y)$ is the electric field distribution in the rectangular aperture.

A similar integral relationship exists between the field in a circular aperture and its far field pattern

$$E(\theta, \varphi) = \int_r \int_\psi F(r, \psi) \exp[-jkr \sin\theta \cos(\varphi - \psi)] dr\, d\psi \qquad (4.88)$$

where $F(r, \psi)$ is the electric field distribution in the circular aperture in cylindrical coordinates.

In most cases, the electric field distribution in the aperture of an antenna is not a simple function, and the integrals in Eqs. (4.87) and (4.88) cannot be evaluated analytically. Many computer programs have been written which calculate far field patterns for antennas with various aperture shapes using numerical calculation of the integrals. In reflector antennas, the field in the aperture is created by a feed system which may include secondary reflectors, as in Cassegrain and Gregorian antennas, and the computer program must trace the energy radiated by the feed through the reflector system to determine the field in the aperture.

Some insight into the gain and far field patterns of antennas can be obtained by using simple functions to model the electric field in the antenna aperture. In nearly all cases a plane wave (uniform phase) is assumed in the aperture. If phase errors exist across the aperture, their effect is treated separately. Prior to the use of computers to evaluate far field integrals, analytical integration was used with simple aperture distribution functions to yield illumination efficiency, 3 dB beamwidth and sidelobe levels in the far field.

Rectangular Apertures

The aperture distributions for a rectangular aperture are assumed to be separable, so that the far field patterns in the principal planes, the x- and y-planes, depend only on the linear aperture field distributions $F(x)$ and $F(y)$ respectively. This reduces the rectangular antenna to two linear antennas for analysis purposes.

The gain of antenna is maximized when the aperture field is uniform in amplitude and phase, so $F(x, y) = 1$ for the rectangular aperture and $F(r, \psi) = 1$ for the circular aperture.

The far field patterns of a rectangular antenna are then given by:

$$E(\theta) = \int_x \exp[-jkx \sin\theta \, dx] = a \sin X / X \qquad (4.89a)$$

for the x-plane with integration limits $\pm a/2$, where $X = \pi a/\lambda$ and

$$E(\varphi) = \int_y \exp[-jky \sin\theta \, dy] = b \sin Y / Y \qquad (4.89b)$$

for the y plane with integration limits $\pm b/2$, and $Y = \pi b/\lambda$.

The shape of the $\sin X/X$ function is well known, since it occurs in Fourier transforms of rectangular functions. The first sidelobe of $\sin X/X$ is at -13.2 dB relative to the main lobe maximum (at 0 dB), and the half power width of the main lobe is $51\lambda/a$ degrees for an aperture width a. The illumination efficiency of this aperture is unity, and the gain is

$$G = 4\pi ab/\lambda^2 \tag{4.90}$$

The distribution of electric field in the aperture of the antenna is rarely uniform, since uniform illumination results in undesirably high sidelobes. Usually, a *tapered distribution* of electric field in the aperture is used. Tapered distributions are typically described by an analytical function for $F(x)$ or $F(y)$, which is cut off at some specific level at the edge of the aperture. The literature of antennas abounds with analyses of far field patterns for analytically integrable aperture field distributions.

A tapered aperture field distribution has lower gain, lower sidelobes, and broader main beam than the uniformly illuminated aperture. Table 4.9 shows a small selection of results for the major characteristics of linear aperture distributions with specific tapers.

The function labeled "cosine to power n" is an aperture electric field of the form

$$F(x) = \cos^n(\pi x/a) \text{ where } -a/2 < x < a/2$$

and the aperture has width a. The function labeled "cosine squared on a pedestal" is an aperture electric field of the form

$$F(x) = A + B\cos^2(\pi x/a) \text{ where } -a/2 < x < a/2$$

and the aperture has width a.

Table 4.9 Far Field Pattern Characteristics of Linear Apertures With Specific Tapered Field Distributions

Illumination Function	Illumination Efficiency	3 dB Beamwidth (degrees)	First Sidelobe Level (dB)
Cosine to power n			
$n = 0$ (uniform illumination)	1.00	$51\lambda/a$	-13.2
$n = 1$	0.81	$69\lambda/a$	-23
$n = 2$	0.67	$83\lambda/a$	-32
$n = 3$	0.58	$95\lambda/a$	-40
Cosine squared on a pedestal			
$A = 0.33, B = 0.67$	0.88	$63\lambda/a$	-25.7
$A = 0.08, B = 0.92$	0.74	$77\lambda/a$	-42.8

Circular Apertures

The aperture distributions for a circular aperture can be simplified by assuming symmetry in the aperture field distribution in the circumferential direction, so that $F(\psi) = 1$. The far field pattern is then the same in all planes, and depends only on the radial field distribution $F(r)$ in the aperture. The integral of Eq. (4.89a) then becomes

$$E(\theta) = \int_r F(r)\exp[-jkr\sin\theta]dr \qquad (4.91)$$

where the integration is carried out from $r = 0$ to $r = r_0$ and r_0 is the radius of the circular aperture.

In the particular case of uniform illumination of the circular aperture, $F(r) = 1$ and

$$E(\theta) = \int_r F(r)\exp[-jkr\sin\theta]dr = \pi r_0^2 2J_1(u)/u \qquad (4.92)$$

where $J_1(u)$ is a Bessel function of the first kind. The shape of the $J_1(u)/u$ function is well known; the first sidelobe of $J_1(u)/u$ is at -17.6 dB relative to the main lobe maximum (at 0 dB). The half-power width of the main lobe is $58\lambda/D$ degrees for a circular aperture diameter D. The illumination efficiency of this aperture is unity, and the gain is

$$G = 4\pi r_0^2/\lambda^2 = (\pi D/\lambda)^2 \qquad (4.93)$$

As with the rectangular aperture antenna, a tapered field distribution in the aperture produces a far field pattern with lower gain, lower sidelobes, and a broader main beam than the uniformly illuminated circular aperture. Some selected results are shown in Table 4.10 for a circular aperture of

Table 4.10 Far Field Pattern Characteristics of Circular Apertures With Specific Tapered Field Distributions.

Illumination Function	Illumination Efficiency	3 dB Beamwidth (degrees)	First Sidelobe Level (dB)
Parabolic to power n			
$n = 0$ (uniform illumination)	1.00	$58\lambda/a$	-17.6
$n = 1$	0.75	$73\lambda/a$	-24.6
$n = 2$	0.56	$84\lambda/a$	-30.6

diameter D with a "parabolic to the power n" aperture field distribution of the form

$$F(r) = (1 - r^2)^n$$

where r is a normalized radius in the aperture with $0 < r < 1$.

Example 4.6

A rectangular waveguide horn with dimensions $a = 0.1$ m and $b = 0.13$ m supports the TE_{10} waveguide mode in its aperture with the electric field polarized in the a direction. The horn is operated at a frequency of 4.0 GHz. Find the on-axis gain, and the 3 dB beamwidths and first sidelobe levels in the principal planes of the horn.

The TE_{10} waveguide mode has uniform electric field distribution $F(x) = 1$ in the direction in which it is polarized (the E-plane), and a distribution $F(y) = \cos(\pi y/b)$ in the orthogonal plane (the H-plane), where $y = \pm b/2$ at the edges of the aperture in the H-plane. The characteristics of the far field pattern can be determined from Table 4.9.

In the E-plane, the field distribution in the aperture is uniform, so illumination efficiency is unity, 3 dB beamwidth of the main beam is $51\lambda/a$ degrees, and first sidelobe level is -13.2 dB. In the H-plane, the distribution is cosine, so the illumination efficiency is 0.81, 3 dB beamwidth of the main beam is $69\lambda/b$ degrees, and first sidelobe level is -23 dB.

At a frequency of 4 GHz, the wavelength is 0.075 m, so $a = 1.33\lambda$ and $b = 1.73\lambda$. The aperture efficiency of the horn is the product of the illumination efficiencies for the E-and H-planes, so $\eta_A = 1.0 \times 0.81 = 0.81$. The gain of the horn antenna is given by Eq. (4.80):

$$G = \eta_A 4\pi A/\lambda^2 = 0.81 \times 4\pi \times 1.33 \times 1.73 = 23.4 = 13.7 \text{ dB}$$

In the E-plane, the beamwidth is $51/1.33 = 38.3$ degrees and in the H-plane the beamwidth is $69/1.73 = 39.9$ degrees. The first sidelobe level in the E-plane is -13.2 dB, and in the H-plane -23 dB. Note that choosing horn aperture dimensions in the ratio 1:1.3 in the E- and H-planes produces approximately equal beamwidths in the two planes.

The results given above are accurate for small horns supporting the TE_{10} waveguide mode. Large horns tend to have a non-uniform aperture phase distribution, which results in a lower gain and wider beamwidth. The phase error is caused by the difference in path length from the throat of the horn to the aperture for wave paths in the center of the aperture and at the edge of the aperture. The maximum achievable gain with a waveguide horn is about 23 dB unless a lens is used to correct the aperture phase error.

Example 4.7

A dish antenna has a circular aperture which has a diameter of 18 inches. The antenna is operated at a frequency of 12.0 GHz, and has a circularly symmetric uniform phase aperture field distribution with a parabolic taper function. Find the gain, beamwidth, and first sidelobe level for this antenna.

The required results are found in Table 4.10. The illumination efficiency of the aperture is 0.75, the main beam 3 dB beamwidth is $73\lambda/D$, and the first sidelobe is at -24.6 dB. The 18 inch dish diameter equates to 0.457 m, and the wavelength at 12 GHz is $\lambda = 0.025$ m, so the 3 dB beamwidth of the antenna beam is given by $\theta_{3\,\text{dB}} = 73 \times 0.025/0.457 = 4.0$ degrees.

The aperture efficiency of the antenna will be lower than the illumination efficiency because there are always additional losses in a reflector antenna. Assuming the ideal case with no losses, the antenna gain from Eq. (4.82) would be:

$$G = \eta_A (\pi D/\lambda)^2 = 0.75 \times (\pi 0.457/0.025)^2 = 2473 \text{ or } 33.9 \text{ dB}$$

A more realistic value for aperture efficiency would be 65%, giving a gain $G = 2143$ or 33.3 dB.

5
Algorithms Used in Control Systems

Hugh F. VanLandingham
Virginia Polytechnic Institute and State University, Blacksburg, Virginia

5.1 INTRODUCTION

In this chapter several algorithmic calculations will be presented. These methods have a broad range of application, but are generally considered to be part of the discipline of *control systems*. As will be seen, the subject matter will fall into the two main categories of *modeling* and *design*. System descriptions are described as state space models, a general form of representation. Variations include continuous time (CT) or analog forms, discrete time (DT) forms, and digital forms. The accepted distinction between DT and digital forms is whether the signal amplitudes are quantized (digital) or not (DT). A significant aspect of modeling involves the conversion of models from one form to another, e.g. from DT to CT, or from state space form to transfer function form (which can be an equivalent form as represented in the appropriate transform domain).

The topics were chosen because of their relevance to control system, and it was thought that to be useful to practicing engineers the algorithms should be augmented with examples. Consequently, for each of the eight topics given a brief discussion is provided for motivation followed by an illustrative example and the MATLAB code used in the example. The use of MATLAB was based on the fact that it is a respected and widely available utility program. This material can be found in reference [2] with more detail.

The order of the specific topics discussed here was selected to lead most naturally from one topic to the next.

5.2 MATRIX FUNCTIONS AND THE MATRIX EXPONENTIAL

As will be seen later in this chapter, an important application of matrix functions to system theory is in the conversion from CT models to DT models. In this context it is necessary to have a robust procedure with which to calculate the exponential function of a square matrix. The extension from ordinary functions $f(x)$ to the same function, but of a (square) matrix A, instead of a scalar variable x, denoted $f(A)$, is done through the Taylor series expansion of the function; this may be done for any function within the range of convergence of the expansion [2]. To illustrate, the *exponential function*, which converges for all real values of x, will be considered:

$$\exp(x) = \sum_{i=0}^{\infty} \frac{x^i}{i!} \tag{5.1}$$

Although defined as an infinite sum of terms, it is clear that for engineering calculations only a finite number of terms is possible, allowing the number of terms to correspond to the required accuracy. This series converges uniformly, so that the accuracy of a partial sum, for a fixed value of x, can be estimated from the number of terms in the sum, usually by estimating the error from the next term in the series.

Given a square numerical matrix, A, the *matrix exponential* of A would be calculated from the finite sum of matrices [1]

$$\exp(A) = \sum_{i=0}^{N} \frac{A^i}{i!} \tag{5.2}$$

where $A^0 = I$, the identity matrix. Depending on the entries of the matrix A, the matrix exponential calculation can have problems converging, i.e. the number of terms N in Eq. (5.2) may have to be very large, which, in turn, may cause inaccuracy due to numerical round-off. This situation occurs when the entries of A are large, or equivalently the norm of A is large ($\gg 1$), causing the numerator of the summand in Eq. (5.2) to increase rapidly with increasing powers n.

To resolve this problem, the following property of the exponential function is used:

$$\exp(x) = [\exp(x/r)]^r \tag{5.3}$$

Algorithms Used in Control Systems

In addition, it is convenient to define the *scaling factor* r as $r = 2^m$, where m is an integer defined so that $\| A/r \| \leq \delta$, a "small" number ($\delta \leq 0.5$), and the norm is any convenient norm, e.g., the Frobenius norm, which is the square root of the sum of the squares of all the elements of A. With this value of m, the series

$$M_1 = \sum_{i=0}^{N} \frac{(A/r)^i}{i!} \tag{5.4}$$

will converge satisfactorily since $\| A/r \| \leq 0.5$. From the property in Eq. (5.3) it may be shown that using the following recursion provides the desired matrix, $M = M_m = \exp(A)$, within some required numerical error, as determined by the selection of N in Eq. (5.4):

$$M_{k+1} = [M_k]^2, \text{ for } k = 1, 2, \ldots, m \tag{5.5}$$

where M_1 is given in Eq. (5.4). The number of terms, N, can be found by ensuring that the value of the "$N+1$" term is less than the allowable error. An example of the procedure is presented below. The MATLAB code at the end of the section will implement the scaled exponential matrix calculation developed here. For additional details the reader is referred to reference [1].

Since N will be known in advance, the truncated series in Eq. (5.4) may be calculated in the following "nested" form for computational convenience.

$$\sum_{i=0}^{N} \frac{B^i}{i!} = \left\{ I + B \left[I + \frac{B}{2} \left(I + \cdots + \frac{B}{N-2} \left[I + \frac{B}{N-1} \left(I + \frac{B}{N} \right) \cdots \right] \right) \right] \right\} \tag{5.6}$$

where B represents the scaled matrix A/r in Eq. (5.4). The computation begins with the center parentheses and expands outward, as illustrated in the flow diagram of Fig. 5.1. Note that the last term in Eq. (5.6) contains an N factorial in the denominator, which gives the series a strong tendency to converge. Considering that $30! > 10^{32}$, there may well be numerical problems if the series requires more than 30 terms.

5.2.1 Example 5.1 (Matrix Function)

Consider the matrix A given by

$$A = \begin{bmatrix} 0 & 1 \\ 0 & -b \end{bmatrix} \tag{5.7}$$

where the parameter $b = 5$. With this simple matrix the exponential matrix can be determined with enough terms; however, it is interesting to compare the direct series with the scaled series in order to see the effectiveness of the

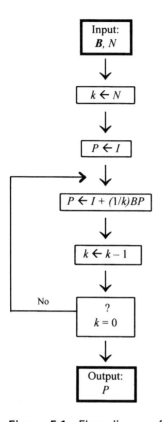

Figure 5.1 Flow diagram for computation of exponential matrix series.

scaling algorithm. Figure 5.2 illustrates the error difference between the scaled and the unscaled matrices as a plot of the logarithm of the Frobenius norm of the error matrix versus the number of terms of the series expansion. In this case convergence can be accomplished in either case, given enough terms; but, if b is selected to be 12, the unscaled series does not converge within 30 terms. Specifically, the second columns of the true $\exp(A)$ and the 31 term (unscaled) approximating series are given by:

$(\text{true})_{21} = 0.083\ 332\ 821\ 315\ 64 \quad (\text{approx})_{21} = 0.136\ 962\ 028\ 222\ 06$
$(\text{true})_{22} = 0.000\ 006\ 144\ 212\ 35 \quad (\text{approx})_{22} = -0.643\ 544\ 338\ 664\ 72$

which clearly shows that the approximation is not good. By comparison the scaled series gives values that are identical to the "true" values shown above, i.e., to at least 14 place accuracy.

Algorithms Used in Control Systems

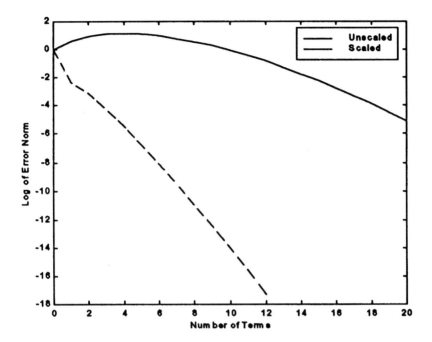

Figure 5.2 Error difference between the scaled and unscaled matrices as a plot of the logarithm of the Frobenius norm of the error matrix versus the number of terms of the series expansion.

5.2.2 Computer Code

The function $p = \text{scalexp}(A, n)$ implements the scaled exponential matrix calculation, $\exp(A)$, using an $(n + 1)$ term expansion.

```
function p = scalexp(A,n);
[dim,dim] = size(A); N = 0; aa = norm(A,'fro');
while aa > .5, A = A/2; N = N + 1; aa = norm(A,'fro'); end
temp = A; k = n; p = eye(dim);
for i = 1:n, p = eye(dim) + temp*p/k; k = k-1; end
for i = 1:N, p = p*; end
```

5.2.3 Summary

The concept of a function of a (square) matrix was defined as the matrix version of the series representation of the function and was illustrated using the exponential function. Numerical calculation of such series typically requires a pre-scaling of the matrix to assist in the convergence of the series, i.e., to provide a good approximation to the function. A simple example was used to illustrate the importance of scaling by comparing the effects of calculating the series with and without scaling. Matrix functions are used extensively in the conversion of continuous time models to discrete time models as well as in the conversion between different types of models, such as between state space models and transfer functions – topics that are discussed later in the chapter.

5.3 CONVERSION FROM CONTINUOUS TO DISCRETE MODELS

In this section specific system models will be considered. The reader will be able to see the material of the previous section applied in the conversion process. We begin with a basic linear, constant coefficient, continuous time (CT) state space model:

$$\dot{x}(t) = Ax(t) + Bu(t)$$
$$y(t) = Cx(t) + Du(t) \tag{5.8}$$

where x is an n-dimensional state vector; u, an m-dimensional input vector, y, a p-dimensional output vector; and the coefficient matrices are constant with compatible dimensions. Along with the system model, if the initial state is given as $x(0)$, the solution for the state vector (given the input vector) can be written as:

$$x(t) = \exp(At)x(0) + \int_0^t \exp[A(t-\alpha)]Bu(\alpha)d\alpha \tag{5.9}$$

The solution for $y(t)$ in terms of $x(t)$ and $u(t)$ is straightforward from Eq. (5.8). We will use this basic solution form to generate a discrete time (DT) model which is suitable for use when the CT system is interfaced with a digital computer.

5.3.1 Zero-Order Hold (ZOH) Equivalent Model

Consider that the CT system in Eq. (5.8) is interfaced as shown in Fig. 5.3. The ideal sampler is a model for a digital-to-analog (D/A) converter, and when combined with a ZOH represents a model for an analog-to-digital (A/D) converter. Figure 5.4 shows a time axis with sample instants indicated. The sampling will be assumed to be uniform, i.e., T is constant.

With reference to Fig. 5.4 let us associate $t = kT$ with the "initial" time and $t = kT + T$ with "present" time. Equation (5.9) then becomes [3]

Algorithms Used in Control Systems

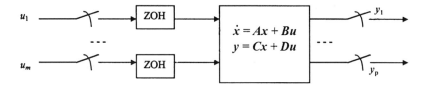

Figure 5.3 Interfacing of the CT system in Eq. (5.8).

$$x(kT + T) = \exp(AT)x(kT) + \int_{kT}^{kT+T} \exp[A(kT + T - \alpha)]Bu(\alpha)d\alpha \qquad (5.10)$$

Now, recognizing that $u(t) = u(kT)$ is constant during the sampling interval, i.e.

$$u(t) = u(kT) \quad \text{for} \quad kT \leq t < kT + T \qquad (5.11)$$

the input vector u can be taken out of the integration in Eq. (5.10). After a change of integration variable, the DT model can be written as

$$\begin{aligned} x(t+1) &= A_d x(t) + B_d u(t) \\ y(t) &= Cx(t) + Du(t) \end{aligned} \qquad (5.12)$$

The time index t now represents *discrete* (normalized) time, which takes on integer values, to be interpreted as multiples of T, the sample interval. In other words the time axis has been changed to *normalized* time, t/T. Note that the dynamics are now in terms of *difference* equations instead of differential equations. The state and input coefficient matrices in the DT model are (from Eq. 5.11)

$$A_d = \exp(AT) \quad \text{and} \quad B_d = \int_0^T \exp(At)B\,dt \qquad (5.13)$$

Again, like any other functions of a matrix, as discussed in the previous section, these new matrices are defined by the series expansion of their right-hand sides. We have seen the expansion for A_d in Eqs. (5.1) and (5.2), i.e.,

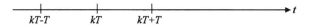

Figure 5.4 Time axis with sample instants indicated.

$$A_d = \exp(AT) = \sum_{i=0}^{\infty} \frac{(AT)^i}{i!} \qquad (5.14)$$

Similarly, by integrating the exponential series term by term,

$$B_d = \int_0^T \exp(At)B\,dt = \sum_{i=0}^{\infty} \frac{(AT)^i}{(i+1)!} BT \qquad (5.15)$$

If A is non-singular,

$$B_d = A^{-1}\exp(AT - I) \qquad (5.16)$$

Since both A_d and B_d must be calculated, we first evaluate a common series:

$$E = \sum_{i=0}^{\infty} \frac{(AT)^i}{(i+1)!} \qquad (5.17)$$

so that

$$A_d = I + EAT \quad \text{and} \quad B_d = EBT \qquad (5.18)$$

The robust scaling method of the previous section must be modified. After E is calculated to the required accuracy, A_d and B_d are determined from Eq. (5.18) using Bingulac's Algorithm to determine the ZOH equivalent DT model from knowledge of the sampling interval T and the original CT model [2].

Bingulac's Algorithm: working with A_d in Eq. (5.18), and recognizing the rescaling process of Eq. (5.5):

$$M_{k+1} = [M_k]^2, \quad \text{for} \quad k = 1, 2, \ldots, m \qquad (5.19)$$

we can write

$$I + E_{k+1}AT_{k+1} = (I + E_k AT_k)^2 \qquad (5.20)$$

Expanding,

$$I + E_{k+1}AT_{k+1} = I + 2E_k AT_k + (E_k AT_k)^2 \qquad (5.21)$$

This last equation leads to the following recursion:

$$\begin{aligned} T_{k+1} &= 2T_k \\ E_{k+1} &= E_k(I + E_k AT_k/2) \end{aligned} \qquad (5.22)$$

which is initialized with

Algorithms Used in Control Systems

$$T_1 = T/r$$

$$E_1 = \sum_{i=0}^{N} \frac{M^i}{(i+1)!} \tag{5.23}$$

where $M = AT/r$ is the initial scaled matrix. The computer code for this algorithm is given at the end of this section. The following example illustrates the conversion method.

5.3.2 Example 5.2 (CT to DT Conversion)

Consider the given CT model representing a state space model of the form in Eq. (5.8)

$$A_c = \begin{bmatrix} -1 & 1 & 0 & -1 \\ -1 & -2 & 1 & 1 \\ 1 & 0 & -2 & 1 \\ 0 & 0 & 0 & -1 \end{bmatrix}, \quad B_c = \begin{bmatrix} 1 \\ -1 \\ 2 \\ 1 \end{bmatrix} \tag{5.24}$$

$$C = \begin{bmatrix} 1 & -1 & 0 & 0 \\ 2 & 0 & 1 & -1 \end{bmatrix}, \quad D = \begin{bmatrix} 1 \\ 0 \end{bmatrix} \tag{5.25}$$

Only A_c and B_c of Eq. (5.24) are involved in the conversion process, the matrices of Eq. (5.25) will be unchanged. A ZOH equivalent (also called a *step-invariant*) model is to be obtained using a sample interval of $T = 2$ s.

Solution: Calculating E, we first note that the (Frobenius) norm of ET is 8.25. This is large enough to require some scaling. Using the scaled algorithm with $N = 20$ terms, we have the following matrices (in addition to C and D which remain unchanged):

$$A = \begin{bmatrix} 0.0722 & 0.0798 & 0.0632 & -0.0159 \\ -0.0167 & -0.0076 & 0.0167 & 0.2061 \\ 0.0798 & 0.0632 & 0.0555 & 0.0805 \\ 0 & 0 & 0 & 0.1353 \end{bmatrix}, \quad B = \begin{bmatrix} 0.4111 \\ 0.3785 \\ 1.5338 \\ 0.8647 \end{bmatrix}$$

5.3.3 Computer Code

The function $[A, B] = \text{zohequiv}(a, b, T)$ calculates the ZOH equivalent system $x(t + 1) = Ax(t) + Bu(t)$ with sampling interval T, using a scaling algorithm

```
function [A,B]=zohequiv(a,b,T);
[n,n]=size(a); at=a*T; I=eye(n); N=0; n=20;
aa=norm(at,'fro')
```

```
temp = at; while aa > .5, temp = temp/2; N = N + 1;
aa = norm(temp,'fro'); end
r = 2^N; k = n + 1; e = I; for i = 1:n, e = I + temp*e/k; k = k-1; end
t = T/r; for j = 1:N, e = e*(I + e*a*t/2); t = 2*t; end
A = I + e*a*T; B = e*b*T;
```

5.3.4 Summary

The concept of a DT equivalent model, which is important when the effect of a CT plant is to be viewed from the perspective of a digital control loop, was introduced. In particular, the ZOH model (also called a step-invariant model) was developed. The development was based on a robust numerical algorithm which uses the unique combination of scaling with a common matrix series [2].

5.4 CONTROLLABILITY AND OBSERVABILITY CALCULATIONS

Once we have an appropriate model, either in discrete time or continuous time, it is important to determine the inherent properties of the model. For example, it is a simple matter to determine the stability of the system by calculating the eigenvalues of the A matrix. For a CT model all eigenvalues must have strictly negative real parts, and for DT models they must have magnitudes strictly less than unity [1]. Other properties deal with the ability of inputs (actuators) to affect the states of the system, *controllability*, and of outputs (measurements) to monitor variations of the states, *observability*. These system properties, although mathematical in nature, have important physical implications. Specifically, an uncontrollable model would suggest poorly designed actuators, requiring relocation of, or additional actuators, whereas an unobservable system indicates a similar need with regard to sensors.

5.4.1 Controllability

By "controlling" a plant we mean to use the system inputs, specifying their time variations, to obtain some desired response. For simplicity, we will only consider DT models here, but the same results, i.e., the identical tests, also apply to CT models. Thus, assuming the given DT system model:

$$\begin{aligned} x(t+1) &= \boldsymbol{A}x(t) + \boldsymbol{B}u(t) \\ y(t) &= \boldsymbol{C}x(t) + \boldsymbol{D}u(t) \end{aligned} \quad (5.26)$$

Algorithms Used in Control Systems

we first note that the solution can be found by direct recursion from a given initial state $x(0)$, as follows (the input sequence $u(t)$ is, of course, known):

$$x(1) = Ax(0) + Bu(0)$$
$$x(2) = Ax(1) + Bu(1) = A^2 x(0) + ABu(0) + Bu(1) \quad (5.27)$$

Continuing this recursion, the *general solution* for the state is

$$x(t) = A^t x(0) + \sum_{i=0}^{t-1} A^{t-i-1} Bu(i) \quad (5.28)$$

Recognizing that it is the internal state of the system, and not just the output, that is of concern, we make the following definition:

Definition: The model in Eq. (5.26) is (*completely state*) *controllable* if it is possible to force the state from any initial state x_0 to an arbitrary "target" state x_f in a finite number of steps.

A simple rank test for the controllability property is easily derived. Assuming k steps, let us expand Eq. (5.28), recognizing that $x(k) = x_f$ is the state after k steps, and reorganize the terms as follows:

$$x_f - A^k x(0) = \begin{bmatrix} B & AB & A^2 B & \cdots & A^{k-1} B \end{bmatrix} \begin{bmatrix} u(k-1) \\ u(k-2) \\ \cdots \\ u(0) \end{bmatrix} \quad (5.29)$$

Since the left-hand side is arbitrary (because the initial and final states are), the coefficient (partitioned) matrix must have full rank. However, we have not yet determined the integer k. Is it possible that additional terms will continue to generate linearly independent columns as k increases? The answer is provided by the Cayley–Hamilton Theorem of matrix algebra [1], which indicates that beyond $k = n$ (where n is the order of the system) there is no new information. (Since A satisfies its own characteristic equation, it follows that A^n can be expressed in terms of A^{n-1} and lesser powers.) For this special case we define the *controllability matrix* for the DT state space model in Eq. (5.26) as

$$Q_c = \begin{bmatrix} B & AB & \cdots & A^{n-1} B \end{bmatrix} \quad (5.30)$$

Controllability Test

The system of Eq. (5.26) is *controllable* if and only if its controllability matrix Q_c given in Eq. (5.30) has rank n, where n is the order of the system.

Controllability is an inherent structural property of the system so that equivalent models will exhibit the same property. Further, this property is crucial to any control method, since without it not all states can be "guided" by input manipulation. Unfortunately, the question of controllability gives rise to a *yes* or *no* answer and does not directly indicate the "degree of controllability," a measure of how close the system is to being uncontrollable.

5.4.2 Observability

As in the previous discussion, the DT model of Eq. (5.26) will be assumed to represent the system under consideration accurately. The property of *observability* is a fundamental property of systems related to how the measurements (outputs) interact with the system states. It has been shown that the simple problem of identifying the initial state $x(0)$ by observing a finite number of outputs is equivalent to knowing that the complete state information is transmitted to the outputs. From Eqs. (5.26) and (5.28) we can write the general solution for the output vector as

$$y(t) = CA^t x(0) + \sum_{i=0}^{t-1} CA^{t-i-1} Bu(i) + Du(t) \qquad (5.31)$$

Only the first term, the zero-input response, need be considered since the inputs are known, i.e. the remaining terms can be precalculated and subtracted from $y(t)$ to simplify the following definition.

Definition: The model in Eq. (5.26) is (*completely state*) *observable* if it is possible to determine the initial state $x(0)$ from a knowledge of $u(t)$ and $y(t)$ over a finite number of time steps.

This definition will be used to develop a simple rank test for the property of observability of a system. Since without loss of generality we can assume that $u(t) = 0$, as discussed above, we can write and arrange $(k+1)$ consecutive measurements as follows:

$$\begin{bmatrix} C \\ CA \\ \cdots \\ CA^{k-1} \end{bmatrix} x(0) = \begin{bmatrix} y(0) \\ y(1) \\ \cdots \\ y(k) \end{bmatrix} \qquad (5.32)$$

We can solve for $x(0)$ given the measurements on the right-hand side of Eq. (5.32) if and only if the n columns of the coefficient matrix on the left are linearly independent. Since the number of linearly independent columns of a matrix is equal to the number of its linearly independent rows, we can collect

Algorithms Used in Control Systems

additional measurements to increase the linearly independent columns, but only up to a certain point. As in the case of the controllability test, the maximal rank of the coefficient matrix is assured when the final partition is CA^{n-1}. For this case the *observability matrix* is defined as follows:

$$Q_o = \begin{bmatrix} C \\ CA \\ \vdots \\ CA^{n-1} \end{bmatrix} \tag{5.33}$$

Observability Test

The system of Eq. (5.26) is *observable* if and only if its observability matrix Q_o given in Eq. (5.33) has rank n, where n is the order of the system.

Like controllability, observability is an intrinsic property of a system. Equivalent state models exhibit identical test results. Again, the question of observability provides a *yes* or *no* answer and does not directly indicate the "degree of observability."

The following example illustrates the two tests.

Example 5.3 (Controllability and Observability of a DT State Space Model)

Consider the given DT model representing a state space model of the form in Eq. (5.26):

$$A = \begin{bmatrix} -1 & 1 & 0 & -1 \\ -1 & -2 & 1 & 1 \\ 1 & 0 & -2 & 1 \\ 0 & 0 & 0 & -1 \end{bmatrix}, B = \begin{bmatrix} 1 \\ 0 \\ -1 \\ 1 \end{bmatrix} \tag{5.34}$$

$$C = \begin{bmatrix} 1 & 2 & 0 & 0 \\ 2 & 1 & 1 & 0 \end{bmatrix}, \quad D = \begin{bmatrix} 1 \\ 0 \end{bmatrix} \tag{5.35}$$

Calculation: Using the methods developed in this section (see the code below), the following matrices were found, indicating both a controllable and observable system. The reciprocal of the condition number is also given.

$$Q_c = \begin{bmatrix} 1 & -2 & 2 & 4 \\ 0 & -1 & 7 & -26 \\ -1 & 4 & -11 & 25 \\ 1 & -1 & 1 & -1 \end{bmatrix} \quad \text{and} \quad r_c = 0.0026$$

$$Q_o = \begin{bmatrix} 1 & 2 & 0 & 0 \\ 2 & 1 & 1 & 0 \\ -3 & -3 & 2 & 1 \\ -2 & 0 & -1 & 0 \\ 8 & 3 & -7 & 1 \\ 1 & -2 & 2 & 1 \end{bmatrix} \quad \text{and} \quad r_o = 0.0665$$

Computer Code

The function **[Qc,r] = tstcont(*A,B*)** calculates the controllability matrix Q_c of the system in Eq. (5.26) and the reciprocal of its condition number.

```
function [Qc,rc] = tstcont(A,B)
n = size(A,2); m = size(B,2); T = B; I = n-m+1; Qc = [];
for i = 1:I
    Qc = [Qc T]; T = A*T; end
rc = 1/cond(Qc);
```

The function **[Qo,r] = tstobsr(*A,C*)** calculates the observability matrix Q_o of the system in Eq. (5.26) and the reciprocal of its condition number.

```
function [Qo,ro] = tstobsr(A,C)
n = size(A,2); p = size(C,1); T = C; I = n-p+1; Qo = [];
for i = 1:I
    Qo = [Qo; T]; T = T*A; end
ro = 1/cond(Qo);
```

Summary

The properties of controllability and observability of a state space model were defined. Rank tests for each property were developed from the definitions; and, even though the developments were in the context of a DT model, the tests are valid for a CT model as well. As pointed out, one of the drawbacks of these tests is the binary nature of their results. As a means of examining to what extent these properties hold, the reciprocal of the condition number of the controllability (or observability) matrices can be useful. The condition number of a matrix is the ratio of its largest singular value to its smallest. Consequently, a "good" number is near unity, and the condition number of a singular matrix is infinite. To avoid extremely large numbers, the reciprocal is used, in which case very small values indicate a poor numerical "condition."

5.5 CONVERSION TO CANONICAL FORMS

We know that state space models are not unique. A simple change of variables will create a new and, perhaps, unfamiliar looking set of equations which is *equivalent* to the original model in that it represents the same underlying system. In this section we will demonstrate equivalent state space models, with particular emphasis on some special models which are referred to as *canonical*. Canonical forms are uniquely related to the structure of the state space model. Consequently, they are useful in comparing different models; but perhaps of more importance is the property that canonical forms exhibit a "minimal" set of parameters for describing the model. For instance, not all $n \times n$ elements of the A matrix are independent parameters! We will concentrate on two forms: a controllability form and an observability form. Although our developments are in the context of DT models, the reader must keep in mind that the same procedures apply equally well to CT models.

5.5.1 Equivalent State Space Models

Let us begin with a given (DT) model of the form in Eq. (5.26), repeated here:

$$\begin{aligned} x(t+1) &= Ax(t) + Bu(t) \\ y(t) &= Cx(t) + Du(t) \end{aligned} \tag{5.36}$$

which has no particular form other than compatibility of matrix dimensions. Now, let us make a change of state variables using

$$x(t) = Pz(t) \tag{5.37}$$

where the *transformation matrix* P must be non-singular in order to be able to invert the transformation, i.e. return to the x variables. For present purposes we will assume that P is also a constant matrix.

Making the indicated substitutions into Eq. (5.36), we arrive at an *equivalent* state model:

$$\begin{aligned} z(t+1) &= P^{-1}APz(t) + P^{-1}Bu(t) \\ y(t) &= CPz(t) + Du(t) \end{aligned} \tag{5.38}$$

Thus, a new set of matrices now describes the system dynamics. In the next discussion specific transformation matrices will be determined to obtain certain "standard" or *canonical* forms.

5.5.2 Controllability Canonical Forms

Recall the controllability matrix defined in Eq. (5.30),

$$Q_c = \begin{bmatrix} B & AB & \cdots & A^{n-m}B \end{bmatrix} \quad (5.39)$$

Note that the number of partitions has been modified to reflect the fact that there are m independent inputs. Since A is an $n \times n$ array and B is $n \times m$, Q_c is $n \times m(n - m + 1)$. The procedure that we will follow is to select n linearly independent columns from Q_c for the construction of the transformation matrix P in Eq. (5.38).

Since it is reasonable to assume that the m inputs are independent, we will always select the columns of the B matrix, the remaining $(n - m)$ columns coming from the other partitions. A convenient visual aid is to create a "table" (called a *crate*), each entry of which corresponds to a particular column of the controllability matrix [4]. Figure 5.5 illustrates this idea for a system of order $n = 6$ with $m = 3$ inputs, i.e. having a (6×6) A matrix and a (6×3) B matrix (3 columns).

As is the convention, the columns of the controllability matrix are selected as "chains" of vectors corresponding to a particular input. In other words, say for the ith input, the chosen columns would be of the form

$$\{ b_i \quad Ab_i \quad A^2 b_i \quad \cdots \quad A^{u-1} b_i \} \quad (5.40)$$

This corresponds to selecting those "blocks" of the crate diagram of Fig. 5.5 which are below b_i, extending down to the A^{u-1} row. We will refer to such a collection of vectors as a *chain*, and selection will be designated by placing a "1" in the corresponding blocks. To emphasize the termination of a chain, the block following a chain will contain a "0." Finally, this information can be summarized by m integers representing the lengths of the chains for each input, i.e. $u = [u_1, u_2, \ldots, u_m]$. Clearly, the only constraint is that the selected

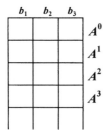

Figure 5.5 Crate diagram for a system of order $n = 6$ with $m = 3$ inputs. Each entry in the crate corresponds to a particular column in the controllability matrix.

vectors are linearly independent, since they will be used as columns of a transformation matrix **P**.

When a transformation matrix **P** is constructed as described above, the equivalent model (5.38) is called a controllable form (corresponding to the "indices" **u**). The **A** and **B** matrices of such a canonical form have a predictable structure. The details can be found in the references [1, 2, 4]. The following example illustrates the method of converting to a controllable form.

Example 5.4 (Conversion to Controllable Canonical Forms)

A pair of matrices, **a** and **b**, was created using random values (between 0 and 1). Two controllable forms were then constructed using different sets of indices: [3, 1] and [2, 2]. Only the first equation of Eq. (5.38) is considered since the second has no particular structure with this form.

Original model matrices:

$$a = \begin{bmatrix} 0.70 & 0.05 & 0.76 & 0.98 \\ 0.91 & 0.74 & 0.99 & 0.72 \\ 0.76 & 0.33 & 0.37 & 0.75 \\ 0.26 & 0.63 & 0.25 & 0.65 \end{bmatrix}, b = \begin{bmatrix} 0.07 & 0.44 \\ 0.63 & 0.77 \\ 0.88 & 0.48 \\ 0.27 & 0.24 \end{bmatrix}$$

Matrices of controllable form with **u** = [3, 1]:

$$A = \begin{bmatrix} 0 & 0.32 & 0 & -0.88 \\ 0 & -0.45 & 0 & 1.79 \\ 1 & 1.05 & 0 & -1.69 \\ 0 & 0.02 & 1 & 2.90 \end{bmatrix}, B = \begin{bmatrix} 1 & 0 \\ 0 & 1 \\ 0 & 0 \\ 0 & 0 \end{bmatrix}$$

Matrices of controllable form with **u** = [2, 2]:

$$A = \begin{bmatrix} 0 & 0 & -17.6 & -19.4 \\ 0 & 0 & 24.8 & 27.4 \\ 1 & 0 & -58.5 & -64.2 \\ 0 & 1 & 55.7 & 61.0 \end{bmatrix}, B = \begin{bmatrix} 1 & 0 \\ 0 & 1 \\ 0 & 0 \\ 0 & 0 \end{bmatrix}$$

Computer Code

(For first conversion.)

```
a=rand(4,4); b=rand(4,2); [q,r]=tstcont(a,b); u=[3 1];
p=q(:,[1 2 3 5]); A=inv(p)*a*p; B=inv(p)*b;
```

5.5.3 Observability Canonical Forms

In a similar manner to the last section we will develop observability canonical forms. (This "similarity" can be formalized as the concept of *duality*). We begin with a slight modification of the observability matrix defined in Eq. (5.33),

$$Q_o = \begin{bmatrix} C \\ CA \\ \cdots \\ CA^{n-p} \end{bmatrix} \quad (5.41)$$

As before the number of partition has been modified to reflect the fact that there are p independent output measurements. Since A is an $n \times n$ array and C is $p \times n$, Q_o is $p(n - p + 1) \times n$. The procedure that we will follow is to select n linearly independent *rows* from Q_o for the construction of the transformation matrix P^{-1} in Eq. (5.38). Note that the role of the transformation matrix is reversed, i.e., "inverted:"

$$x(t) = P^{-1} z(t) \quad (5.42)$$

We will always include the rows of the C matrix, since the p outputs are (assumed to be) independent; the remaining $(n - p)$ rows coming from the other partitions. Again, the selected rows of the observability matrix can be represented as in Fig. 5.5. The only change necessary is that the B matrix columns b_i are replaced by the rows c_i of the C matrix. The same idea of chains of vectors (in this case *row* vectors) holds. As before, this information can be summarized by p integers representing the lengths of the chains for each output, i.e., $v = [v_1, v_2, \cdots, v_p]$. The selected (row) vectors must be linearly independent, since they will be used as rows of the inverse of the transformation matrix P.

When a transformation matrix P is constructed as described above, the equivalent model (5.38) is called an observable form (corresponding to the "indices" v). The A and C matrices of such a canonical form have a predictable structure. The details can be found in the references [1, 2, 4]. The following example illustrates the method of converting to an observable form.

Example 5.5 (Conversion to Observable Canonical Forms)

The pair of matrices a and c, was generated randomly as in Example 5.4. Two different observable forms, corresponding to sets of indices [3, 1] and [2, 2], are constructed. Again, only the A and C matrices are shown since the other matrices have no particular structure.

Original model matrices:

Algorithms Used in Control Systems 243

$$a = \begin{bmatrix} 0.05 & 0.13 & 0.63 & 0.89 \\ 0.76 & 0.02 & 0.74 & 0.23 \\ 0.77 & 0.69 & 0.73 & 0.31 \\ 0.83 & 0.87 & 1.00 & 0.35 \end{bmatrix}, c = \begin{bmatrix} 0.51 & 0.85 & 0.84 & 0.42 \\ 0.59 & 0.41 & 0.27 & 0.54 \end{bmatrix}$$

Matrices of observable form with $v = [3, 1]$:

$$A = \begin{bmatrix} 0 & 0 & 1 & 0 \\ -0.22 & -0.16 & -0.24 & 0.49 \\ 0 & 0 & 0 & 1 \\ 0.78 & 0.13 & 1.73 & 1.30 \end{bmatrix}, C = \begin{bmatrix} 1 & 0 \\ 0 & 1 \\ 0 & 0 \\ 0 & 0 \end{bmatrix}^T$$

Matrices of observable form with $v = [2, 2]$:

$$A = \begin{bmatrix} 0 & 0 & 1 & 0 \\ 0 & 0 & 0 & 1 \\ 0.45 & 0.33 & 0.48 & 2.03 \\ 0.57 & 0.20 & 0.82 & 0.66 \end{bmatrix}, C = \begin{bmatrix} 1 & 0 \\ 0 & 1 \\ 0 & 0 \\ 0 & 0 \end{bmatrix}^T$$

Computer Code
(for first conversion)

```
a=rand(4,4); c=rand(2,4); [q,r]=tstobsr(a,c); v=[3 1];
p=q([1 2 3 5],:); A=p*a*inv(p); C=c*inv(p);
```

Summary

A general method of constructing controllable and observable forms was developed and illustrated. These standard forms are uniquely related to the given system and the specified structure, which is described by controllability (or observability) indices. They are useful when comparing two different state space models. In addition, their property of exhibiting a minimal number of parameters can be important for problems in system identification.

5.6 INTERMODEL CONVERSION

There are several different methods of describing a linear time-invariant system. This chapter emphasizes state space models, but others are equally useful, making it necessary to have a means of conversion between two different types. In this section we will investigate such conversions involving state space models, transfer function matrices, and a closely related variation of transfer function matrices, called matrix fraction descriptions.

5.6.1 Transfer Function Matrices

State space models, which were introduced in the early 1960s, are *time domain* models in that they describe a dynamic system either in terms of differential or difference equations. Prior to the advent of inexpensive digital computation, it was much easier to work with *frequency domain* models, which are Fourier or Laplace transformed variations of the time domain equations. Frequency domain models have been in use since the 1940s, and from experience provide a great deal of insight into the behavior of a system. The insight gained from transfer functions is, however, severely reduced as the systems become more complex, such as with many input and output variables.

Given a system defined with an input vector $u(t)$ and output vector $y(t)$, the transfer function matrix is the $p \times m$ matrix $G(s)$ of rational functions in s which satisfies

$$y(s) = G(s) u(s) \qquad (5.43)$$

where $u(s)$ and $y(s)$ represent (component by component) Laplace transforms of $u(t)$ and $y(t)$, respectively. (This misuse of the functional notation is, nevertheless, convenient. Misinterpretation can be avoided with careful use of the argument notation.)

Conversion from state space to transfer function matrix: beginning with the CT model given in Eq. (5.8),

$$\begin{aligned}\dot{x}(t) &= Ax(t) + Bu(t) \\ y(t) &= Cx(t) + Du(t)\end{aligned} \qquad (5.44)$$

the transfer function matrix of Eq. (5.43) is obtained by taking Laplace transforms of these equations and solving for $y(s)$ in terms of $u(s)$:

$$\begin{aligned}sx(s) &= Ax(s) + Bu(s) \\ y(s) &= Cx(s) + Du(s)\end{aligned} \qquad (5.45)$$

With some simple algebraic manipulations we arrive at

$$y(s) = [C(sI - A)^{-1}B + D]u(s) \qquad (5.46)$$

For multiple input, multiple output (MIMO) systems there are several options for representing the transfer function matrix $G(s)$, shown in the brackets of Eq. (5.46). We will consider two representations, which we will call *transfer function* (TF) and *matrix fraction description* (MFD).

From Eq. (5.46) it is clear that $a(s) = \det(s\mathbf{I} - \mathbf{A})$ will be the common denominator of the matrix $\mathbf{G}(s)$, in which case the "numerator" will be a matrix polynomial. Specifically, we can write

$$\mathbf{G}(s) = \frac{\mathbf{W}(s)}{a(s)} \tag{5.47}$$

where $\mathbf{W}(s)$ is a $p \times m$ matrix of polynomials in s. One way of representing $\mathbf{W}(s)$ is as a matrix polynomial where the coefficients are $p \times m$ constant matrices, and another is as an array of individual (scalar) polynomials. We will illustrate these options in a later example. There are several algorithms that can be used for state space to transfer function conversion. Even though prone to accumulation of errors, due to its recursive nature, the Leverrier algorithm is, perhaps, the best known; others are available in the literature [2].

Leverrier's Algorithm: the *resolvent matrix*, $(s\mathbf{I} - \mathbf{A})^{-1}$, is the most difficult calculation of Eq. (5.46). Since an inverse matrix is equal to its adjoint matrix divided by its determinant, we can write

$$\text{adj}(s\mathbf{I} - \mathbf{A}) = \det(s\mathbf{I} - \mathbf{A}) = a(s)\mathbf{I} \tag{5.48}$$

where the *characteristic polynomial*, $a(s)$ is of the form

$$a(s) = s^n + a_{n-1}s^{n-1} + a_{n-2}s^{n-2} + \ldots + a_1 s + a_0 \tag{5.49}$$

The adjoint matrix can be expanded as

$$\text{adj}(s\mathbf{I} - \mathbf{A}) = \mathbf{I}s^{n-1} + (\mathbf{A} - a_1\mathbf{I})s^{n-2} + (\mathbf{A}^2 - a_1\mathbf{A} - a_2\mathbf{I})s^{n-3} \\ + \cdots + (\mathbf{A}^{n-1} - a_1\mathbf{A}^{n-2} - \cdots - a_{n-1}\mathbf{I}) \tag{5.50}$$

To see that the expansion (5.50) is valid, the reader should take time to multiply Eq. (5.50) by $(s\mathbf{I} - \mathbf{A})$, thereby checking Eq. (5.48). Note that the Cayley–Hamilton Theorem requires that a (square) matrix satisfy its own characteristic equation, i.e. $a(\mathbf{A}) = 0$. Let us formally write

$$(s\mathbf{I} - \mathbf{A})^{-1} = \frac{1}{a(s)}(\mathbf{R}_1 s^{n-1} + \mathbf{R}_2 s^{n-2} + \cdots + \mathbf{R}_n) \tag{5.51}$$

Since the $\mathbf{R}$ coefficients in Eq. (5.51) should match up with the corresponding coefficients in Eq. (5.50), we observe that

$$\mathbf{R}_1 = \mathbf{I}, \mathbf{R}_2 = \mathbf{R}_1\mathbf{A} - a_1\mathbf{I}, \mathbf{R}_3 = \mathbf{R}_2\mathbf{A} - a_2\mathbf{I}, \ldots, \mathbf{R}_n = \mathbf{R}_{n-1}\mathbf{A} - a_{n-1}\mathbf{I} \tag{5.52}$$

Leverrier's algorithm is the following recursion process:

$R_1 = I$, $a_1 = -\text{trace}(A)$
$R_2 = R_1A + a_1I$, $a_2 = -\text{trace}(R_2A)/2$
$\vdots$
$R_n = R_{n-1}A + a_{n-1}I$, $a_n = -\text{trace}(R_nA)/n$

As a test of accuracy, $R_nA - a_nI$ should be a zero matrix.

Example 5.6 (Conversion From State Space to Transfer Function Matrix [2])

Given the state space model:

$$A = \begin{bmatrix} 0 & 1 & 0 \\ 0 & 0 & 1 \\ -2 & -4 & -3 \end{bmatrix}, B = \begin{bmatrix} 0 & 1 \\ 1 & 0 \\ 1 & 0 \end{bmatrix}$$

$$C = \begin{bmatrix} 1 & 0 & 0 \\ 0 & 0 & 2 \end{bmatrix}$$

Solution: Applying the Leverrier algorithm using the MATLAB code at the end of this section, we obtain

$$W = \begin{bmatrix} 1 & 0 & 3 & 1 & 5 & 3 & 6 & 4 \\ 0 & 0 & 2 & 0 & -8 & -4 & -4 & 0 \end{bmatrix}, a = \begin{bmatrix} 1 & 3 & 4 & 2 \end{bmatrix}$$

which we directly interpret as

$$G(s) = \frac{\begin{bmatrix} 1 & 0 \\ 0 & 0 \end{bmatrix} s^3 + \begin{bmatrix} 3 & 1 \\ 2 & 0 \end{bmatrix} s^2 + \begin{bmatrix} 5 & 3 \\ -8 & -4 \end{bmatrix} s + \begin{bmatrix} 6 & 4 \\ -4 & 0 \end{bmatrix}}{s^3 + 3s^2 + 4s + 2}$$

or alternatively, the numerator may be written as

$$W(s) = \begin{bmatrix} s^3 + 3s^2 + 5s + 6 & s^2 + 3s + 4 \\ 2s^2 - 8s - 4 & -4s \end{bmatrix} \tag{5.53}$$

Computer Code

```
function [W,a] = lever(A,B,C,D)
[n,m] = size(B); p = size(C,1); I = eye(n); a = -trace(A);
R = I; t = a; T = R; W = [D (C*R*B+t*D)];
for i = 2:n, T = T*A + t*I; t = -trace(T*A)/i; a = [a t]; R = [R T];
W = [W (C*T*B+t*D)]; i = i+1; end
a = [1 a];
```

Algorithms Used in Control Systems

5.6.2 Conversion from Transfer Function Matrix to State Space

This conversion will be accomplished in two steps. First, the conversion is made to a non-minimal state space model; then the second step is to use the Jordan canonical form to obtain a minimal-order model. The first step requires that the transfer function matrix be strictly proper, i.e., the equivalent of a zero D matrix state model.

In Eq. (5.47) the numerator expression can be written as

$$W(z) = \sum_{i=0}^{n} W_i z^i \tag{5.54}$$

where $W_n = D$. Since the denominator coefficients in Eq. (5.49) are known, the strictly proper transfer function is obtained by removing the effect of the D matrix. This can be done by eliminating W_n and modifying the remaining W matrices as follows

$$W_i := W_i - a_i W_n \tag{5.55}$$

for $i = [0, n-1]$, using the coefficients as shown in Eq. (5.49).

We will use a conversion to a controllable model as the first step. Assuming that the system is nth order with m inputs and p outputs, the matrices are established as follows. The A and B matrices are block diagonal with m blocks of the form:

$$a = \begin{bmatrix} 0 & 1 & \cdots & 0 \\ 0 & 0 & \ddots & 0 \\ \vdots & \vdots & \ddots & \vdots \\ -a_0 & -a_1 & \cdots & -a_{n-1} \end{bmatrix}, \quad b = \begin{bmatrix} 0 \\ 0 \\ \vdots \\ 1 \end{bmatrix} \tag{5.56}$$

i.e. A is $nm \times nm$ and B is $nm \times m$. The C matrix contains the coefficients of W in the form of the coefficients of the polynomials in Eq. (5.53). For example, using Example 5.6, and eliminating the effect of the D matrix, the C matrix, representing the polynomials:

$s + 4$, $s^2 + 3s + 4$, $2s^2 - 8s - 4$, and $-4s$

would be

$$C = \begin{bmatrix} (0 & 1 & 4) & (1 & 3 & 4) \\ (2 & -8 & -4) & (0 & -4 & 0) \end{bmatrix} \tag{5.57}$$

The C matrix is, therefore, a $p \times nm$ array. Of course, the system must be reduced from this nmth order system to an nth order system using, e.g., the Jordan form to eliminate non-controllable and non-observable models, or the MATLAB built in function "minreal." Additional details and more approaches can be found in [2].

Summary

In this section some different methods to describe a system model mathematically were presented along with some discussion of how algorithms perform the computation to convert between different forms. Definitions and structural forms were given as background for the algorithms. Although other forms are used for system representation, our discussion was limited to methods of conversion between state space models and transfer function matrices. An example was used to illustrate these techniques.

5.7 TIME-DOMAIN RESPONSE CALCULATIONS

It is important to be able to check any system design that might be derived. Computer-aided design techniques rely a great deal on the accurate simulation of a system to certain "test" inputs. If we have a CT model with given functions (either specified analytically or in tabular form), the material in Sect. 5.2 can be used to convert to a DT model. Once the problem is in the DT model form, the time domain solution is readily carried out using the natural recursion of the DT state space model. Recall Eq. (5.12), repeated here:

$$x(t+1) = Ax(t) + Bu(t)$$
$$y(t) = Cx(t) + Du(t) \tag{5.58}$$

The argument t is in "normalized time," i.e., integer values which represent multiples of uniform sample intervals. The recursive solution is easily derived. Starting with a given "initial state vector" $x(0)$, and knowing the input function values $u(t)$ for $t = [0, t_f]$, where t_f is the final time of the solution interval, we have that

$$x(1) = Ax(0) + Bu(0)$$
$$x(2) = Ax(1) + Bu(1) = A^2 x(0) + ABu(0) + Bu(1)$$
$$\vdots \tag{5.59}$$
$$x(t) = A^t x(0) + \sum_{n=0}^{t-1} A^{t-n-1} Bu(n)$$

Finally, the output can be written as

Algorithms Used in Control Systems

$$y(t) = CA^t x(0) + \sum_{n=0}^{t-1} CA^{t-n-1} Bu(n) + Du(t) \tag{5.60}$$

After the model matrices, initial state vector and input vectors have been established, finding $y(t)$ is just a matter of programming Eq. (5.60).

Example 5.7 (Simulated System Response)

For the DT system below, we will calculate the unit step resonse:

$$A = \begin{bmatrix} 0 & 1 & 0 \\ 0 & 0 & 1 \\ 0.01 & -0.17 & 0.80 \end{bmatrix}, B = \begin{bmatrix} 0 \\ 0 \\ 1 \end{bmatrix}, x(0) = \begin{bmatrix} 1 \\ 1 \\ 1 \end{bmatrix}$$

$$C = [0.1 \quad 0.2 \quad 0.3] \quad D = [0]$$

Figure 5.6 illustrates the resulting step response output. The following function provides the state and output history of a DT system given the state space model $\{A, B, C, D\}$, the initial state $x(0)$ and the input history using

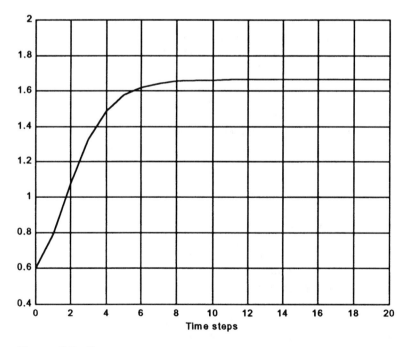

Figure 5.6 Step response output.

simple recursion. The input *u* is assumed to be an $m \times N$ array representing the time sequence of the input(s), i.e. $u(0), u(1), \ldots, u(N-1)$.

Computer Code

```
function [x,y] = response(A,B,C,D,x0,u)
x = x0; y = C*x + D*u(:,1);
for k = 1:length(u)-1
    x = [x A*x(:,k) + B*u(:,k)]; y = [y C*x(:,k+1) + D*u(:,k+1)]; end
```

Summary

In this section the generation of a time domain response from a DT system was explained. The method also applies to CT systems if the CT to DT conversion of Sect. 5.2 is first performed. A simple example and the corresponding computer code were also given.

5.8 ROOT LOCUS CALCULATIONS

Often in the design of control systems an important single parameter must be set, for example, the forward-path gain of the system (following some initial design). For this situation a technique is available to determine the locus of possible *closed-loop* system poles (eigenvalues) as a function of the parameter.

This method, called the (*Evans*) *root-locus* method, is based on the closed-loop characteristic equation, including the variable parameter which we will denote as K. The characteristic polynomial is the denominator of the corresponding transfer function matrix, as indicated in Eq. (5.47). For a closed-loop system some feedback loop (or combination of feedback loops) must already be established. Having this structure, the closed-loop characteristic equation can then be written as

$$a(z) = D(z) + KN(z) = 0 \qquad (5.61)$$

where those terms of $a(z)$ which do not depend on K are collected together in the polynomial $D(z)$ and those which have K as a factor, in $N(z)$. Here we are using the z notation which is conventional for DT systems, but a simple change of notation from z to s would make the method applicable to CT systems, i.e., there is no difference in the technique, only in the interpretation. Finally, the *root-locus equation* is obtained by dividing Eq. (5.61) by $D(z)$:

Algorithms Used in Control Systems

$$1 + K\frac{N(z)}{D(z)} = 0 \tag{5.62}$$

thus creating an equivalent (unity feedback) "open-loop function" $G(z) = N(z)/D(z)$ which is a single input, single output transfer function (even though the original system might be a multiple input, multiple output system). The key idea is to "construct" the locus of the roots of Eq. (5.61), i.e., the closed-loop poles, from knowledge of the *zeros* (roots of $N(z) = 0$) and the *poles* (roots of $D(z) = 0$) of $G(z)$. In the early days of control system design, going back to the 1950s, several "rules of construction" were developed and used to construct the root locus (RL). With the increased computational power available to present day designers these rules have, to a large extent, become obsolete.

We will approach RL construction from the numerical side by calculating the roots for a sequence of values for K. Typically, the range of K is from zero to some upper limit. The number and spacing of the values of K, as well as this range, depend on what the designer is interested in. For example, there would be no interest in that part of the RL for which the system is unstable, i.e., where one or more segments of the locus lie in the right-hand half of the *s*-plane.

Assuming that a polynomial root finding algorithm is available, which is the case in most utility software, all that is necessary to implement the root locus plot for a specific system is to specify the $G(s)$ discussed previously, the desired range of K and call the root-finding algorithm for each value of K. A simple example will be presented to illustrate the method, using MATLAB's root finding capability (even though MATLAB also has a root locus function also).

Example 5.8 (Root Locus Calculation)

The DT system of Example 5.7, repeated here, represents $G(z)$ in a unity (negative) feedback configuration. Hence Eq. (5.62) holds with $G(z)$ as the transfer function of the following state space model:

$$A = \begin{bmatrix} 0 & 1 & 0 \\ 0 & 0 & 1 \\ 0.01 & -0.17 & 0.80 \end{bmatrix}, B = \begin{bmatrix} 0 \\ 0 \\ 1 \end{bmatrix}$$

$$C = \begin{bmatrix} 0.1 & 0.2 & 0.3 \end{bmatrix}, D = \begin{bmatrix} 0 \end{bmatrix}$$

To illustrate the root locus method, a plot of the closed-loop roots is presented in Figure 5.7 for K varying from 0 to 10 in unit increments. The increments between pole locations can vary quite nonlinearly with K. Figure 5.8 shows how connecting the poles together can aid in the visualization, by

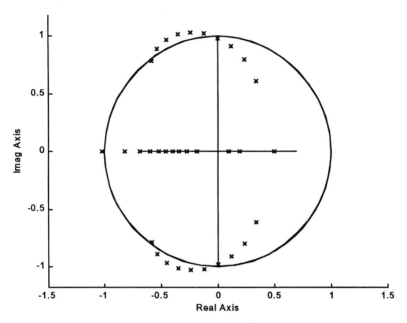

Figure 5.7 Plot of closed-loop poles for *K* varying from 0 to 10 unit increments.

creating a continuous locus for each closed-loop pole; this figure is a standard function in MATLAB, called "rlocus." The unit circle is included in both figures for reference.

Computer Code

```
[n,d] = ss2tf(A,B,C,D); figure(1); hold on;
for k = 0:10, p = d + k*n; P = roots(p); pzmap(1,p); end
figure(2); rlocus(n,d); zgrid(0,0)
```

Summary

A brief description of the root-locus method was given in this section to illustrate how a root-locus plot can provide insight into the closed-loop system response as a function of a single parameter. This method, while extremely important in past decades, has been made virtually obsolete by (other) computer-aided design techniques, one of which is discussed in the next section.

Algorithms Used in Control Systems

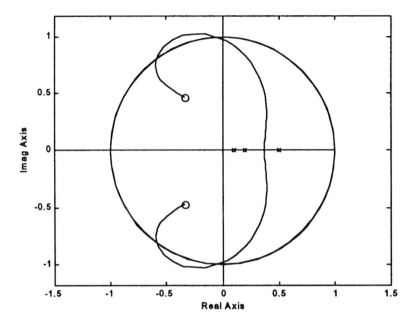

Figure 5.8 Connecting the poles creates a continuous locus for each closed-loop pole, called a "root locus."

5.9 STATE FEEDBACK DESIGN (POLE PLACEMENT)

A standard design method of modern control theory is state (variable) feedback. This method assumes that the plant has an accurately known (linear) model in state space form, and that all variables in the state vector are "measurable." Let us assume that the plant model is given by the following state space equations, repeated here from Eq. (5.12)

$$x(t+1) = Ax(t) + Bu(t)$$
$$y(t) = Cx(t) + Du(t) \quad (5.63)$$

State feedback is defined as the generation of the (control) inputs(s) $u(t)$ as a function of the state variable measurements, i.e.,

$$u(t) = -Kx(t) + Gr(t) \quad (5.64)$$

The matrix K is the *feedback gain matrix*. The second term can be used as an external reference signal for generality. Introducing Eq. (5.64) into the first equation of Eqs. (5.63), we obtain

$$x(t+1) = (A - BK)\, x(t) + BGr(t) \tag{5.65}$$

Given that the plant model of Eqs. (5.63) is controllable (see Sect. 5.3), a key result of control theory says that for *any* desired set of eigenvalues of $(A - BK)$, there exists a (real-valued) matrix K which makes it so [1–4]. For simplicity of development we will work with only a single input system. (The number of outputs is arbitrary.)

Knowing that any controllable state space model can be converted into a controllable canonical form, we can assume that the A and B matrices have the forms

$$A_c = \begin{bmatrix} 0 & 1 & \cdots & 0 \\ 0 & 0 & \ddots & 0 \\ \vdots & \vdots & \ddots & 1 \\ -a_0 & -a_1 & \cdots & -a_{n-1} \end{bmatrix}, \quad B_c = \begin{bmatrix} 0 \\ 0 \\ \vdots \\ 1 \end{bmatrix} \tag{5.66}$$

which can be obtained by using $P = QQ_c^{-1}$ for a similarity transformation matrix, as in Eq. (5.37), where Q is the controllability matrix for the given system (arbitrary A and B matrices) and Q_c is the controllability matrix for the system having the form given in Eq. (5.66) [3]. Note that Eq. (5.66) can be constructed by knowing (only) the characteristic polynomial of the original A matrix. In the following we will develop Ackermann's formula.

Ackermann's Formula: first, to define some notation, we have the following row vectors:

$$\begin{aligned} K_c &= \begin{bmatrix} k_0 & k_1 & \cdots & k_{n-1} \end{bmatrix} \\ a &= \begin{bmatrix} a_0 & a_1 & \cdots & a_{n-1} \end{bmatrix} \\ d &= \begin{bmatrix} d_0 & d_1 & \cdots & d_{n-1} \end{bmatrix} \end{aligned} \tag{5.67}$$

where K_c is the feedback gain matrix, a is the negative of the array of characteristic polynomial coefficients, and d is the corresponding array of desired values. From the structure of Eq. (5.66) it can be seen that the required gain values to achieve the desired values are

$$K_c = d - a \tag{5.68}$$

Expanding the right-hand side into its components,

$$-K_c = a - d = (a_0 - d_0)e_1 + (a_1 - d_1)e_2 + \cdots + (a_{n-1} - d_{n-1})e_n \tag{5.69}$$

where e_j is the jth unit row vector (i.e., a row of zeros with a single unit in the jth position).

The (open-loop) characteristic equation is

$$a_{\mathrm{OL}}(z) = z^n - a_{n-1}z^{n-1} - a_{n-2}z^{n-2} - \cdots - a_1 z - a_0 \tag{5.70}$$

Algorithms Used in Control Systems

We also know by the Cayley–Hamilton Theorem of matrix theory that a matrix satisfies its own characteristic equation; therefore,

$$A_c^n = -a_0 I - a_1 A_c - \cdots - a_{n-1} A_c^{n-1} \tag{5.71}$$

To complete the development of Ackermann's formula, let us establish the matrix expression

$$a_{CL}(A_c) = A_c^n + d_{n-1} A_c^{n-1} + \cdots + d_1 A_c + d_0 I \tag{5.72}$$

Although this expression does not equal zero, we may substitute for $(A_c)^n$ from Eq. (5.71) and collect terms to obtain

$$-a_{CL}(A_c) = (a_0 - d_0) I + (a_1 - d_1) A_c + \cdots + (a_{n-1} - d_{n-1}) A_c^{n-1} \tag{5.73}$$

It follows from the structure of A_c that by pre-multiplying by the first unit row vector,

$$-e_1 a_{CL}(A_c) = (a_0 - d_0) e_1 + (a_1 - d_1) e_2 + \cdots + (a_{n-1} - d_{n-1}) e_{n-1} \tag{5.74}$$

Note that the "companion" form of A_c gives us the equations:

$$e_1 A_c = e_2;\; e_2 A_c = e_3;\; \cdots e_{n-1} A_c = e_n \tag{5.75}$$

Now let us begin to consolidate some of the previous steps. Comparing Eqs. (5.74) and (5.69), we have

$$K_c = e_1 a_{CL}(A_c) = e_1 a_{CL}(P^{-1} A P) \tag{5.76}$$

Factoring out the transformation matrix P,

$$K_c = e_1 P^{-1} a_{CL}(A) P \tag{5.77}$$

Since the state feedback for the original model is $K = K_c P^{-1}$, where $P = Q Q_c^{-1}$,

$$K = (e_1 Q_c) Q^{-1} a_{CL}(A) \tag{5.78}$$

From Eq. (5.78) we write the final Ackermann's formula for state variable feedback as

$$K = \begin{bmatrix} 0 & \cdots & 0 & 1 \end{bmatrix} \begin{bmatrix} B & AB & \cdots & A^{n-1} B \end{bmatrix}^{-1} a_{CL}(A) \tag{5.79}$$

Equation (5.79) takes advantage of the special structure of Q_c, namely that $e_1 Q_c = e_n$. From Eq. (5.79) the state feedback gains can be calculated using only matrix operations from a given (controllable) system model and knowledge of the desired pole locations, p_j for $j = [1, n]$ so that the desired characteristic polynomial is constructed from

$$a_{CL}(z) = \prod_{j=1}^{n} (z - p_j) = z^n - d_{n-1} z^{n-1} - \cdots - d_1 z - d_0 \tag{5.80}$$

A simple example will be used to illustrate the method.

To remove the "single input" restriction, the input vector of a multiple input system can be represented as

$$u(t) = Tu_s(t) \tag{5.81}$$

where $u_s(t)$ is an effective "single input." For example, if T is a vector of ones, then Eq. (5.81) implies that each input of the system receives the same feedback control signal. On the other hand, the designer can use the components of T to weight the effect of the feedback control signal on the actual system inputs (actuators). More details on this method can be found in [3]. In addition, various related topics, such as "output" feedback (where only a strict subset of the states are measured), are discussed in [2].

Example 5.9 (State Variable Feedback – Ackermann's Formula)

A random DT system is used to represent the plant (entries between 0 and 1):

$$A = \begin{bmatrix} 0.0535 & 0.0077 & 0.4175 \\ 0.5297 & 0.3834 & 0.6868 \\ 0.6711 & 0.0668 & 0.5890 \end{bmatrix} \quad B = \begin{bmatrix} 0.9304 \\ 0.8462 \\ 0.5269 \end{bmatrix}$$

After testing to ensure that the system is controllable, let us specify that the desired closed-loop poles are located at 0.2, 0.4, and 0.6. Thus, the desired characteristic equation is $a_{CL}(z) = z^3 = 1.2z^2 + 0.44z - 0.48$.

Applying Ackermann's formula the feedback gain matrix is calculated to be

$$K = \begin{bmatrix} 0.1272 & -2.0080 & 2.6694 \end{bmatrix} \tag{5.82}$$

which provides the gains to be used on the three (measured) state variables of the given system as in Eq. (5.64). The reference gain G in Eq. (5.64) can be used to adjust the d.c. gain of the closed-loop system.

Computer Code

```
A=rand(3,3); B=rand(3,1); pd=[1 -1.2 .44 -.48];
Pcl=polyvalm(pd,A); [Q,r]=tstcont(A,B); K=inv(Q)*Pcl;
K=K(size(K,1),:); Acl=A-B*K; poly(Acl);
```

Summary

The basic state variable feedback design was discussed in this section. The presentation assumed that the model was only a single input system and Ackermann's formula was developed. The single input assumption is, however, not a constraint for using this method. As discussed above, an *equivalent single input* can be defined for the multiple input system. This method constitutes the basic design technique of state space systems, but the designer must decide in advance what the desired pole locations are. For high-order systems this may not be obvious. One solution in this situation is to choose the pole locations required for Butterworth or Bessel filters, or some combination to achieve the required specifications on e.g. step response [5].

REFERENCES

1. W. L. Brogan (1991), *Modern Control Theory*, 3rd edn, Prentice-Hall, Inc., Englewood Cliffs, NJ.
2. S. Bingulac and H. F. VanLandingham (1993), *Algorithms for Computer-Aided Design of Multivariable Control Systems*, Marcel Dekker, New York.
3. H. F. VanLandingham (1985), *Introduction to Digital Control Systems*, Macmillan, New York.
4. T. Kalaith (1980), *Linear Systems*, Prentice-Hall, Inc., Englewood Cliffs, NJ.
5. L. Weinberg (1962), *Network Analysis and Synthesis*, McGraw-Hill, New York.

6
Computer Engineering

Peter Athanas and Yosef Tirat-Gefen
Virginia Polytechnic Institute and State University, Blacksburg, Virginia

6.1 TRANSMISSION LINES, REFLECTIONS, AND TERMINATIONS

In digital logic design terminology, an *event* is a transition of a signal either from a logic 0 to a logic 1 or from a logic 1 to a logic 0. As a digital signal propagates through a transmission medium (a cable, a trace on a printed circuit board, or a metal wire on an integrated circuit, for example), the signal may become distorted, and interact with other nearby signals. The behavior of the signal can be examined in a similar fashion to a wavefront in traditional transmission line analysis. In the physical implementation of the logic circuit, an event adopts the properties of rise time and fall time. The *length* of an event is defined as the physical distance that would be covered during the transition period of the event, and is calculated by multiplying the event duration by the speed of propagation of the signal in the transmission medium. The length of an event, t_{event}, becomes significant with respect to the length of the transmission medium, t_{wire}, when $t_{event} \leq t_{wire}$. When this condition is true, then it may be necessary to consider the effects of *reflections* on signal transmission. This criterion is becoming ever more significant as clock speeds of digital systems increase to well over 100 MHz.

When a wavefront encounters an abrupt change in the characteristic impedance of the transmission medium or a mismatched termination network, a portion of the wavefront continues to propagate away from the source in the new medium, while the remainder is reflected back towards the wavefront source. The reflected wave obeys the principles of superposition, along with Kirchhoff's voltage and current laws. The magnitude of the reflected wave is determined by the relative magnitudes of the characteristic impedance of the materials at the transmission line boundary. A complete analysis can be found in several books on the topic [1]. A summary of lossless transmission line reflections pertaining to digital systems follows.

The incident voltage wave, denoted as $v_0(t, x)$, travels in the positive x-direction along a wire or cable having a characteristic impedance of Z_0 ohms ($Z_0 = \sqrt{L_0/C_0}$, where L_0 is the inductance per unit length of wire and C_0 is the capacitance per unit length of wire). The velocity of the wave is equal to $1/\sqrt{L_0 C_0}$. The incident wave is accompanied by a current wave, $i_0(t, x)$, equal to $v_0(t, x)/Z_0$. When the incident wave encounters an impedance change to Z_L (such as a mismatched termination), a reflected wave, denoted as $v_r(t, x)$, is formed and propagates in the negative x-direction. The magnitude of the reflected voltage waveform is equal to $v_r(x, t) = \rho v_0(x, t)$, where ρ is referred to as the *reflection coefficient*, and computed by $\rho = (R_L - Z_0)/(R_L + Z_0)$. Note that the magnitude of ρ is limited to be less than or equal to 1. When $R_L = 0$ (the incident impedance is shorted to ground), $\rho = -1$, and the reflected wave is equal in magnitude but opposite in sign of the incident wave. When $R_L = \infty$ (the incident impedance is an open circuit), the reflected wave equals the incident wave. No reflection occurs when $R_L = Z_0$. If the impedance of the signal source, Z_s, is not equal to Z_0, a wave reflected back to the source will once again reflect, with a reflection coefficient equal to $\rho = (R_s - Z_0)/(R_s + Z_0)$.

Reflections in digital systems can cause false transitions and glitches, which in turn may result in adverse behavior. Consider the following example.

A high-speed computer peripheral is to be driven by a CMOS inverter as shown in Fig. 6.1. The output of the logic gate is connected to the peripheral load through a 10 meter cable with a characteristic impedance of 100 ohms. The peripheral input consists of another CMOS logic gate. Input V_{IN} is driven from 0 to 10 volts in a negligible amount of time at $t = 0$ seconds. The logic for all gates is 5 volts, and the rise/fall times are negligible. The input impedance of the load CMOS gate is to be considered as infinite (essentially an open circuit).

Determine the waveforms for the points X, Y, and Z along the cable for the period from 0 to 600 ns.

Construct a series termination network at the peripheral input for the above circuit.

Construct a parallel termination network at the peripheral input for the above circuit.

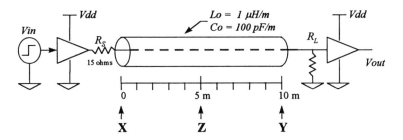

Figure 6.1 Transmission line example.

6.1.1 Solution

In this example, the characteristic impedance of the cable, Z_0, is $\sqrt{1\ \mu\text{H}/100\ \text{pF}} = 100\ \Omega$, and the wavefront velocity is $\sqrt{1\ \mu\text{H}100\ \text{pF}} = 10^8$ m/s. At this rate, it takes $T = 100$ ns for a signal to propagate from end to end of the cable. Since the load gate has essentially an infinite input impedance, the reflection coefficient at the load, ρ_L, is 1. The reflection coefficient at the source, ρ_s, is equal to $(15\ \Omega - 100\ \Omega)/(15\ \Omega + 100\ \Omega) = -0.74$. Initially, the system is at rest, and all node voltages are at 0 volts. After a long time has passed, the steady state voltage at the load end is equal to the voltage source, 10 volts. The magnitude of the initial wave is determined by the voltage division between the source resistance and the cable characteristic impedance: $V_x(0+) = 10\ Z_0/(Z_0 + R_s) = 8.7$ volts. The initial current wave is equal to $V_x(0+)/Z_0 = 87$ mA. This wave crosses point Z at time $0.5T$ seconds, and arrives at point Y at $t = T$ seconds. Since $\rho_L \neq 0$, a reflected wave is produced at the load end of the cable and travels back to the source end. The reflected wave is equal to $\rho_L \times V_r(T-) = 8.7$ volts. The resultant voltage at node Y at time $t = T+$ seconds can be determined using the principle of superposition. There are three voltage components to consider: the initial voltage at node Y, the incident wave voltage, and the reflected wave voltage. For this point, the initial voltage is 0 volts and the incident wave voltage and the reflected wave voltage are both 8.7 volts. Therefore, $V_r(T+)$ is equal to the sum of these three, which is 17.4 volts.

The computation of $V_x(2T)$ proceeds in a similar manner as the computation for $V_r(T)$. The 8.7 volt reflected wave passes node Z at $t = 1.5T$, and reaches the source node at $t = 2T$. A portion of this wave is in turn reflected back to the load end of the cable. The new reflected wave is equal to the incident wave times the source node reflection coefficient, which is equal to $\rho_s \times 8.7$ volts $= -6.43$ volts. Once again, applying the principle of superposition, the resulting voltage, $V_x(2T+)$, is equal to the sum of the initial node X voltage $[V_x(2T-)]$, the incident wave voltage, and the reflected wave voltage: $V_x(2T+) = 8.7 + 8.7 - 6.4 = 11.0$ volts. The reflected wave of -6.43 volts travels to the load end of the cable, where it is once again reflected. This process repeats until a steady state voltage of 10 volts along the cable is asymptotically reached. Figure 6.2(a) illustrates the voltage waveform at node X and Fig. 6.2(b) depicts the voltage waveform at node Y. The signal at node Y is applied to the input of the cable receiver buffer shown on the right-hand side of Fig. 6.1. The response of the receiver

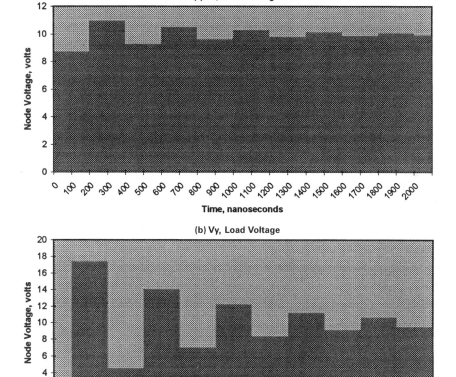

Figure 6.2 (a) Voltage waveform at node X. Node X is at 0 volts before $t = 0$ seconds. (b) The voltage waveform at node Y.

buffer is shown in Fig. 6.3 (the buffer delay is assumed to be negligible). Note that the buffer output rises to a logic 1 after 100 ns, but then returns to a logic 0 after 300 ns for a brief period. This 200 ns glitch is due to the reflection received at node Y at $t = 200$ ns, which brings the node voltage below 5 volts (the threshold of the receiver buffer). In some situations, a glitch like this could cause unwanted logical behavior. Another effect may be observed at the source end of the transmission line. Note in the above example that the voltage at node X occasionally rises above 10 volts and, in one case, to nearly 18 volts for a period of 200 ns. Overshoots such as these could stress the transmission line driver output circuitry. For example, if the cable driver were CMOS, overshoots may introduce parasitic currents, which in turn may lead to failure modes, such as latch-up.

Reflections can be greatly attenutated by the application of an appropriate *termination network*. A termination network is a passive or active circuit that attempts to compensate for mismatches in impedance at transmission line boundaries. There are a number of strategies in deriving termination networks, yet all share a common goal: to modify the boundary reflection coefficient so that it nears zero. This is achieved by changing the effective load resistance to match the transmission line characteristic impedance. Two strategies are illustrated in Fig. 6.4. A common method used to reduce the load impedance is to insert a parallel termination network (labeled R_{TL} in he right-hand shaded region in Fig. 6.4). The values of the

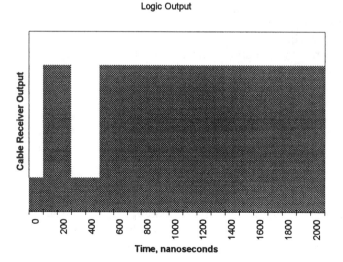

Figure 6.3 Logic transitions on the output of the transmission line receiver.

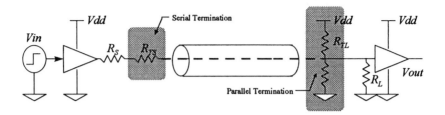

Figure 6.4 Transmission line example illustrating two termination strategies: serial termination and parallel termination.

pull-up resistor and the pull-down resistor are chosen to make the Thévenin resistance of the right-hand circuit equal to Z_0. The two resistors in R_{TL} are chosen so that

$$\frac{1}{R_{TLpull\text{-}up}} + \frac{1}{R_{TLpull\text{-}down}} + \frac{1}{R_L} = \frac{1}{Z_0}$$

In addition, the pull-up resistor value should be chosen large enough not to exceed the current sinking ability (I_{OL}) of the line driver. Likewise, the pull-down resistor should be large enough so that it doesn't draw more current than what the line driver can supply (I_{OH}).

The second method shown is to insert a serial termination network at the source. In Fig. 6.4, this is accomplished by a resistor, R_{TS}, on the output of the line driver. The value of R_{TS} is chosen so that the effective source resistance increases to Z_0; hence $R_{TS} + R_s = Z_0$. When this is accomplished, the source reflection coefficient becomes zero.

Note that the parallel termination network will consume d.c. power and may impose fan-out restrictions on the line driver. The serial termination network has the advantage that it consumes no power during steady state conditions; however, it has the disadvantage that the source node will not stabilize for a period of $2T$ seconds, which may be detrimental in some instances.

6.2 COMPUTER ORGANIZATION

Broadly speaking, the building blocks within a microprocessor can be categorized as either control elements or data path elements. Data path elements provide a means of transferring data within the processor (through busses), providing intermediate storage (in registers), and performing arithmetic and logic operations (ALUs and pipeline stages). The control element provides a means of interpreting instructions, sequencing data movement, scheduling coordination with past and future instructions. Data path functionality is often expressed in terms of register transfer language (RTL) or by graphical flow diagrams. Control aspects of a processor are often best expressed in terms of state diagrams and Boolean equations. The boundary between these units is not black and white; there are microprocessor functions that are not so readily classified (such as discrete event and property flags, often associated with the ALU of a microprocessor).

There are three primary levels of abstraction used to describe the functionality of a microprocessor. The highest level is the *instruction set architecture* (ISA). The ISA provides a complete behavioral view of the instruction repertoire and the processor resources that are visible to the user (or compiler writer). Because of this, the instruction set architecture is sometimes referred to as the *programmer's model*. The most popular ISA definition to date is the Intel ×86 architecture [2]. In the case of the Intel ×86, the ISA of the latest generation processor has been intentionally created to be a superset of the first generation ×86 machine (the 8086 or, to some extent, the 8080).

A given ISA can often be satisfied by a number of *organizations*, which is the next lower level in the abstraction hierarchy. The organization of a computer, often depicted graphically, denotes the arrangement and interconnectivity (the number of internal busses) of processor internals. A given organization of an ISA often determines the temporal properties of the machine (how many clock cycles are required by each instruction, the degree of pipelining and latency, and input/output data movement capabilities).

The last layer in the abstraction hierarchy usually relies somewhat on a given fabrication technology and is referred to as the *physical design* or *implementation* of the processor. Like the many-to-one relationship between the ISA and the organization, the organization of a processor can be satisfied by several possible implementations. Decisions made at the implementation level determine the processor clock speed, the physical size, power dissipation, and cost.

In the remainder of this section, an illustrative example will be used to expound upon the relationship between the control and data implementation of the organization of a very simple accumulator-based processor (shown in Fig. 6.5). This processor is organized using a single bus (labeled A-BUS in the figure), which is shared with all of the resources within the processor. All of the rectangular objects in the figure denote registers that, when selected by the control unit, capture data from the A-BUS on the rising edge of a clock signal (which is not explicit in Fig. 6.5, yet is implied). When directed by the controller, the registers can transfer data onto the A-BUS by using tri-state drivers or multiplexors. The ISA for this processor specifies

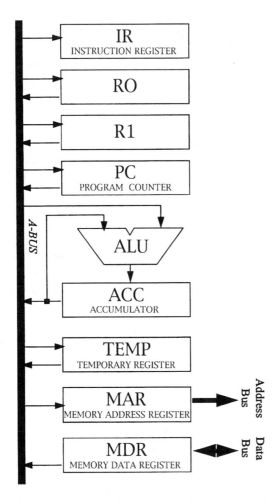

Figure 6.5 Organization of a simple single-bus processor.

the existence of an accumulator (which registers the output of the ALU) and two general-purpose registers, R0 and R1. The instruction register (IR) contains the current instruction being executed, and the program counter contains an address that points to the next instruction (in memory) to be executed. Two memory support registers are also explicit in this organization: MDR and MAR. MAR (the memory address register) contains the address of the current location being referenced in memory. The MDR (memory data register) serves two functions in this organization. When data are transferred from the A-BUS to the MDR, the processor initiates a *memory write cycle*, where the new data in MDR are moved to the location specified by MAR. When a transfer is requested from MDR to the A-BUS, a *memory read cycle* is initiated (data from the location referenced by MAR are transferred to MDR and A-BUS). The register labeled TEMP is not part of the ISA specification, but is provided in the organization to facilitate the execution of ISA-specified instructions.

The ALU is assumed to be built with combinational gates only; hence it contains no latches or flip-flops. The ALU in this example is assumed to be capable of performing the functions ADD A B, INCREMENT A, NOT A, and PASS A. The B input of the ALU is fixed to the accumulator output. The ALU is assumed to have control inputs that determine the function which emanates from the control unit.

Given the above processor organization, construct the sequence of events to form the following instructions, which are assumed to be part of the instruction set architecture:

(a) add r1; add the contents of register R1 to the accumulator
(b) sub r0; subtract the accumulator from R0, and store the result into the accumulator
(c) branch r0; unconditionally branch to the location indexed by r0

6.2.1 Solution

When implementing the behavior for each of these instructions, it is notable to mention that in addition to the instruction behavior, each instruction is responsible for incrementing the program counter and fetching the next scheduled instruction. Since this simple architecture is fairly restrictive, several clock cycles may be necessary to fully execute each instruction. RTL notation will be used to describe data movement within the processor. The transfer operator, ←, implies that the data source on the right-hand side of the operator drives onto the A-BUS at the beginning of the cycle, and the destination captures the data at the cycle's end. To eliminate contention, only one source may drive a bus at a time.

The table below summarizes the sequence of events for the ADD instruction. Note that five clock cycles are necesary to perform this operation.

Sequence of Events for the ADD Instruction

Clock	RTL Expression	Comment
Cycle 1	ALU ← R1 f(ALU) ← ADD AB Acc ← ALUout	; Present R1 to A input of ALU ; Set ALU for ADD function ; Perform add, store result into ; accumulator
Cycle 2	TEMP ← Acc	; Prepare to fetch next instruction. ; Temporarily store away addition ; result.
Cycle 3	ALU ← PC f(ALU) ← INC Acc ← ALUout	; Present PC to A input of ALU ; Select INC A function for the ALU ; Increment PC, store in accum
Cycle 4	MAR ← Acc PC ← Acc < MDR READ >	; Enable ACCM onto ABUS, and capture ; into both the MAR and PC. ; Initiate memory read operation (next ; instruction).
Cycle 5	ALU ← TEMP f(ALU) ← PASS A	; Restore addition result back to the ; accumulator
Cycle 6	IR ← MDR Acc ← ALUout	; When read complete, transfer new ; instruction into IR. ; Instruction execution is now ; complete.

The table below lists the sequence of events for the SUB R1 instruction, which subtracts the content of R1 from the contents of the accumulator. Note that the ALU in this organization does not have a native SUBTRACT instruction; however, this operation can be emulated with the existing operations. In the table below, this is accomplished by the equivalence expression,

Acc-R1 = -[R1 + -Acc)]

where the negation operation is performed by computing the 2's complement, which is accomplished by first taking the 1's complement (NOT A) and then incrementing (INC A). This imposes several cycles of overhead: the implementation below requires a total of 10 clock cycles.

Sequence of Events for the SUB R1 Instruction

Clock	RTL Expression	Comment
Cycle 1	ALU ← Acc	; Construct 2's complement of the accumulator.
		; Present Accumulator to A input of ALU.
	f(ALU) ← NOT A	; Set ALU for NOT (1's complement) function.
	Acc ← ALUout	; Negation complete.
Cycle 2	ALU ← Acc	; Increment the negated accumulator
	f(ALU) ← INC A	; and finish the 2's complement function.
	Acc ← ALUout	
Cycle 3	ALU ← R1	; Present R1 to A input of ALU.
	f(ALU) ← ADD AB	; Set ALU for ADD function.
	Acc ← ALUout	; Perform ADD, store result in accumulator.
Cycle 4	ALU ← Acc	; Construct 2's complement of the accumulator.
		; Present Accumulator to A input of ALU.
	f(ALU) ← NOT A	; Set ALU for NOT (1's complement) function.
	Acc ← ALUout	; Negation complete.
Cycle 5	ALU ← Acc	; Increment the negated accumulator
	f(ALU) ← INC A	; and finish the 2's complement function.
	Acc ← ALUout	
Cycle 6	TEMP ← Acc	; Prepare to fetch next instruction.
		; Temporarily store away addition result.
Cycle 7	f(ALU) ← INC	; Select INC A function for the ALU.
	Acc ← ALUout	; Increment PC, store in accumulator.
Cycle 8	MAR ← Acc	; Enable ACCM onto ABUS, and capture into both the MAR and PC.
	PC ← Acc	
	< MDR READ >	; Initiate memory read operation (new
		; instruction).
Cycle 9	ALU ← TEMP	; Restore addition result back to the accumulator
	f(ALU) ← PASS A	
Cycle 10	IR ← MDR	; When read complete, transfer new instruction into IR.
	Acc ← ALUout	; Instruction execution is now complete.

This last example illustrates the implementation of a program control instruction: BRANCH R0. On completion, this instruction sets the program counter to PC = PC+R0+1. This instruction requires 6 clock cycles to execute.

Clock	RTL Expression	Comment
Cycle 1	TEMP ← Acc	; BRANCH instruction should not destroy ; the contents of the accumulator. Save it ; TEMP.
Cycle 2	ALU ← R0 f(ALU) ← INC A Acc ← ALUout	; Perform the "+1" portion of the branch. ; R0 contains the offset for the branch.
Cycle 3	ALU ← PC f(ALU) ← ADD AB Acc ← ALUout	; Present PC to A input of ALU. ; Set ALU for ADD function. ; Perform add, store result into ; accumulator.
Cycle 4	PC ← Acc MAR ← Acc < MDR READ >	; Restore new program counter value. ; Prepare to fetch new target instruction ; Read target instruction.
Cycle 5	ALU ← TEMP f(ALU) ← PASS A	; Restore addition result back to the ; accumulator.
Cycle 6	IR ← MDR Acc ← ALUout	; When read complete, transfer new ; instruction into IR. ; Instruction execution is now complete.

6.3 CMOS LOGIC GATE DESIGN

The most prevalent integrated circuit fabrication technology is CMOS (complementary MOS). CMOS technology has a number of advantages over bipolar fabrication technologies, including higher device density, lower power consumption, easier design creation, and simpler fabrication (fewer fabrication masks). Furthermore, the complementary MOSFET transistors that form the basis of this technology (the PFET and the NFET) are well suited for high-speed switching operations. This section will present a brief review of the behavior of enhancement mode MOSFETs used in complementary digital logic applications.

Computer Engineering

Contemporary texts on CMOS digital integrated circuits present a more complete account of device properties and mathematical models of CMOS transistors [3].

For a zero-order approximation of their behavior in complementary digital gate design, N-channel MOSFETs (or NFETs) provide continuity between the *source* and *drain* terminals when the *gate* terminal is a logic 1.* When the gate terminal is a logic 0, the drain and source terminals become electrically isolated. The complementary device, the P-channel MOSFET (or PFET), provides continuity between the source and drain terminals when the gate terminal is connected to a logic 0. When considering a slightly more realistic first-order model of the MOSFETs, the ideal switching behavior must be modified to account for the actual voltage–current behavior of devices. As an additional complication, NFETs pass the logic 0 value between the source and drain terminals with little degradation; however, when attempting to pass a logic 1 between the source and drain terminals, the logic 1 value is "diminished." Likewise, PFETs do well in passing the logic 1 value, yet impair the transmission of the logic 0 value (refer to Fig. 6.6).

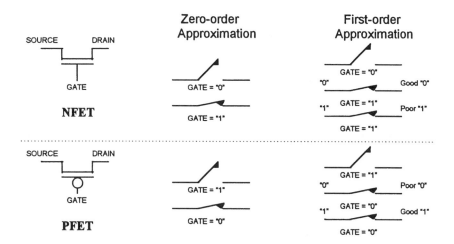

Figure 6.6 The simplified logical behavior of MOSFETs for complementary logic gate design.

---

*For the purposes of this discussion, a logic 1 is equal to V_{dd}, and logic 0 is equal to ground. MOSFETs are four-terminal devices; however, the connection of *body* terminal is implied in this discussion (the body is assumed to be connected to ground for NFETs and V_{dd} for PFETs).

A complementary CMOS logic gate is a fundamental building block for static digital logic circuits. A transistor-level complementary logic gate design can be readily formulated for any arbitrary combinational logic function. While fully complementary gate structures are but one of several implementation styles available to a logic designer, they offer a few considerable advantages. One advantage is that switching structures are fairly simple to design and can be readily automated for logic synthesis tools. Furthermore, complementary structures consume virtually no d.c. power, and have high immunity to switching noise because of their rail-to-rail output swing voltages.

A generalized complementary logic gate is shown in Fig. 6.7. A complementary CMOS gate provides continuity between the *Vdd* terminal and the output terminal for input combinations to the gate that are to evaluate to logic 1. Likewise, the gate provides continuity between ground and the output terminal for logic 0s in the function's truth table. Because of the first-order limitations depicted in Fig. 6.6, PFETs are chosen to provide continuity between the *Vdd* terminal and the function output (the pull-up network). Therefore, the PFETs within the pull-up network are responsible for covering the logic 1s in the function truth table. Similarly, the pull-down network is constructed with NFETs to furnish continuity for the logic 0s in

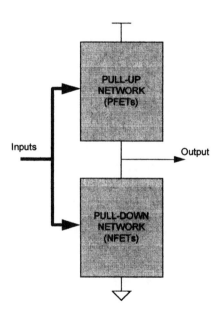

Figure 6.7 Generalized model of an inverting complementary gate.

Computer Engineering

the logic function. The process of designing a complementary gate consists of deriving two switching networks: the pull-down network, consisting of NFETs, to cover the logic 0s; and the pull-up network, consisting of PFETs, to cover the logic 1s. For a given input to a fully complementary design, either the pull-up network will provide continuity or the pull-down network will provide continuity – there are no input sequences where both networks are on or off simultaneously.

To illustrate the design process for a complementary gate, consider the following problem: given the four input signals $[a, b, c, d]$, implement the Boolean function $f = \overline{a + d(c+b)} + \bar{b}\bar{d}$ at the transistor level using fully complementary logic with the fewest number of NFET/PEET pairs.

6.3.1 Solution

The method of deriving a transistor-level complementary logic gate structure from an arbitrary combinational expression is to derive the switching network separately for the pull-up structure and for the pull-down structure, and piece them together in the form illustrated in Fig. 6.7. The solution presented here will provide a transistor-level circuit that satisfies the logical behavior of the function f. Additional improvements might be necessary if one were optimizing this switching function for, say, high speed, low power, or minimal layout area.

The pull-down structure for this function will be designed first. The pull-down network should provide continuity between the ground terminal and the output terminal whenever f evaluates to a logic 0. This condition is true when the expression $\bar{f} = \overline{a + d(c+b)} + \bar{b}\bar{d}$, is true. The most intuitive way to construct a switching function for a complex logical expression is to break the expression down into simpler subexpressions and then piece together the components. A switching network for implementing a logical OR operation of two variables, X and Y, can be constructed with two switches connected in parallel (refer to Fig. 6.8(a)). Each switch provides continuity when the

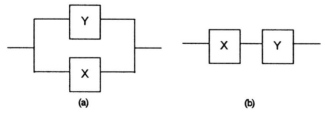

Figure 6.8 Continuity organization for switching circuits: (a) a parallel connection of $x + y$, and (b) the series connection of xy.

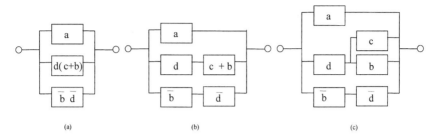

Figure 6.9 Factoring the pull-down network into a sum of three product terms.

control variable evaluates to a logic 1 (note that for the final circuit, the control variables will be connected to the gate terminals of the transistors – NFETs for the pull-down network – which provide continuity with a logic 1 input). In Fig. 6.8(a), continuity is provided between the two end points of the network when either Y is true or X is true. In general, to implement an OR function of n variables requires n switches assembled in parallel.

A switching network for implementing an AND operation of two variables, X and Y, can be constructed with two switches connected in series. Referring to Fig. 6.8(b), continuity between the two end points will be provided only when both of the control variables, X and Y, evaluate to be true. Once again, this can be generalized to an n-input AND function by assembling n switches in a series topology.

Given the procedure for assembling AND and OR structures, arbitrarily large functions can be synthesized using a top-down divide-and-conquer strategy. For the problem at hand, the function $a + d(c + b) + \bar{b}\bar{d}$ is the sum (OR) of three product terms: a, $d(c + b)$, and $\bar{b}\bar{d}$. A top-level switching network that satisfies this OR behavior is illustrated in Fig. 6.9(a). Each of these product terms can in turn be decomposed into subnetworks in a recursive fashion (Figs. 6.9(b) and (c)). The first product term, a, needs no further reduction. The second product term, $d(c + b)$, is itself a product of two expressions, namely d and $c + b$. The second sub-term can be further

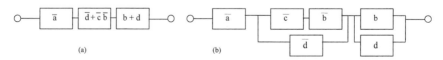

Figure 6.10 Forming the pull-up is done by constructing the dual of the pull-down network. Implementation of the switching function (a) as a fully complementary gate and (b) as a pseudo-NMOS gate.

expressed as a sum of the terminals c and b. Similarly, the third product term $\bar{b}d$ can be decomposed into a series network to produce the product between the terms $\bar{b}$ and d. Note that in this case, the inverse of both b and c must be constructed and conveyed to the control input of the switch (NFET). This can be done simply with an inverter. The final decomposition of the pull-down network is given in Fig. 6.9(c).

The pull-up network is responsible for providing continuity when the function f evaluates to be true, which is when $f = a + d(c + b) + \bar{b}d = (\bar{a})(\bar{d} + \overline{cb})(b + d)$. The derivation of the pull-up network can proceed in a similar manner as the pull-down network. Since PFETs will be used exclusively in the pull-up network, and PFETs provide continuity when the control input is a logic 0, the control structure will have to be derived considering active-low control variables. An alternative method for constructing the pull-up network is to exploit the symmetry properties of the complementary gate. The pull-up network can be created by constructing the *dual graph* of the pull-down network. Classical methods can be used to create the dual graph, but the simplest method is to restructure the graph so that the element in series in the pull-down network are parallel networks in the pull-up network. Similarly, elements in parallel in the pull-down network become series structures in the pull-up network. Figure 6.10 illustrates the process for creating the pull-up network. Note that since PFETs will be used for implementation, the elements that have control variables which do not indicate inversion will require inverters in the final circuit.

The final gate can be formed by connecting these two networks in the manner illustrated in Fig. 6.7. The completed transistor-level circuit is provided in Fig. 6.11. Note that since the subnetworks could have been interconnected in a number of equivalent ways, this solution is not unique.

6.4 SEQUENTIAL STATE MACHINE DESIGN

Construct a one-input, one-output clocked sequential circuit that will recognize all possible occurrences of the sequence of 01101 in a serial input data stream, where overlap occurrences of the pattern are allowed. For example, if the binary stream

00110110110101000101101000111

were presented to the machine (from left to right), the output of the machine should produce the binary stream

0000001001001000000000010000

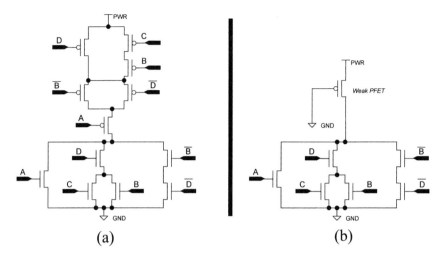

Figure 6.11 Completed transistor-level circuit.

Notice that in this machine definition, a logic 1 is produced, indicating that a valid symbol was detected. This occurs on the clock cycle following the last bit of the template symbol pattern. Since overlap occurrences are allowed in the 5-bit pattern, 1s in the output stream may be produced in less than 5-bit periods. In the example above, four occurrences of the test pattern were detected – three of which were produced from overlapping sequences.

6.4.1 Solution

One method of solving this problem is to apply classic state machine design. The solution presented here is a Moore machine – a machine whose outputs are strictly a function of the state of the machine. A Moore output of a sequential machine transitions only on the change of state of the machine. In contrast, a machine with Mealy outputs can be a direct function of the principal inputs as well; thus, it can change value when either the principal inputs change or when the state of the machine changes. The general procedure for solving state machine problems is first to construct a state diagram or state transition table for the problem, then derive the next-state equations, and finally to construct the necessary logic.

The first step in the process is describing the behavior of the state machine by a state transition table or state diagram. The state of the system serves as a means of remembering where the machine is throughout the symbol decoding process. For this problem, a total of six states are selected – one state will serve as an initialization state (referred to as *IDLE*), and five states are chosen to remember each bit in the five-symbol search (referred to as *S1* through *S5*). When the last bit in the sequence is detected, the final state (*S5*) is entered and the machine produces a "1" to indicate that the symbol is found. Transitions from each of these states into another state are recorded in the state transition table (Table 6.1) or the state diagram (Fig. 6.12) – both are equivalent and supply the same information.

Table 6.1 State Transition Table for the Sequence Detector

Current State	Next State Input = 0	Next State Input = 1	Machine Output	Comment
IDLE	S1	*IDLE*	0	Searching for the beginning of symbol ("0")
S1	S1	S2	0	First "0" found, looking for a "1"
S2	S1	S3	0	Bit 2 found ("1"), looking for Bit 3
S3	S4	*IDLE*	0	Bit 3 found ("0"), looking for Bit 4
S4	S1	S5	0	Bit 4 found ("0"), looking for Bit 5
S5	S2	*IDLE*	1	Final bit found

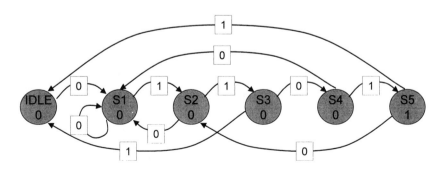

Figure 6.12 State transition diagram for the sequencing detector.

The column labeled *Current State* in Table 6.1 enumerates all of the possible states for this symbol detector circuit. The column labeled *Next State* lists the destination states allowed from the *Current States* column. There are two sub-columns listed under *Next State*, indicating the allowed transition for the given value of the machine input (labeled *Input*). For example, when the machine is in state *IDLE*, it can either remain in state *IDLE* (when the input remains a "1") or it can transition into state *S1* (when the first "0" of the sequence is detected). Note that the state sequences frequently branch to intermediate states to allow the detection of overlapping symbols in the input stream. To illustrate this, refer to the case when the current state is state *S2*. In this state, the first two bits of the sequence have been found. If the third bit is a "1," then the conditions are satisfied to indicate that the third bit has been recognized; thus, it moves into third state, *S3*. If the input should be "0" while in *S2* – which is not the proper sequence for the symbol – then the search begins again; yet this "0" may be the first bit of a new sequence. Therefore, when the input is "0" while in state *S2*, a transition to state *S1* is allowed. The machine output remains a "0" for all states except *S5*. When the machine changes into this state, a valid symbol has been detected.

In the state diagram, the legal transitions from states are indicated by the directed arcs from the nodes – which represent the six states of the machine. Each node has two arcs leaving – one for when the input is a "0" and one for when the input is a "1." The machine output is noted as the value immediately below the state label within each of the nodes (circles).

The next step in the process is *state assignment*, or the process of associating a binary vector with each of the six states. There are a number of strategies that one can use for state assignment, depending upon an assortment of criteria, such as minimal power, minimal decoder logic, fastest implmentation, and so on. Optimizing for any of these criteria is beyond the scope of this problem; hence, a rather arbitrary state assignment strategy is chosen. With six states, a minimum of $\lceil \log_2(6) \rceil = 3$ flip-flops are needed to identify uniquely all of the states of the system. Three D flip-flops are chosen for this design, the outputs of which are referred to as $[Q3, Q2, Q1]$. The state assignment is summarized in Table 6.2. State *IDLE* was given the assignment of "000," which is the value in which all flip-flops are initialized in the implementation. All other state assignments have been made arbitrarily. Two of the eight possible states of the three flip-flops remain unassigned.

With the state assignments made, the next step in the process is to derive the logic equations for the next state logic. If it has not been done yet, a decision must be made about the underlying structure of the state machine. In this case, the organization of the machine is assumed to have

Computer Engineering

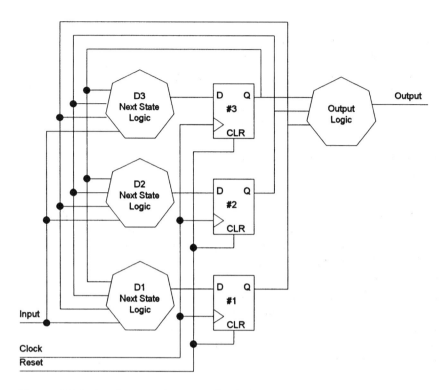

Figure 6.13 Implementation organization.

the structure shown in Fig. 6.13. It is the purpose of this step to derive the logic expressions for each of the four unknown functions in Fig. 6.14.

Table 6.3 is a copy of Table 6.1 in which the state variable names have been replaced with the state assignments made in Table 6.2. From this table, a truth table can be derived for each of the four functions.

Table 6.2 State Assignments for Symbol Detector Circuit

State	State Assignment ($Q3$, $Q2$, $Q1$)
IDLE	000
S1	001
S2	010
S3	011
S4	100
S5	101

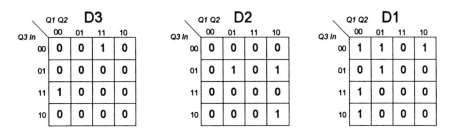

Figure 6.14 Karnough maps for the next state equations.

Each of the three functions for the D inputs to the flip-flops depends on the three Q outputs (the *Current State* column) and the dependent input, *Input*. The values in each of the columns under *Next State* are to be considered as the outputs of the three functions. For example, when the current state is $S3$ (011) and *Input* = 0, then $D3$ should evaluate to "1," $D2$ to "0," and $D1$ to "0." Repeating this process for all states and input values, complete truth-tables can be formed. Karnaugh maps for the three functions, constructed from the data in Table 6.3, are provided in Fig. 6.14. From these tables, it is a simple matter to derive sum-of-products or products-of-sums expressions for the next state computations for each of the flip-flops:

Table 6.3 State Table with Bound State Assignments

Current State	Next State		Machine Output	Comment
	Input = 0	Input = 1		
000	001	000	0	Searching for the beginning of symbol ("0")
001	001	010	0	First "0" found, looking for a "1"
010	001	011	0	Bit 2 found ("1"), looking for Bit 3
011	100	000	0	Bit 3 found ("0"), looking for Bit 4
100	001	101	0	Bit 4 found ("0"), looking for Bit 5
101	010	000	1	Final bit found

$$D3 = Q_1 Q_2 \overline{Q_3} \, \overline{In} + Q_1 Q_2 \overline{Q_3} \, In$$
$$D2 = Q_1 \overline{Q_2} \, \overline{Q_3} \, In + \overline{Q_1} Q_2 \overline{Q_3} In + Q_1 \overline{Q_2} Q_3 \overline{In}$$
$$D1 = \overline{Q_1} \, \overline{Q_2} Q_3 + Q_1 Q_2 \overline{In} + \overline{Q_1} \, \overline{Q_3} \, In + \overline{Q_1} \overline{Q_2} Q_3 + \overline{Q_2} \, \overline{Q_3} \, \overline{In}$$

The fourth expression to derive is the equation for the state machine output. This is a simple expression to derive, since by design, the output is to be a "1" when state $S5$ is entered. Thus,

$$Output = \text{"1" when State} == S5$$
$$= Q_1 \overline{Q_2} Q_3$$

The final implementation for the symbol detector can be realized by the instantiation of the above equations into the appropriate points in Fig. 6.13.

6.5 INFORMATION CODING

When data are transferred from one point to another in a harsh electrostatic or electromagnetic environment, or through a medium with non-ideal transport properties, the electrical signal may be degraded to the point where occasionally it may be improperly received. We assume that a bit is improperly received when either we transmit a "1" and receive a "0," or if a transmitted "0" is interpreted as a "1." In this problem we assume that data are passed through a degraded channel which may introduce random noise pulses to the signal. The channel is assumed to be robust enough that the occurrence of errors is relatively rare; however, the errors which do occur should be detected and, if possible, corrected. The input data sequence is coded for the purpose of not only detecting errors at the receive side, but also correcting single bit errors. In this case, *Hamming coding* is used.

Determine the even-parity Hamming-coded word for the eight-bit message 10011010.

The Hamming-coded message 010001111010 is received using the same scheme. Determine whether any errors occurred. If so, provide the corrected message if possible.

6.5.1 Solution

The Hamming code is a self-correcting code that can detect and correct a single error. To accomplish both error detection and error correction, additional redundant data bits are appended to each transmitted word. At the receiver end, the redundant bits are used to determine whether an error has occurred (whether any one of the bits in the received word has flipped). If so, the redundant data can be used to determine which bit is faulty. Once determined, the faulty bit can be corrected and the word restored to its original transmitted state. The first step in this process is referred to as *error detection*, and the second is referred to as *error correction*.

The most common way to perform error detection is through parity circuits. In a typical parity error detection circuit, each data word is augmented with (at least) one redundant bit, or *parity bit*. The parity bit is chosen so that the modulo-1 sum of all bits within the word (including the parity bit) is a constant. The constant can either be a "1" or a "0." For an even parity scheme, the constant is "0"; hence the sum of all of the 1s in the coded word is an even number. In an odd parity scheme, the modulo-1 sum of all bits in the encoded word is "1"; thus, the encoded word would have an odd number of 1.

At the transmitter side of the channel, the parity bits can be computed by performing the modulo-1 sum on the unencoded word. Depending on whether even or odd parity is used, the modulo-1 sum is either passed through or inverted, and augmented to the unencoded word. Parity generators can be easily generated using XOR gates to perform the modulo-1 sum.

At the receiver side of the channel, the modulo-1 of the encoded word is computed. This is performed in a similar manner as in the parity generator circuit in the transmitter, except the parity bit is included in the modulo-1 computation. If the word was transmitted using even parity, then the resulting sum should be "0." If the modulo-1 sum of the received word were not "0," then it could be assumed that an error had occurred. In fact, a parity checker circuit will provide sufficient data to indicate that an odd number of errors had been introduced to the encoded word. Note, however, that a single parity bit cannot determine whether an even number of errors have been introduced to the encoded word.

Parity checking circuitry can determine whether an error occurred during transmission; however, with a single parity bit, there is insufficient information to determine which bits are in error. Hamming coding is a common scheme for single-error detection and correction. In a Hamming coder, the n unencoded data bits are augmented with p parity bits placed in specific locations prior to transmission. Each of the parity bits is computed to provide even parity or odd parity on an overlapping subset of the encoded word. The receiver recomputes the parity for each of the subfields within the encoded word using a parity checking circuit. From the syndrome produced at the receiver by the parity checker, the exact position of the single error can be determined. With this information, the error can be corrected.

The number, position, and computation of each of the p parity bits depend on n, the size of the data word. The procedure for the Hamming encoder is as follows.

For n unencoded data bits, the minimum number of parity bits, p, satisfies the expression

$p \geq \log_2(p + n + 1)$

The encoded word is created by placing parity bits at bit positions $1, 2, 4, 8, \ldots, 2^k$, where bit position 1 is the least significant bit, and $k \leq \log(m + p)$. The remining bits are filled with n bits of the unencoded data. Then, each bit position of the encoded word is expressed in binary format using p bits. For example, for $p = 5$, bit position 1 is expressed as 00001, bit position 2 as 00010, and so on. Each of the parity bits is computed to cover a subset of the encoded word in the following manner: Construct p sets of numbers: $P_1, P_2, P_3, \cdots, P_p$. In set P_1, include all of the bit position numbers that have a "1" in bit position 1. Therefore, set P_1 would consist of $[1, 3, 5, 7, 9, 11, \ldots]$. Set P_2 would consist of all of the bit position numbers that have bit position 2 set to a "1," which are $[2, 3, 6, 7, 10, \ldots]$. This process is continued for all p sets. By construction, each of these sets will contain one of the p parity bits. The parity bit for each set is computed so that all of the bits within the set exhibit even parity (modulo-1 sum). Once computed for all sets, the encoded word is ready for transmission.

Upon reception, the data are first checked with a Hamming checking circuit. Within this circuit, the parity is computed for each of the p sets. This is accomplished by computing the modulo-1 sum for all bits in a given set. For each of the p computations, since even parity was used in the encoder, a "1" will indicate that an error exists in one of the data bits within the set. For discussion purposes, let $E_p, \cdots, E_2, E_1$ denote the computed (even) parity for each of the p sets, and let S be the binary number formed by the concatenation of these bits, i.e., $S = [E_p, \cdots, E_2, E_1]$. If no error had occurred during transmission, all of these bits would be zero: $S = 0000\ldots0$. If a single error had occurred in transmission, one or more of the E values would be "1". Because of this, S would be non-zero. By construction, the binary number formed by S would reflect the index of the bit position that has been received incorrectly. The vector S can be sent to a Hamming corrector circuit that would rectify the faulty bit. Figure 6.15 illustrates the overall Hamming coding transmitter/receiver system.

- For the problem at hand, the 8-bit number 10011010 ($n = 6$) is to be encoded using Hamming coding and then transmitted. From the formula above, p is chosen to be 4. The total number of bits in

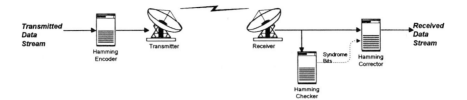

Figure 6.15 Transmitter receiver system using Hamming coding.

the encoded sequence will be $8 + 4 = 12$. For an eight-bit unencoded data sequence, the four parity bits are to cover the bit positions listed in Table 6.4. The bit assignments in the encoded word are summarized in Table 6.5. The computed parity for each of the groups listed in Table 6.4 is also provided in Table 6.5.

- As a result, the Hamming coded message for the input [10011010] is [100101011011].

For the second part of the problem, if the message received was 010001111010, we need to determine the corrected (if necessary) received message. Table 6.6 summarizes the operations performed by the receiver. The receiver computes the parity for each of the four groups of bits in the encoded message as determined by Table 6.6. The column labeled *Computed Parity* in Table 6.6 lists the operations performed to compute the check bits (note that the $\otimes$ operator refers to modulo-1 addition). Since one or more of the computed parity bits are "1," then an error had occurred during transmission. The bit in error is determined by the S value, which in this case is 1001, which is binary for bit position 9 (note that 9 is common to

Table 6.4 Bit Positions Assigned to Each of the Parity Bits for a 12-Bit Encoded Sequence

Parity Bit	Bit Positions Covered
$P1$	1, 3, 5, 7, 9, 11
	[$P1$, $D1$, $D2$, $D4$, $D5$, $D7$]
$P2$	2, 3, 6, 7, 10, 11
	[$P2$, $D1$, $D3$, $D4$, $D6$, $D7$]
$P3$	4, 5, 6, 7, 12
	[$P3$, $D2$, $D3$, $D4$, $D8$]
$P4$	8, 9, 10, 11, 12
	[$P4$, $D5$, $D6$, $D7$, $D8$]

Table 6.5 Bit Assignments and Computed Parity for the Word [10011010]

Bit Position	Bit Assignment ($P \to$ Parity, $D \to$ Data)	Data	Parity
1	0001	P1		1
2	0010	P2		1
3	0011	D1	0	
4	0100	P3		1
5	0101	D2	1	
6	0110	D3	0	
7	0111	D4	1	
8	1000	P4		0
9	1001	D5	1	
10	1010	D6	0	
11	1011	D7	0	
12	1100	D8	1	

both $P1$ and $P4$ in Table 6.4). Bit 9 was mistakenly received as a "0." The corrected encoded message is, therefore, 010101111010. The decoded data word, which is the corrected encoded word stripped of the parity bits, is, therefore, 01011110.

Table 6.6 Computed Parity for the Received (Corrupted) Encoded Word 010001111010

Bit Position	Bit Assignment ($P \to$ Parity, $D \to$ Data)	Received Data	Computed Parity
1	0001	P1	0	$0 \otimes 0 \otimes 1 \otimes 1 \otimes 0 \otimes 1 = 1$
2	0010	P2	1	$1 \otimes 0 \otimes 1 \otimes 1 \otimes 0 \otimes 1 = 0$
3	0011	D1	0	
4	0100	P3	1	$1 \otimes 1 \otimes 1 \otimes 1 \otimes 0 = 0$
5	0101	D2	1	
6	0110	D3	1	
7	0111	D4	1	
8	1000	P4	0	$0 \otimes 0 \otimes 0 \otimes 1 \otimes 0 = 1$
9	1001	D5	0	
10	1010	D6	0	
11	1011	D7	1	
12	1100	D8	0	

6.6 COMPUTER MEMORY HIERARCHY AND PAGING

It is quite common for a computer program, such as a spreadsheet or video game, to have enormous run-time random access memory (RAM) requirements. In fact, it is often the case that computer programs on contemporary personal computers have run-time memory requirements that exceed the physical random access memory in the machine by a factor of 10 or more. Furthermore, in many circumstances it is not economically plausible to stock a general-purpose computer with enough RAM to satisfy the requirements of the most demanding application. Instead, contemporary machines are designed to emulate a much larger *logical* RAM using a relatively small physical RAM and a secondary store (such as a hard disk drive). Portions of memory (or *pages*) are rapidly swapped between the physical RAM and the secondary store, thus presenting portions of the logical RAM to the processor in a time-sharing manner. With this strategy, accesses to the physical RAM are unhindered. When access is needed to data that currently reside in the secondary store, the data are transferred from the secondary store to the local physical RAM. If there is no physical RAM free to accommodate the new data, the processor must decide which page of physical RAM to replace temporarily with the new secondary store data. Once the new data is restored, the application continues. Note that if the physical RAM page targeted for replacement has been recently updated by the processor, it must be saved before it is overwritten. Saving the page is done by writing it back to the secondary store.

Through careful system programming, hardware design, and address remapping, a large logical RAM and the swapping process can be accomplished in a way that is transparent and independent of the application program. Therefore, an application programmer can assume the existence of an enormous logical RAM, or *virtual memory*, and not be limited by the constraints of the physical RAM resources of the underlying platform. The automatic demand-driven memory management process greatly reduces the complexity of program development.

Page swapping in a virtual memory environment does not come without a cost. The access time of the secondary store (the time between the request for data and when the data are delivered to the processor) is usually much longer than the access time of the local physical RAM, usually by several orders of magnitude. In conventional computers, the secondary store is on hard disk drives, which have access times of the order of several milliseconds, while RAMs have access times of the order of nanoseconds. Because of this, an application that requires excessive swapping (or *page faults*), can incur a substantial performance penalty. Factors that determine the frequency of page faults are the amount of available physical RAM, physical page size, and the page replacement policy. The page replacement policy determines which page of physical memory to supplant with a page from the secondary store. It is desirable to devise a page replacement policy that has low complexity yet minimizes the numbr of page faults.

To illustrate the different page replacement strategies, consider the following problem: a given hypothetical machine has four physical pages of memory, and a sufficiently large secondary store. Memory references made by an application program span a total of 12 pages – each page is labeled with a letter [A through L]. It is the responsibility of the operating system to swap in these pages as needed by the application. During execution, the processor makes the following sequence of references to the pages of the application program:

ABCBDEDBCFCGCHICBDEJKCLACBCFHICBDEJKDCLA.

Determine the number of times a page needs to be replaced for each of the following page replacement algorithms:

First in, first out (FIFO)
Least recently used (LRU)
Ideal paging (considering future knowledge)

Solution

In this sample trace above, the processor makes a total of 40 references to the 12 pages throughout the execution lifetime of the program. The performance of the mythical four-page processor is examined for each of the paging strategies. The first strategy, *first in, first out paging*, is the simplest to implement of all the strategies examined here. This page replacement policy makes no use of past history or any property of the application in determining the page to replace during a page fault. Pseudocode for this strategy is as follows:

Initialize *index* = 0
Loop always {
    If a page fault occurs {
    Backup (write) *PAGE[index]* to secondary store if necessary.
    Physical *PAGE[index]* is replaced with *NEWPAGE* from secondary store.
    index < = *(index + 1) MOD 4* // Increment INDEX to point
    }
}

In this algorithm, *index* points to the next page to replace when a page fault occurs. After the page has been swapped, *index* is incremented (in a modulo-4 or *round-robin* fashion) to point to the next adjacent page to swap out during the next page fault. Table 6.7 summarizes the execution and paging history for the test program. The column labeled *working set* lists the pages that currently reside in physical RAM. Using round-robin paging, a total of 31 page faults occurred.

The second strategy, *Least Recently Used (LRU)*, utilized past history information as a statistic to pick the page to replace. In this page replacement policy, the operating system monitors the memory references made by a program or set of programs over a period of time. The pages that have not been referenced for a long time are presented as candidates for swapping. The reasoning behind this strategy ties in closely with the Principle of Locality, which states that computer programs tend to reference memory locations that are the same as or nearby past memory references. Because of this stationary behavior, LRU relies on the loose statistic that if the time from the last reference to a page is relatively long, then the likelihood of a future reference is diminished. The performance benefit gained from LRU depends upon the statistical properties (past history) of the program behavior. Table 6.8 illustrates the paging behavior when LRU is used on the sample trace above. For this trace, 31 page faults are made when LRU is applied.

The best way to determine which page within the working set to swap out on a page fault is to have future knowledge of the behavior of the program. With future knowledge, one could pick the page replacement sequence to minimize overall page faults. As a result, with this optimal paging strategy, the application performance is highest since only the minimum of page faults are performed. Unfortunately, the operating system rarely (if at all) can determine the future behavior of programs. Because of this, *optimal page replacement policy* is an unimplementable strategy. Nonetheless, it is useful way to determine the upper-bound of performance that could be achieved for a given program trace.

Table 6.9 summarizes the optimal paging policy for the given trace. In this trace, the page picked for swapping is the one which has the next reference made to it most distant in the future (or no further references). In this manner, pages that will have the most immediate future references are kept within physical RAM. With optimal paging on this trace, 22 page faults occur.

In summary, the goal of a paging policy is to reduce the number of page faults and page replacements that occur during program execution. Paging strategies that use past history tend to perform better than those strategies that do not. The actual performance of a paging strategy often depends

Computer Engineering

Table 6.7 Reference History for FIFO Paging

Time	Page	Page Fault?	Working Set	Notes
1	A	Y	A__	Page 0 loaded
2	B	Y	AB_	Page 1 loaded
3	C	Y	ABC_	Page 2 loaded
4	B		ABC_	
5	D	Y	ABCD	Page 3 loaded
6	E	Y	EBCD	Page 0 replaced
7	D		EBCD	
8	B		EBCD	
9	C		EBCD	
10	F	Y	EFCD	Page 1 replaced
11	C		EFCD	
12	G	Y	EFGD	Page 2 replaced
13	C	Y	EFGC	Page 3 replaced
14	H	Y	HFGC	Page 0 replaced
15	I	Y	HIGC	Page 1 replaced
16	C		HIGC	
17	B	Y	HIBC	Page 2 replaced
18	D	Y	HIBD	Page 3 replaced
19	E	Y	EIBD	Page 0 replaced
20	J	Y	EJBD	Page 1 replaced
21	K	Y	EJKD	Page 2 replaced
22	C	Y	EJKC	Page 3 replaced
23	L	Y	LJKC	Page 0 replaced
24	A	Y	LAKC	Page 1 replaced
25	C		LAKC	
26	B	Y	LABC	Page 2 replaced
27	C		LABC	
28	F	Y	LABF	Page 3 replaced
29	H	Y	HABF	Page 0 replaced
30	I	Y	HIBF	Page 1 replaced
31	C	Y	HICF	Page 2 replaced
32	B	Y	HICB	Page 3 replaced
33	D	Y	DICB	Page 0 replaced
34	E	Y	DECB	Page 1 replaced
35	J	Y	DEJB	Page 2 replaced
36	K	Y	DEJK	Page 3 replaced
37	D			
38	C	Y	CEJK	Page 0 replaced
39	L	Y	CLJK	Page 1 replaced
40	A	Y	CLAK	Page 2 replaced

Table 6.8 Least Recently Used (LRU) Paging History

Time	Page	Page Fault?	Working Set	Notes
1	A	Y	A__	Page 0 loaded
2	B	Y	AB__	Page 1 loaded
3	C	Y	ABC_	Page 2 loaded
4	B		ABC_	
5	D	Y	ABCD	Page 3 loaded
6	E	Y	EBCD	Page 0 last referenced at $T=1$
7	D		EBCD	
8	B		EBCD	
9	C		EBCD	
10	F	Y	EBFD	Page 2 last referenced at $T=3$
11	C		EFCD	
12	G	Y	GFCD	Page 0 last referenced at $T=6$
13	C		GFCD	
14	H	Y	GFCH	Page 3 last referenced at $T=7$
15	I	Y	GFIH	Page 2 last referenced at $T=9$
16	C	Y	GCIH	Page 1 last referenced at $T=10$
17	B	Y	BCIH	Page 0 last referenced at $T=12$
18	D	Y	BDIH	Page 1 last referenced at $T=13$
19	E	Y	BDIE	Page 3 last referenced at $T=14$
20	J	Y	BDJE	Page 2 last referenced at $T=15$
21	K	Y	KDJE	Page 0 last referenced at $T=17$
22	C	Y	KCJE	Page 1 last referenced at $T=18$
23	L	Y	KCJL	Page 3 last referenced at $T=19$
24	A	Y	KCAL	Page 2 last referenced at $T=20$
25	C		KCAL	
26	B	Y	BCAL	Page 0 last referenced at $T=21$
27	C		BCAL	
28	F	Y	BCAF	Page 3 last referenced at $T=23$
29	H	Y	BCHF	Page 2 last referenced at $T=24$
30	I	Y	BIHF	Page 1 last referenced at $T=25$
31	C	Y	CIHF	Page 0 last referenced at $T=26$
32	B	Y	CIHB	Page 3 last referenced at $T=28$
33	D	Y	CIDB	Page 2 last referenced at $T=29$
34	E	Y	CEDB	Page 1 last referenced at $T=30$
35	J	Y	JEDB	Page 0 last referenced at $T=31$
36	K	Y	JEDK	Page 3 last referenced at $T=32$
37	D		JEDK	
38	C	Y	JECK	Page 0 last referenced at $T=33$
39	L	Y	JLCK	Page 1 last referenced at $T=34$
40	A	Y	ALCK	Page 0 last referenced at $T=35$

Computer Engineering   291

Table 6.9 Optimal Scheduling Using Future Paging Knowledge

Time	Page	Page Fault?	Working Set	Notes
1	A	Y	A___	Page 0 loaded
2	B	Y	AB__	Page 1 loaded
3	C	Y	ABC_	Page 2 loaded
4	B		ABC_	
5	D	Y	ABCD	Page 3 loaded
6	E	Y	EBCD	Data A not referenced until $T=24$
7	D		EBCD	
8	B		EBCD	
9	C		EBCD	
10	F	Y	FBCD	Data E not referenced until $T=19$
11	C		FBCD	
12	G	Y	GBCD	Data F not referenced until $T=28$
13	C		GBCD	
14	H	Y	HBCD	Data G never referenced again
15	I	Y	IBCD	Data H not referenced until $T=28$
16	C		IBCD	
17	B		IBCD	
18	D		IBCD	
19	E	Y	IBCE	Data D not referenced until $T=33$
20	J	Y	IBCJ	Data E not referenced until $T=34$
21	K	Y	IBCK	Data A not referenced until $T=35$
22	C		IBCK	
23	L	Y	IBCL	Data K not referenced until $T=36$
24	A	Y	IBCA	Data L not referenced until $T=39$
25	C		IBCA	
26	B		IBCA	
27	C		IBCA	
28	F	Y	IBCF	Data A not referenced until $T=40$
29	H	Y	IBCH	Data F never referenced again
30	I		IBCH	
31	C		IBCH	
32	B		IBCH	
33	D	Y	DBCH	Data I never referenced again
34	E	Y	DECH	Data B never referenced again
35	J	Y	DECJ	Data H never referenced again
36	K	Y	DKCJ	Data E never referenced again
37	D		DKCJ	
38	C		DKCJ	
39	L	Y	DKLJ	Data C never referenced again
40	A	Y	DKLJ	Data J never referenced again

upon the actual behavior and statistical properties of the memory references made by application programs. There is a minimum number of page replacements that must occur for a given program trace, which can be identified by an optimal paging strategy.

6.7 INSTRUCTION SET ARCHITECTURES

A mythical RISC processor, called *ZAP-1*, is a pipelined machine that operates at 20 MHz. LOADs and STOREs on the *ZAP-1* produce one delay slot. A number of benchmark programs are compiled and executed on *ZAP-1*, and some observations are made. Twenty percent of the instructions executed are LOADs and STOREs, each of which has one delay slot. In the compiled benchmark, only half of the LOADs and STOREs could make use of the delay slots; the others are lost to NO-OPs.

(a) A new version is produced, *ZAP-1A*. In this machine, the operating frequency is boosted to 55 MHz. To compensate for the increased clock rate, LOADs and STOREs now produce two delay slots. Once again, 50% of the LOADs and STOREs in the benchmarks can utilize the first delay slot; however, only 18% can utilize the second. Compare the performance of *ZAP-1A* with *ZAP-1*.

(b) *ZAP-1* had no MULTIPLY instruction. Multiplication was emulated with a number of SHIFT/ADD and related instructions. Approximately 25 cycles were required to perform the emulated multiplication. Eight per cent of the execution time in the benchmarks was consumed with the execution of multiplication. *ZAP-1B* was introduced. It featured a hardware (non-pipelined) multiplier that could multiply two numbers in a single cycle. Unfortunately, the multiplier increased the critical timing path (and hence the operating clock period) of the processor by 5 ns. Compare the performance of *ZAP-1B* with *ZAP-1*.

6.7.1 Solution

It is not always straightforward to compare the performance between two different machines. Many times, changes in the micro-architecture of a processor may allow improvements in the memory organization and support software (compilers and operational system), i.e., collateral effects that may make it difficult to compare different versions of a processor. We may, however, compare the relative performance of *ZAP-1*, *ZAP-1A*, and *ZAP-1B* by using the following equations:

$$TPI(M) = \frac{CPI(M)}{f_{CK}(M)} \qquad CPI(M) = \sum_{i=1}^{N_1} F_i \times CPI_i$$

where M is a particular machine (in this exercise M = ZAP-1, ZAP-1A, or ZAP-1B); $CPI(M)$ = average number of clock cycles per instruction on a machine M; $CPI_i(M)$ = average number of clock cycles for an instruction of type i on a machine M; F_i = frequency of occurrence of instructions of type i:

$$0 \le F_i \le 1$$

$$\sum_{i=1}^{N_i} F_i = 1;$$

N_i = number of instruction types; $TPI(M)$ = average instruction time for machine M; $f_{CK}(M)$ = clock frequency for machine M:

$$f_{CK}(M) = \frac{1}{T_{CK}(M)};$$

$T_{CK}(M)$ = clock period for machine M; and $MIPS(M)$ = rate of execution of machine M in million of instructions per second (MIPS).

$MIPS(M)$ is a possible performance measure for machine M. By comparing $MIPS(ZAP\text{-}1)$ and $MIPS(ZAP\text{-}1A)$, we can have a good idea of the relative gains in performance of ZAP-1A over ZAP-1.

$MIPS(M)$ is not a good performance figure, however, to compare machines with different instruction sets. It is more realistic in this situation to compare the overall *execution time* (CPU time) of a set of benchmarks **B** in each machine:

$$CPU \text{ time } (M, B) = \frac{NI(M, B)}{10^6 \times MIPS(M)}$$

where $NI(M, B)$ is the overall number of instructions of the compiled code of a set of benchmarks **B** running on machine M. $NI(M, B)$ may also be a function of the compiler (code optimizer) used in machine M.

Estimating CPI$_i$

Typically, an instruction is executed in stages, which are usually of the order of three to five in most commonly used scalar processors. For the sake of simplicity, it is assumed that each instruction (Fig. 6.16) has four stages for all versions of the ZAP machines:

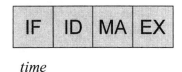

time ⟶

Figure 6.16 Instruction stages.

Instruction fetch (IF)
Instruction decode and operand fetch (ID)
Memory access (MA) – for load and store and other instructions accessing external memory.
Execution and result write-back (EX)

As *ZAP-1* and its other versions (*ZAP-1A*, *ZAP-1B*) are RISC machines, it can be assumed that all stages are performed in one clock cycle with the exception of the *memory access* stage that may take zero, one, or more cycles, depending on the instruction type. If *ZAP-1* and its sequels were not pipelined (Fig. 6.17), the average number of clock cycles $CPI_i(M)$ for an instruction type i on a processor $M = ZAP\text{-}1, ZAP\text{-}1A$, or $ZAP\text{-}1B$ would be given by:

$$CPI_i(M) = \#cycles_i(IF, M) + \#cycles_i(ID, M)$$
$$+ \#cycles_i(MA, M) + cycles_i(EX, M),$$

where $\#cycles_i(X, M)$ is the number of cycles used by stage X of an instruction of type i on a machine M.

As $\#cycles_i(IF, M) = \#cycles_i(ID, M) = \#cycles_i(MA, M) = 1$, M is a RISC machine. For a non-pipelined machine, $CPI_i(M) = 3 + \#cycles_i(MA, M)$.

The *ZAP* machines are assumed to be pipelined. Therefore, the calculation of CPI_i has to consider the effect of latency hiding due to pipelining (Fig. 6.18).

Figure 6.17 Non-pipelined stream of instructions.

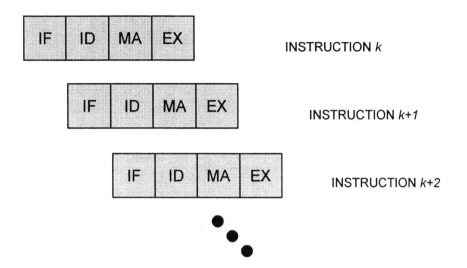

Figure 6.18 Pipelined stream of instructions.

In the pipelined case, a lower bound for $CPI_i(M)$ is one clock cycle for all instructions types, i.e., given a stream S of N instructions, the number of cycles to execute the stream S is at least equal to $N + 3$ cycles.

$$CPI = CPI_i \geq \lim_{N \to \infty} \left(\frac{N+3}{3}\right) = 1$$

However, the occurrence of data and control dependencies (conflicts) may not allow CPI_i to be equal to 1 for some of the instruction types, i.e., delay slots may appear (Fig. 6.19).

The number of delay slots between two instructions INS_k and INS_{k+1} on a stream S can be greater than or equal to zero. The compiler (code optimizer) for the processor should be able to rearrange the order of execution of the instructions of a program in such a way that the occurrence of delay slots is minimized (Fig. 6.20).

Therefore, $CPI_i(M)$ is also a function of the particular compiler (code optimizer) C used by the pipelined machine M. It is possible to evaluate $CPI_i(M)$ by means of benchmarking. In this case,

$$CPI_i(M) = 1 + \sum_{j=0}^{\infty} j \times r_j^i(M, C)$$

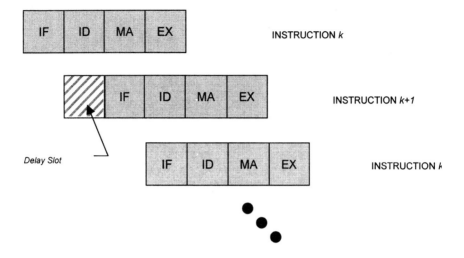

Figure 6.19 Pipelined stream of instructions with a delay slot.

$$\sum_{j=0}^{\infty} r_j^i(M, C) = 1$$

$$0 \leq r_j^i(M, C) \leq 1$$

where $r_j^i(M, C)$ is the frequency of occurrence of j delay slots after an instruction of type i, executing on a pipelined processor M, executing code optimized by a compiler C. Properly designed processors and compilers should have

$$r_j^i(M, C) \geq r_{j+1}^i(M, C)$$

For instruction type i, the occurrence of $j + 1$ delay slots is less likely than the occurrence of j delay slots.

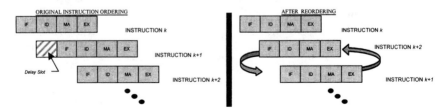

Figure 6.20 Delay slot minimization by reordering.

Computer Engineering

Part (a): Comparing the Performance of ZAP-1 and ZAP-1A

For example, in *ZAP-1* 50% *of the load/store* operations have no delay slots and 50% have one delay slot:

$$CPI_{\{LOAD, STORE\}}(ZAP\text{-}1) = 1 + 0 * 0.5 + 1 * 0.5 = 1.5 \text{ cycles}$$

In comparison, in the *ZAP-1A*, 18% of the *load/store* operations have no unused delay slots, while 32% of the *load/stores* use only the first delay slot, and 50% are not able to fill first delay slot:

$$CIP_{\{LOAD, STORE\}}(ZAP\text{-}1A) = 1 + 0 * 0.18 + 1 * 0.32 + 2 * 0.5$$
$$= 2.32 \text{ cycles}$$

In both processors, $CPI_i(ZAP\text{-}1) = CPI_i(ZAP\text{-}1A) = 1$ for all instruction types other than the *load/stores* (i.e. delay slots can only be inserted after *load/store* operations). Considering complete instruction mixes for both machines, 20% instructions are *load/stores*.

$$CPI(ZAP\text{-}1) = 0.8 * 1.0 + 0.2 * 1.5 = 1.1 \text{ cycles}$$
$$CPI(ZAP\text{-}1A) = 0.8 * 1.0 + 0.2 * 2.32 = 1.264 \text{ cycles}$$

Since

$$MIPS(M) = \frac{f_{CK}(M)}{10^6 * CPI(M)}$$

then,

$$f_{CK}(ZAP\text{-}1) = 20 \text{ MHz}$$
$$f_{CK}(ZAP\text{-}1A) = 55 \text{ MHz}$$

The following performance figures are derived for *ZAP-1* and *ZAP-1A*:

$$MIPS(ZAP\text{-}1) = \frac{20 \times 10^6}{10^6 \times 1.1} = 18.1 \text{ } MIPS$$

$$MIPS(ZAP\text{-}1A) = \frac{55 \times 10^6}{10^6 \times 1.264} = 43.5 \text{ } MIPS$$

Because *ZAP-1* and *ZAP-1A* have the same instruction set, it is straightforward to evaluate the speed-up ratio S_{up} as

$$S_{up} = \frac{(ZAP\text{-}1A)}{MIPS(ZAP\text{-}1)} = 3.4$$

Conclusion

A speed-up in performance of $S_{up} = 3.4$ may be achieved by increasing the clock frequency from 20 MHz to 55 MHz in the *ZAP-1* processor. $S_{up} = 3.4$ is actually an optimistic value because the assumption of no delay slots after instructions that are not LOAD/STORES may not be realistic.

Part (b): Comparing the Performances of ZAP-1 and ZAP-1B

As *ZAP-1* and *ZAP-1B* have different instruction sets, it is more realistic to compare the CPU time needed to execute a benchmark *B* in both processors instead of using the approach of Part (a). Furthermore,

$$CPI_{(MULTIPLY)}(ZAP\text{-}1B) = 1 \text{ cycle}$$

and

$$CPI_{(\text{simulated MULTIPLY})}(ZAP\text{-}1) = 25 \text{ cycles.}$$

It is a good approximation to say that a MULTIPLY operation in *ZAP-1* is simulated by means of about 25 SHIFT/ADD and related instructions. Therefore, given a benchmark, *B*, if the expected number of instructions (instruction count) of the compiled code on *ZAP-1B* is *NI(ZAP-1B, B)*, then the instruction count on *ZAP-1* should be

$$NI(ZAP\text{-}1, \mathbf{B}) = NI(ZAP\text{-}1B, \mathbf{B}) * (0.92 + 0.08 * 25)$$
$$= NI(ZAP\text{-}1B, \mathbf{B} * 2.92)$$

where 8% of the operations are MULTIPLY instructions.

Assuming that the code for *ZAP-1B* fills the delay slots in the same fashion as *ZAP-1*, both processors have the same clocks per instruction average (*CPI*):

$$CPI(ZAP\text{-}1) = CPI(ZAP\text{-}1B) = 0.8 * 1.0 + 0.2 * 1.1 = 1.22 \text{ cycles}$$

20% of the operations are *load/stores*–see Part (a). However, the clock periods are different for the two machines:

$TCK(ZAP\text{-}1) = 50$ ns

$TCK(ZAP\text{-}1B) = 55$ ns

The overall speed-up, S_{up}, can be evaluated as

Computer Engineering

$$S_{UP} = \frac{CPUtime(ZAP\text{-}1, \beta)}{CPUtime(ZAP\text{-}1B, \beta)} = \frac{MIPS(ZAP\text{-}1B)}{MIPS(ZAP)} * \frac{NI(ZAP\text{-}1, \beta)}{NI(ZAP\text{-}1B)}$$
$$= \frac{CIP(ZAP\text{-}1B)}{CPI(ZAP\text{-}1)} * \frac{T_{CK}(ZAP\text{-}1)}{T_{CK}(ZAP\text{-}1B)} * \frac{NI(ZAP\text{-}1, \beta)}{NI(ZAP\text{-}1B, \beta)}$$
$$= \frac{50}{55} * 2.92 = 2.8$$

Conclusion

Although *ZAP-1B* has a 10% longer clock cycle than *ZAP-1*, the former will execute 2.8 times faster than the latter on average because it has a dedicated MULTIPLY unit.

6.8 MULTIPLIERS

Compare the approximate speed and area of the following unsigned integer 5×5 multipliers:

(a) an array multiplier
(b) a Wallace tree multiplier
(c) a 2-stage pipelined array multiplier

Assume that the above multipliers are built with the following parameterized cells:

Cell Name	Normalized Propagation Delay (Any Output w.r.t. Any Input)	Normalized Area
AND gate	1	1
Half adder	3	6
Full adder	3	8

6.8.1 Solution

In each case of the three cases (a), (b), and (c), the following notation is used to denote the inputs, output, and product terms of a 5×5 multiplier:

Inputs:

$$X = X_4^*16 + X_3^*8 + X_2^*4 + X_1^*2 + X_0$$
$$X = Y_4^*16 + Y_3^*8 + Y_2^*4 + Y_1^*2 + Y_0$$

Output:

$$M = M_9^* 512 + M_8^* 256 + M_7^* 128 + M_6^* 64 + 32^* M_5 + M_4^* 16 + M_3^* 8 \\ + M_2^* 4 + M_1^* 2 + M_0$$

Product terms:

					X_4	X_3	X_2	X_1	X_0	
				×	Y_4	Y_3	Y_2	Y_1	Y_0	
$S_0 =$					X_4Y_0	X_3Y_0	X_2Y_0	X_1Y_0	X_0Y_0	
$S_1 =$				X_4Y_1	X_3Y_1	X_2Y_1	X_1Y_1	X_0Y_1		
$S_2 =$			X_4Y_2	X_3Y_2	X_2Y_2	X_1Y_2	X_0Y_2			+
$S_3 =$		X_4Y_3	X_3Y_3	X_2Y_3	X_1Y_3	X_0Y_3				
$S_4 =$	X_4Y_4	X_3Y_4	X_2Y_0	X_1Y_4	X_0Y_4					
$M =$	M_9	M_8	M_7	M_6	M_5	M_4	M_3	M_2	M_1	M_0

The cells shown in Fig. 6.21 are the building blocks in all cases.

Part (a)

An array multiplier has organization shown in Fig. 6.22.
    The overall area and propagation delay are given by

$$Area = 25 \times Area(AND) + 12 \times Area(Full\ adder) \\ + 4 \times Area(Half\ adder) = 145.$$

$$Propagation\ delay = Delay\ (critical\ path) \\ = 7 \times Delay\ (Full\ adder) + Delay\ (Half\ adder) \\ + Delay\ (AND) = 24$$

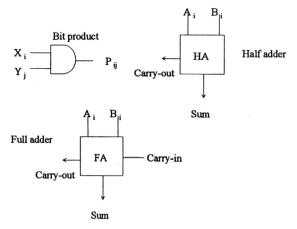

Figure 6.21 Basic cells.

Computer Engineering

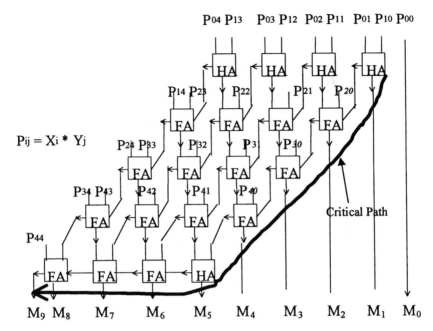

Figure 6.22 5 × 5 unsigned integer array multiplier.

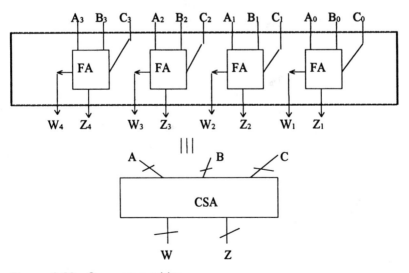

Figure 6.23 Carry save adder.

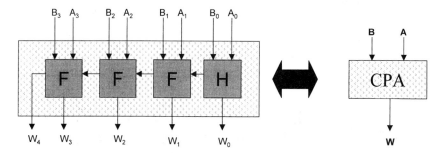

Figure 6.24 Carry propagate adder.

Part (b)

A Wallace tree multiplier uses two major building blocks: carry save adders (Fig. 6.24) and carry propagate adders (Fig. 6.25).

The overall structure of a 5×5 Wallace tree multiplier is shown in Fig. 6.26.

The approximate area and delay are given by:

$$Area = 25 \times Area(AND) + (7 + 8 + 9)$$
$$\times Area(Full\ adder) + 9 \times Area(Full\ adder) = 217$$

$$Delay = Delay(AND + Delay(CSA) + Delay(CPA\ of\ 9\ bits)$$
$$= 1 + 3 + 9 \times 3 = 31$$

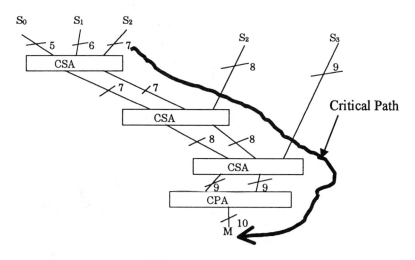

Figure 6.25 Wallace tree.

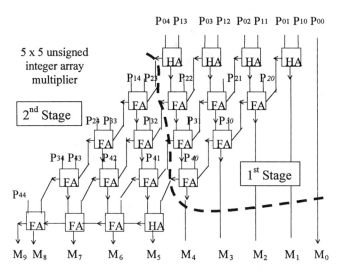

Figure 6.26 Deriving a two-stage pipelined array multiplier.

Part (c)

A two-stage pipelined array multiplier can be derived by inserting latches in each signal line cutting the dashed line in Fig. 6.27.

Therefore,

$Delay(stage) = 4 \times Delay(Full\ adder) = 4 \times 3 = 12$

Estimating the area of a latch as approximately the area of a half adder:

$Overhead\ in\ area = 10 \times Area(Latch) = 10 \times 6 = 60.$

$Area = 145 + 60 = 205.$

Conclusion

An array multiplier offers a better area versus speed trade-off than the Wallace tree for the particular case of a 5 × 5 unsigned integer multiplier. The proposed pipelined multiplier can potentially be less expensive than the Wallace tree in terms of area, i.e., it depends on the accuracy of area estimates.

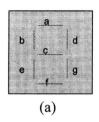 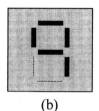

(a) (b)

Figure 6.27 A typical 7-segment display: (a) the location of each of the seven segments and their corresponding control signals [a,b,c,d,e,f,g]: (b) an example showing the illumination of the number nine, where all segments except [e,f] are on (logic 1).

6.9 LOGIC DESIGN

Derive a PLA NOR-NOR structure for each of the terms of a BCD-to-7-segment display device.

Solution

A 7-segment device is shown in Fig. 6.27.

The truth table of the BCD device is given below, where a "1" denotes that the segment is used in the representation of the respective BCD number, and "X" is a *don't care*.

A direct mapping NOR-NOR implementation of a Boolean function W can be easily derived from the product of implicates representation of W, as seen below:

$$W = \prod_i S_i$$

where S_i is an implicate, e.g., $\overline{S_k} = A + B$.

Applying DeMorgan's Theorem, it can be derived that

$$W = \overline{\overline{\prod_i S_i}} = \overline{\sum_i \overline{S_i}}$$

The *product of implicates* form of a function W can be directly derived from the truth table of W.

In the case of a BCD-to-7-segment display device, each implicate will be one of the following maxterms:

Computer Engineering

BCD to 7-Segment Truth Table

BCD Number	ABCD	a	b	c	d	e	f	g
0	0000	1	1	0	1	1	1	1
1	0001	0	0	0	1	0	0	1
2	0010	1	0	1	1	1	1	0
3	0011	1	0	1	1	0	1	1
4	0100	0	1	1	1	0	0	1
5	0101	1	1	1	0	0	1	1
6	0110	1	1	1	0	1	1	1
7	0111	1	0	0	1	0	0	1
8	1000	1	1	1	1	1	1	1
9	1001	1	1	1	1	0	1	1
—	1010	X	X	X	X	X	X	X
—	1011	X	X	X	X	X	X	X
—	1100	X	X	X	X	X	X	X
—	1101	X	X	X	X	X	X	X
—	1110	X	X	X	X	X	X	X

$$S_0 = A + B + C + D$$
$$S_1 = A + B + C + \overline{D}$$
$$S_2 = A + B + \overline{C} + D$$
$$S_3 = A + B + \overline{C} + \overline{D}$$
$$S_4 = A + \overline{B} + C + D$$
$$S_5 = A + \overline{B} + C + \overline{D}$$
$$S_6 = A + \overline{B} + \overline{C} + \overline{D}$$
$$S_7 = A + \overline{B} + \overline{C} + \overline{D}$$
$$S_9 = A + \overline{B} + \overline{C} + \overline{D}$$

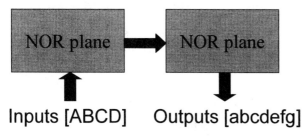

Figure 6.28 General structure for NOR-NOR PLA.

By inspection,

$$a = \overline{\overline{S_1} + \overline{S_4}}$$
$$b = \overline{\overline{S_1} + \overline{S_2} + \overline{S_3} + \overline{S_7}}$$
$$c = \overline{\overline{S_0} + \overline{S_1} + \overline{S_7}}$$
$$d = \overline{\overline{S_5} + \overline{S_6}}$$
$$e = \overline{\overline{S_1} + \overline{S_3} + \overline{S_4} + \overline{S_5} + \overline{S_7} + \overline{S_9}}$$
$$f = \overline{\overline{S_1} + \overline{S_7}}$$
$$g = S_2$$

A NOR-NOR PLA implementing these equations is seen in Fig. 6.29.

6.10 SYSTEM DYNAMICS

The computation time of a functional unit depends on the value of the input, V, and is given by:

$$T_{\text{compute}} = (50 + V/2) \text{ nanoseconds}$$

Assume that V is an 8-bit unsigned quantity, and V takes on values randomly and uniformly throughout the full 8-bit range. The functional unit outputs are valid simultaneously. Neglect wire routing delays between functional units.

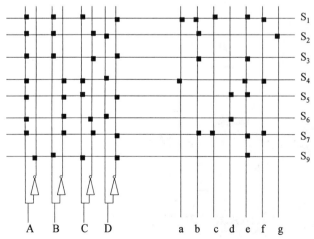

Figure 6.29 Inputs (left) and outputs (right) to a 7-segment display.

Computer Engineering

(a) If this unit was used as a functional unit in a pipeline of a synchronous system and was the slowest unit in the pipeline, what would be the highest allowable clocking frequency for the pipeline?

(b) If the pipeline was changed from synchronous to a self-timed pipeline (with no additional computational delays), what would be the average increase or decrease in throughput (in calculations per second)?

6.10.1 Solution

Part (a)

Assuming that the unit is the slowest in the pipeline, the maximum possible frequency of operation in synchronous mode is given by

$$f_{max} = \frac{1}{T_{compute}^{max}} = 5.6 \text{ MHz},$$

where

$$T_{compute}^{max} = 50 + \frac{255}{2} \text{ ns} = 173.5 \text{ ns} = \text{maximum possible computation time.}$$

Part (b)

A self-timed pipeline would operate at an average frequency given by the following equation, assuming that the unit would still be the slowest unit in the average case.

$$\bar{f} = \frac{1}{\overline{T_{compute}}} = 8.9 \text{ MHz},$$

where

$$\overline{T_{compute}} = 50 + \frac{\overline{V}}{2} \text{ ns} = 113 \text{ ns}$$

and $\overline{X}$ is the average value of random variable X. As V is a random variable between 0 and 255 (i.e., V is an 8-bit quantity), its average value is 127.5.

Conclusion

A self-timed pipeline using the specified unit would be on average 56% faster than a synchronous pipeline.

REFERENCES

1. S. Rosenstark (1995), *Transmission Lines in Computer Engineering*, McGraw-Hill, New York.
2. Intel Corporation (1995), *The Intel Pentium Progarmmer's Manual*, Intel Corporation, Santa Clara, CA.
3. S. M. Kang and Y. Leblebici (1995), *CMOS Digital Integrated Circuits: Analysis and Design*, McGraw-Hill, New York.

Index

A/D conversion, 124, 167
Abramowitz and Stegun, 69
Ackermann's formula, 254
Acton, 69
Almost periodic signals, 126
Amplitude shift key, 173
Analog modulation (AM), 157
Antenna, 90, 92, 214
   aperture efficiency, 214
   beamwidth, 214
   gain, 214
   patterns, 218
Arithmetic logic unit (ALU), 265
Asymptotic methods, 62
Athanas, 259
Autocorrelation, 151

Back-up relays, 32
Balanis, 75
Baseband, 167
Basis function, 71
Bessel functions, 65

Bilinear transformation, 140
Bingulac, 232
Biot–Savart law, 77
Bit error rate, 170, 171
Boolean function, 273
Broadside far field, 105
Bus admittance matrix, 11
   data for load flow, 25
   impedance matrix, 11
   injection current, 12
   voltage, 12

Canonical forms, 239
Carson's rule, 161
Cayley–Hamilton theorem, 245
Characteristic polynomial, 245
Charging current, 21
Circular apertures, 221
Circular characteristic, 38
Clarke transformation, 8
Classical model of generator, 44
CMOS, 270

Communication systems, 157
Compact discs (CD), 169
Computer engineering, 259
  organization, 264
Continuous models, 230
Controllability, 234
Control systems, 225
Cross-correlation, 151
Cumulative distribution function, 153
Curl, 64
Current transformer, 32
  multi-ratio, 32

D/A conversion, 124
Decimation, 133, 134
Difference equations, 231
Digital radio link, 173
Digital signal processing (DSP), 124
Dimensionality, 66
Directional overcurrent relays, 30
Discrete Fourier transform (DFT), 132
Discrete models, 230
Distance relays, 36
  ground, 38
  phase, 37
Distribution factor, 13
Downlink, 184
Dual graph, 275

Earth–satellite link, 195
Eigensystems, 146
Electric charge, 61
  power, 1
Electromagnetics, 61
Equivalent admittance matrix, 50, 53, 54
Error
  correction, 282
  detection, 282

Far field, 62
Fast Decoupled Load Flow, 19
Fast Fourier transform (FFT), 123
Fault resistance, 38

Faults, 1, 8
  ground, 13
  three-phase, 12
  unbalanced, 1
FDMA, 190
Field effect transistor (FET), 270
FIFO, 287
Filters, 139
  Finite impulse response (FIR), 142,144
Flat conductors, 6
Fortescue, 1
Fourier series, 130, 136
  transform, 137
Frequency domain, 62, 129,
Frequency modulation (FM), 157, 161
Frequency shift key, 173

Galerkin's method, 72, 73
Gaussian pulse, 108
Generator buses, 18
Geostationary satellite
  communications, 181
Green's function, 62, 67

Hamming coding, 281
Hamming window, 145
Hankel functions, 87
Harrington, 68
Helix antenna, 218
Hertzian dipole, 75
Hub station, 193

Impedance matrix, 1
  bus, 9
  negative sequence, 14
  positive sequence, 14
  zero sequence, 14
Inertia constant, 44
Information coding, 281
Instantaneous relays, 30
Integral equation, 98
Intermodel conversion, 243
Iterations, 26

Index

Jacobian matrix, 20
Jordan form, 146

Kernel function, 73
Kimbark transformation, 8
Kirchhoff, 68, 260

Large carrier (LC), 158
Least recently used (LRU), 288
Legendre polynomials, 65
Leibniz series for π, 114
Leverrier, 245
Line-of-sight (LOS), 177
Load buses, 18
Load flow, 18
Logic design, 305
Look angles, 202, 203
Loop systems, 35
Low earth orbit satellites, 191

Matrix
  exponential, 226
  fraction description, 244
  functions, 226
Maxwell equations, 61
Memory, 286
  address register (MAR), 266
  data register (MDR), 266
Method of moments, 70, 99
Mho characteristic, 38
Miller, 62, 100
MIMO, 244
Mismatch vector, 26
Modal expansions, 66
Multiple access (MA) techniques, 190
Multipliers, 300
Multi-port, 70
Mutual impedance, 6, 7

Negative sequence, 2
Newton–Raphson, 19
NTSC, 162
Nyquist, 175

Observability, 234, 236
  matrix, 236
  test, 237
Optical link, 172
Orthogonal, 62
Overcurrent relays, 30

Paging, 286
Parity bit, 282
Park transformation, 8
Per unit, 4
Periodic signals, 126
Personal communications, 191
Phadke, 1
Phased array antennas, 215
Pick-up setting, 31
Pipelined stream, 296
Pocklington, 104
Poggio, 100
Pole placement, 253
Positive sequence, 2
Post-fault system, 49
Power, 3
  per unit, 4
  phase, 3
  symmetrical component, 4
Poynting vector, 77
Pratt, 157
Pre-fault system, 49
Press, 70
Prewarping, 140
Propagator, 64, 67
Protection, 30
Pseudo random numbers, 127

Quadrilateral characteristic, 38, 40

Radar design, 206
Radiation, 66
Radio frequency (RF), 158
Radio links, 177
Rain attenuation, 187
Random signals, 127, 151
Rayleigh–Ritz, 68
Reactance characteristic, 40

Rectangular apertures, 219
Reed Solomon, 169
Reflections, 259, 260
Reflector antennas, 215
Regenerative transponder, 181
Register transfer language (RTL), 265
Relays, 30
  characteristics, 31
  coordination, 32
Resolvent matrix, 245
Richardson extrapolation, 116
Romberg quadrature, 116
Root locus, 250
Rotating machine, 7
Rotor equations of motion, 44
R–X diagram, 37

Sampling functions, 69
Satellite links, 179
Scalar potential, 76
Scattering, 66
Secondary ohms, 37
Sequence impedances, 2
Sequential state machine, 275
Shanks' method, 113
Short-circuit, 9
Side-band, 157
  double, 158
  single, 157
  vestigial, 157
Signal analysis, 123
Signal convolution, 127
Signal-to-noise (S/N) ratio, 159
Singular value decomposition, 149
Sinusoidal signals, 125, 126
Sparse matrix, 9
Sparse vector, 9
State
  assignment, 279
  transition diagram, 278
  vector, 127
Static electric field, 61
Static magnetic field, 62
Steady state, 9
  stochastic, 152

Superposition, 12
Suppressed carrier (SC), 158
Swerling case, 213
Swing bus, 18
Swing curves, 57
Symmetrical components, 1
Synchronous speed, 44
System dynamics, 307

Tapered field distributions, 220
Target glint, 213
TDMA, 190
TE case, 83
Terminations, 259
  parallel, 264
  serial, 264
Time correlated signals, 152
Time dial setting, 33
Time domain, 62, 248
Tirat-Gefen, 259
TM case, 82
Transfer function, 63, 64, 110
Transfer function matrices, 244
Transformation, 2
  matrix, 239
Transient stability, 43
Transmission line, 5, 259
  transposed, 5
  untransposed, 6
Trapezoidal rule, 116

Unbalanced voltages, 8
Ungrounded neutral, 13
Unit pulse, 128
Uplink, 184

VanLandingham, 123, 225
Vector potential, 76
Vegetative shadowing, 199
Vertical electric dipole, 93, 97
Video satellite signal, 189
Virtual memory, 286
Voltage angles, 24
Voltage magnitudes, 24

Voltage transformer, 37

Wallace tree multiplier, 300
Waveguide, 80
Wavelength, 62
Weight function, 71, 72
Window functions, 143

Yagi antenna, 218

Zero-order hold, 230
Zero sequence, 2
Zones of protection, 36, 37
z-transform, 130